Drug Repositioning

Approaches and Applications for Neurotherapeutics

FRONTIERS IN NEUROTHERAPEUTICS SERIES

Series Editors
Diana Amantea, Laura Berliocchi, and Rossella Russo

Drug Repositioning: Approaches and Applications for Neurotherapeutics
Joel Dudley, Mount Sinai School of Medicine, New York, New York, USA
Laura Berliocchi, Magna Græcia University, Catanzaro, Italy

Mapping of Nervous System Diseases via MicroRNAs
Christian Barbato, Institute of Cell Biology and Neurobiology (IBCN), Rome, Italy
Francesca Ruberti, Institute of Cell Biology and Neurobiology (IBCN), Rome, Italy

Rational Basis for Clinical Translation in Stroke Therapy
Giuseppe Micieli, IRCCS, Pavia, Italy
Diana Amantea, University of Calabria, Rende, Italy

Drug Repositioning
Approaches and Applications for Neurotherapeutics

Edited by
Joel Dudley and Laura Berliocchi

CRC Press
Taylor & Francis Group
Boca Raton London New York

CRC Press is an imprint of the
Taylor & Francis Group, an **informa** business

CRC Press
Taylor & Francis Group
6000 Broken Sound Parkway NW, Suite 300
Boca Raton, FL 33487-2742

First issued in paperback 2019

©2017 by Taylor & Francis Group, LLC
CRC Press is an imprint of Taylor & Francis Group, an Informa business

No claim to original U.S. Government works

ISBN-13: 978-1-4822-2083-4 (hbk)
ISBN-13: 978-0-367-86907-6 (pbk)

Library of Congress Cataloging-in-Publication Data

Names: Dudley, Joel T. | Berliocchi, Laura.
Title: Drug Repositioning: Approaches and Applications for Neurotherapeutics
/ [edited by] Joel Dudley and Laura Berliocchi.
Description: Boca Raton: CRC Press, [2017] | Series: Frontiers in
Neurotherapeutics series | Includes bibliographical references and index.
Identifiers: LCCN 2016059349| ISBN 9781482220834 (hardback: alk. paper) |
ISBN 9781315373669 (ebook)
Subjects: LCSH: Drug development. | Neuropharmacology.
Classification: LCC RM301.25 .D786148 2017 | DDC 615.1/9--dc23
LC record available at https://lccn.loc.gov/2016059349

Visit the Taylor & Francis Web site at
http://www.taylorandfrancis.com

and the CRC Press Web site at
http://www.crcpress.com

Contents

SECTION I   The Rationale and Economics of Drug Repositioning

SECTION II   Repositioning Approaches and Technologies: From Serendipity to Systematic and Rational Strategies

SECTION III   Drug Repositioning for Nervous System Diseases

Preface

A better understanding of many nervous system disorders and their effective treatment represents an important scientific challenge and an increasing concern for health systems worldwide, due to the chronic nature of some of these conditions and their high incidence especially in the increasing aging population. In spite of significant financial and professional investments, and great advances made during the past two decades, the fundamental etiology and pathophysiology of many diseases affecting the nervous system remain unclear, and effective disease-modifying therapies are still lacking. The reasons for such a failure in developing new effective therapeutics for nervous system disorders are several and of different nature. The intricate biology of the nervous system itself, together with the complexity and slow progression of these specific pathologies, made it difficult to understand the basic disease mechanisms and to identify appropriate end points and biomarkers, essential in achieving an accurate stratification of patients' populations. Indeed, patients' heterogeneity, lack of reliable biomarkers for both diagnosis and treatment, and slow progression are some of the factors responsible for the failure of many clinical trials. Furthermore, limitations related to the uncertain predictive validity of animal models seem to have interfered with the successful identification of safe and/or effective new candidate drugs, and contributed to the high rate of late-stage clinical trial failures, for instance in the case of drugs acting on the central nervous system.

In addition to the existing biological reasons, regulatory barriers seem to have contributed to make *de novo* drug discovery and development for nervous system diseases a lengthy, costly, and risky process. Although this is particularly true for drug development in the field of neuroscience, in recent years it has become more and more clear that, in general, the whole traditional paradigm of R&D needed some rethinking. Over the last decade, increasing pharma R&D costs were not paralleled by increased productivity. On the contrary, the relationship between the investments to develop new innovative drugs and the outcome in terms of the resulting medical and financial benefits dramatically decreased, and only very few new drugs were approved.

Among the possible alternative approaches to *de novo* drug discovery, drug repositioning seems to be one of the most promising strategies to develop therapeutic options for currently unmet medical needs. Drug repositioning or repurposing or reprofiling (the terms are sometimes used interchangeably) refers to a designed way to identify new applications for existing drugs, at any stage of their long developmental or clinical path. This also includes drugs that have been shown to be safe but not effective for the indication they were originally developed for or, to the extreme, drugs investigated but not further developed or even removed from the market for safety reasons (drug rescue).

Thanks to the most recent advancements in technologies, including *in vitro/in vivo* screening approaches and computational tools such as bioinformatics, chemoinformatics, network biology, and system biology, the drug repositioning concept has flourished and moved from casual discoveries to targeted strategies.

Repositioning shows several advantages over traditional *de novo* drug discovery, such as reduced development costs and shorter time to approval and launch, and is emerging as a particularly attractive approach for several pathologies including rare and neglected diseases. Although with some challenges, the recovery of failed compounds for new indications clearly represents an interesting business opportunity for the industry, as also shown by the creation of *ad hoc* partnerships between big pharma, academia, and governments. Also from a social standpoint, conveying existing data and knowledge toward new therapeutic applications stands as a highly ethical way to maximize the use of patient information, and several nonprofit organizations have launched programs specifically aimed to support drug repurposing projects and initiatives.

It is clear that collaboration between different entities is key to the success of this attractive and complex new strategy in improving and accelerating therapeutic development for nervous system disorders.

Renowned experts from different settings (academia, industry, nonprofit organizations) will discuss all these aspects in the present volume of the series Frontiers in Neurotherapeutics. The book aims to provide an overview of drug repositioning applications specific to neurotherapeutics and is organized in three sections, each composed of several chapters. *Section I* introduces the concept and rationale of drug repositioning, illustrates the different possible challenges in repurposing by analyzing the cases of Alzheimer's and Parkinson's diseases, and describes the contribution of nonprofit research organizations. *Section II* illustrates the evolution of drug repositioning from a serendipitous advance to a precise strategy, providing some examples of techniques and tools used for the identification of new applications for existing compounds. *Section III* focuses on drug repositioning relevance specifically for nervous system diseases, providing some historical examples and analyzing in individual chapters the status of some of the main nervous system conditions (Alzheimer's, Parkinson's, and Huntington's diseases; amyotrophic lateral sclerosis; spinal muscular atrophy; ischemic stroke; and psychiatric disorders).

We thank all the authors for their participation and their valuable contributions and the reviewers for their critical comments. We are particularly grateful to Hilary LaFoe for her constant support, to Natasha Hallard for her skilled help, and to all CRC Press and Taylor & Francis Group staff for their professional assistance during all phases of book production.

Editors

Dr. Joel Dudley is associate professor of genetics and genomic sciences and director of biomedical informatics at Mount Sinai School of Medicine, New York (NY, USA). Prior to Mount Sinai, he held positions as cofounder and director of informatics at NuMedii, Inc., and consulting professor of systems medicine in the Department of Pediatrics at Stanford University School of Medicine (CA, USA), where he participated in leading research to incorporate genome sequencing into clinical practice. Dr. Dudley's current research is focused toward solving key problems in genomics and precision medicine through the development and application of translational and biomedical informatics methodologies. His publications cover the areas of bioinformatics, genomic medicine, personal and clinical genomics, as well as drug and biomarker discovery. He received a BS in microbiology from Arizona State University and an MS and a PhD in biomedical informatics from Stanford University School of Medicine (CA, USA).

Dr. Laura Berliocchi is associate professor of pharmacology at the Department of Health Sciences, Magna Græcia University (Catanzaro, Italy). She is leading the Pain Unit at the Center of Preclinical and Translational Pharmacology (University of Calabria, Italy), whose research activity is focused on a better understanding of the neurobiology of pain for more effective clinical treatments. Her latest research activity is focused on the role of autophagy in neuronal dysfunction and on the identification of new strategies for pain management. She received a BSc (Hons) in biology from the University of Rome "Tor Vergata" (Italy), a specialization degree in biotechnologies from the same university, and a PhD in molecular toxicology from the University of Konstanz (Germany). She was a research associate at the Medical Research Council (MRC; Leicester, UK), working on the effects of synaptic and axonal damage on neuronal function and survival, then a London Pain Consortium (LPC) senior research fellow and a scientific visitor at University College London (UCL; London, UK), where she trained in experimental models of pain and worked on mechanisms of pain control. She is a *Deutscher Akademischer Austauschdienst* (DAAD) alumna.

Contributors

Cecilio Álamo
Department of Biomedical Sciences
(Pharmacology Area)
University of Alcalá
Madrid, Spain

Diana Amantea
Department of Pharmacy, Health and
  Nutritional Sciences
University of Calabria
Rende, Italy

Giulia Ambrosi
Center for Research in
  Neurodegenerative Diseases
Casimiro Mondino National
  Neurological Institute
Pavia, Italy

Christos Andronis
Biovista Inc.
Charlottesville, Virginia

Giacinto Bagetta
Department of Pharmacy, Health and
  Nutritional Sciences
University of Calabria
Rende, Italy

Fabio Blandini
Center for Research in
  Neurodegenerative Diseases
Casimiro Mondino National
  Neurological Institute
Pavia, Italy

Bruce Bloom
Cures within Reach
Chicago, Illinois

Aleksandra Caban
Creativ-Ceutical
Kraków, Poland

David Cavalla
Numedicus Limited
Cambridge, United Kingdom

Silvia Cerri
Center for Research in
  Neurodegenerative Diseases
Casimiro Mondino National
  Neurological Institute
Pavia, Italy

Alexander W. Charney
Department of Neuroscience
and
Department of Psychiatry
Icahn School of Medicine at Mount
  Sinai
New York, New York

Dennis S. Charney
Department of Psychiatry
Icahn School of Medicine at Mount
  Sinai
New York, New York

John R. Ciallella
Melior Discovery Inc.
Exton, Pennsylvania

Spyros N. Deftereos
Biovista Inc.
Charlottesville, Virginia

Maria P. del Castillo-Frias
Manchester Institute of Biotechnology
The University of Manchester
Manchester, United Kingdom

Andrew J. Doig
Manchester Institute of Biotechnology
The University of Manchester
Manchester, United Kingdom

Alba Esposito
Department of Neuroscience,
    Reproductive and
    Odontostomatological Sciences
"Federico II" University of Naples
Naples, Italy

Faraz Farooq
Molecular Biomedicine Program
Children's Hospital of Eastern Ontario
    Research Institute
Ottawa, Ontario, Canada

and

Mathematics & Science Department
Emirates College for Advanced
    Education
Abu Dhabi, United Arab Emirates

Francesca Romana Fusco
Laboratory of Neuroanatomy
Santa Lucia Foundation
Rome, Italy

Silvia E. García-Ramos
Hospital Pharmacy Service
Principe de Asturias University Hospital
Madrid, Spain

Szymon Jarosławski
Public Health Department
Aix-Marseille University
Marseille, France

Anna Kapuśniak
Creativ-Ceutical
Kraków, Poland

Eftychia Lekka
Biovista Inc.
Charlottesville, Virginia

Christopher A. Lipinski
Melior Discovery Inc.
Exton, Pennsylvania

Francisco López-Muñoz
Faculty of Health Sciences and Chair of
    Genomic Medicine
Camilo José Cela University
and
Faculty of Medicine and Health
    Sciences
University of Alcalá
and
Neuropsychopharmacology Unit
Hospital 12 de Octubre Research
    Institute (i+12)
Madrid, Spain

Christine M. Macolino-Kane
Melior Discovery Inc.
Exton, Pennsylvania

Hermann Mucke
H.M. Pharma Consultancy
Vienna, Austria

Francesco Napolitano
Systems and Synthetic Biology
    Laboratory
Telethon Institute of Genetics and
    Medicine
Pozzuoli, Italy

Nichole Orr-Burks
Department of Infectious Disease
University of Georgia
Athens, Georgia

Emanuela Paldino
Laboratory of Neuroanatomy
Santa Lucia Foundation
Rome, Italy

Andreas Persidis
Biovista Inc.
Charlottesville, Virginia

Aris Persidis
Biovista Inc.
Charlottesville, Virginia

Olivia Perwitasari
Department of Infectious Disease
University of Georgia
Athens, Georgia

Tiziana Petrozziello
Department of Neuroscience,
   Reproductive and
   Odontostomatological Sciences
"Federico II" University of Naples
Naples, Italy

Andrew G. Reaume
Melior Discovery Inc.
Exton, Pennsylvania

Cecile Rémuzat
Creativ-Ceutical
Paris, France

Douglas M. Ruderfer
Department of Medicine, Psychiatry
   and Biomedical Informatics
Vanderbilt University School of Medicine
Nashville, Tennessee

Joseph R. Scarpa
Department of Genetics and Genomic
   Sciences
Icahn School of Medicine at Mount Sinai
New York, New York

Agnese Secondo
Department of Neuroscience,
   Reproductive and
   Odontostomatological Sciences
"Federico II" University of Naples
Naples, Italy

Byoung-Shik Shim
Immunology and Microbial Sciences
The Scripps Research Institute
Jupiter, Florida

Valentina Tedeschi
Department of Neuroscience,
   Reproductive and
   Odontostomatological Sciences
"Federico II" University of Naples
Naples, Italy

Mondher Toumi
Public Health Department
Aix-Marseille University
Marseille, France

Ralph A. Tripp
Department of Infectious Disease
University of Georgia
Athens, Georgia

Vassilis Virvillis
Biovista Inc.
Charlottesville, Virginia

Section I

The Rationale and Economics of Drug Repositioning

1 Scientific and Commercial Value of Drug Repurposing

David Cavalla

CONTENTS

1.1   INTRODUCTION

Drug repurposing is a directed strategy to identify new uses for existing drugs, to be embarked upon at any stage in their developmental or clinical life. For pharmaceutical R&D, the benefits are clear: alongside reduced risk of developmental failure, there is demonstrable reduced cost and time of development. While historically many examples of repurposing arose from serendipitous clinical findings, modern repurposing has other skills in its toolbox; it may also derive from literature-based methods, deliberate *in vitro* or *in vivo* screening exercises, or *in silico* computational techniques to predict functionality based on a drug's gene expression effects, interaction profile, or chemical structure.

From the earliest times of medicine, doctors have sought further uses for available treatments. Traditional folk medicines are often proposed for the treatment of a bewilderingly wide range of purposes. The keystone in the process of new uses for existing drugs is the physician; they approach the issue using the principle of "clinical relatedness," whereby if a drug is useful for condition A, it is likely to be useful for a related condition B.

As distinct from the historical interest in new use for existing medicines, the modern strategy of drug repurposing involves a much fuller evaluation of a drug-like compound, including its chemistry, its medical use, and the biological target through which its effect is derived. This diverges from the traditional discovery approach, which is focused on a particular disease-related target. Instead, drug repurposing starts with the drug, looks at its complete biological profile, and ends with the identification of a number of new diseases for which it might be useful. These hypotheses are then tested experimentally, in preclinical and clinical trials.

The nomenclature in the field of drug repurposing has been rather confusing: other terms such as drug repositioning, reprofiling, and therapeutic switching have been suggested by some authors to relate to subtly different aspects. In this chapter, they will be taken to mean the same thing, broadly, the "concept of branching the development of an active pharmaceutical ingredient, at any stage of the life cycle and regardless of the success or misfortune it has encountered so far, to serve a therapeutic purpose that is significantly different from the originally intended one" (Mucke 2014).

There are three main categories of drug repurposing: the identification for a new indication of a developmental compound, a launched proprietary product, and a generic drug. In addition, relative to the primary indication, repurposing may involve a different dose, a different route of administration, a different formulation, or none of these, in which case it may represent more of a product line extension. Each of these alternatives differs substantially from the other in terms of developmental, regulatory, and commercial prospects. The change of indication may also involve further optimization of the active principle, on the basis that a very good way of discovering a new drug is to start with an old drug.

1.2   CASE HISTORIES

Recent attention to the deliberate strategy of drug repurposing has arisen partly because of the observed frequency with which this has happened by chance in the past. In other words, as success in pharmaceutical R&D becomes evermore challenging, investigators have been attracted to this strategy because pharmaceutical products with secondary uses are known from previous, serendipitous experience. These findings, although serendipitous, have revealed more than just another use for an existing drug. Their frequency has also revealed that a single biological mediator is usually involved in many different diseases, and this pleiotropy makes repurposing (of a modulator of such a mediator) a promising strategy.

The discovery of the use of thalidomide for the treatment of leprosy is an instructive example. Before it was banned by WHO (World Health Organization) for its teratogenicity in 1962, and withdrawn from the market in Europe and Canada, thalidomide was used for the treatment of insomnia and morning sickness. By 1964, almost no one believed that it might be reintroduced after its infamous history.

But at this time, a critically ill patient with erythema nodosum leprosum (ENL), a complication of multibacillary leprosy, was referred to Dr. Jacob Sheskin, who was at Hadassah University in Jerusalem. The patient was originally from Morocco and was being treated by the University of Marseilles, France. Leprosy (Hansen's disease) is a chronic, infectious human disease caused by a bacillus similar to that which causes tuberculosis.

The patient was on the verge of death—for months, the pain of his condition had prevented him from sleeping for more than 2 or 3 hours in any 24-hour period. Sheskin had no available therapy for his patient and as a last resort administered thalidomide because he thought that its original indication for insomnia would allow him to sleep better. Rather to Sheskin's surprise, one day after administering two pills of thalidomide, the patient slept continuously for about 20 hours. After 2 days, the pain, which had been so severe, had disappeared almost entirely. After another 3 days, Sheskin decided to withdraw treatment, and the condition rapidly worsened.

Sheskin was unable to replicate his discovery in Israel, because leprosy was almost unknown. So he traveled to Venezuela, where leprosy was endemic and thalidomide was still available. In clinical trials in subsequent years, he treated 173 patients and symptomatically cured over 90%. The development was taken up by the U.S. pharmaceutical company Celgene, who engaged with the FDA and finally secured their approval in 1997 to use thalidomide for the treatment of erythema nodosum leprosum; in due course, it was also approved for multiple myeloma.

The case of thalidomide represents perhaps the most remarkable of all examples of drug repurposing. If a product that is globally recognized as having terrible effects when prescribed for a certain indication can induce an almost Lazarus-like effect in a life-threatening disease, and then become approved for such use from one of the world's most exacting regulatory agencies, surely are there effectively no existing drugs for which an alternative use cannot be posited? The constraint in this analysis is revealed by the following thought experiment: if thalidomide can be approved for these serious conditions despite its appalling safety record in the context of the original indications, it must equally be the case that an existing drug, deemed safe in an original serious indication, is not necessarily acceptably "safe" in a much less serious secondary indication. The product needs to be subjected to a new regulatory review, and a new safety/efficacy assessment, specifically for this new indication. Thus, it is difficult to countenance the new use of, say, an existing cancer chemotherapeutic for a condition significantly less severe than cancer (unless there are ameliorating factors, such as a lower dose).

It is surprising how new uses can be found even for well-known drugs long after their therapeutic birth. Think of aspirin, which derives from the bark of the willow tree; its use to relieve headaches, pains, and fevers was known to Hippocrates in ancient Greece around 2500 years ago. It was isolated in the early nineteenth century and introduced as a pharmaceutical by Bayer in 1899. It took a further 70 years for the British pharmacologist and Nobel Laureate John Vane to discover that aspirin could disrupt a pathway needed for platelet aggregation (Vane 1971). Further studies in the 1980s showed that this effect could be used for the prevention of heart attacks and stroke; low-dose aspirin is now widely used for this effect. A further 30 years passed while its role in cancer was unraveled, and in December 2010 important clinical information was reported supporting the ability of aspirin to prevent colorectal and other cancers (however, crucially, this preventative effect on colorectal cancer is based on data from 25,000 patients but is published with the caveat that "further research is needed") (Rothwell et al. 2011). Over a century has passed since aspirin was first commercialized as a painkiller, which goes to show how long it can take for therapeutic uses to be discovered even in modern scientific times and with a well-known drug. The main reason for this very long time interval is the lack of commercial incentive to develop a generic drug for a new indication, since the existence of a generic substitution removes any commercial exclusivity that might reward a successful innovator. Clinical trials of aspirin in cancer are currently being financed from the public purse, which results in far longer time frames than if commercial investment were available.

The widespread adoption of a deterministic approach to the identification of new indications for developmental drugs followed the approval of the use of sildenafil for erectile dysfunction in 1998. The commercial success of this product introduction by a large pharmaceutical company, and the prospective identification and pursuit of a secondary indication of an incompletely developed drug, attracted significant interest in drug repurposing as a business strategy.

The discovery of sildenafil began in 1985 at Pfizer in a discovery program focused on inhibitors of cGMP phosphodiesterase type V (PDE5) enzyme as novel antihypertensives. The project changed direction toward angina after test compounds, which were shown to inhibit PDE5 activity, resulted in vasodilatation and platelet inhibition. Human trials began in the United Kingdom, which were disappointing for their primary end point, but some patients reported the unexpected side effect of penile erections, which ultimately led to the development of sildenafil (Viagra™) as a treatment for erectile dysfunction. However, research continued into pulmonary hypertension; as the role of PDE5 within this condition became better understood, sildenafil was repurposed again. Pulmonary hypertension is the general term for a progressive increase in pressure in the vessels supplying the lungs, particularly the pulmonary artery. It can be idiopathic, familial, or secondary to conditions such as rheumatoid arthritis or HIV. Symptoms often include right heart failure, shortness of breath, dizziness, fainting, and leg swelling. With a median survival of 2–3 years from the time of diagnosis, it is a life-threatening disease, unlike erectile dysfunction. In its idiopathic form, pulmonary arterial hypertension is a rare disease with an incidence of about 2–3 per million per year; however, it is far more common as a condition secondary to other diseases.

Sildenafil works by relaxing the arterial wall, which leads to a reduction in pulmonary arterial resistance and pressure. This, in turn, reduces the workload of the right ventricle of the heart and improves symptoms of right-sided heart failure. Pfizer conducted three trials on sildenafil in pulmonary arterial hypertension, the largest being an international, multicenter, randomized, blinded, controlled study involving 278 patients with the disease. Conclusions were drawn from the data produced, which showed improvements in exercise capacity, and the company submitted an additional registration for this indication of sildenafil to the FDA, for which it was approved in 2005. The dose of sildenafil required to treat pulmonary hypertension was as low as one-fifth of the dose for erectile dysfunction.

Finally, we have an example of a determinate development of a product for a new use in modern times, which came to fruition in 2013, when a compound we had known for 200 years was approved for the treatment of relapsing multiple sclerosis (MS) in both Europe and the United States. The product manufacturer was the large biotech company Biogen-Idec, who had licensed it from a small German company called Fumapharm.

The product, codenamed BG-12, is more commonly called dimethyl fumarate, known since the early days of organic chemistry and first synthesized as early as 1819. It therefore took nearly two centuries for the use in MS to be approved. For at least 150 years, dimethyl fumarate was considered as an organic chemical without conceivable therapeutic effects, rather than as a pharmaceutical. For a long while, its primary function was as a mould inhibitor and accordingly was added to leather items such as sofas during storage. However, at very low concentrations (down to 1 part per million), it is an allergic sensitizer: it produces extensive, pronounced eczema that is difficult to treat. This came to the fore in 2007, when 60 Finnish users of leather sofas into which dimethyl fumarate had been incorporated suffered serious rashes; as a consequence, the importation of products containing dimethyl fumarate has been banned in the European Union since 2009.

As a pharmaceutical, dimethyl fumarate (and other fumarate esters) was first used to treat psoriasis, and a product called Fumaderm™ had been approved in Germany for this use since 1994. Biogen was interested in these wider uses of this product in conditions similar to psoriasis. Given that the pathophysiology of psoriasis is based on various immune and inflammatory mechanisms that are shared with other conditions, Biogen undertook an investigation of the product's biochemical pharmacology, during which it was discovered that the mechanism of action involves upregulation of nuclear factor (erythroid-derived 2)-like 2 (Nrf2) protein, followed by induction of an antioxidant response. This is achieved through modification of the cysteine groups of a protein called KEAP1, which normally tethers Nrf2 in the cytoplasm. Once modified by fumarate, the KEAP1/Nrf2 complex dissociates and Nrf2 migrates to the nucleus, where it activates various antioxidant pathways. Armed with this knowledge, Biogen chose to develop dimethyl fumarate for MS.

While Fumaderm was commonly prescribed for psoriasis within Germany under Fumapharm, its use was geographically constrained. Once licensed to Biogen-Idec, far greater funds and priority were allocated to the longer-term studies necessary for MS. They sponsored two main trials to prove the efficacy of the product involving 1200 and 1430 patients, respectively, with relapsing remitting MS and conducted over 2 years.

The extra resources at the campaign's disposal were not wasted; they culminated in the regulatory approval of an oral product, Tecfidera™, which contained a slightly different composition of fumarate esters at a higher dose than Fumaderm. As many of the existing products for MS required intramuscular or subcutaneous injection, Tecfidera as an oral product offers distinct advantages; in addition, when measuring up to other oral MS products, it poses a lower risk of adverse cardiac events relative to fingolimod and a lower risk of liver toxicity compared to teriflunomide. It is now approved in both the United States and Europe and has recently been allocated 10 years of regulatory exclusivity in the latter territory. This remarkable story, concerning the introduction of a valuable new therapeutic option in a very serious disease, that had lain unappreciated for nearly 200 years, should not be underestimated: it proves that major improvements in therapy can derive from evaluating existing compounds in ways that had not previously been anticipated. From the patient's perspective, therefore, drug repurposing offers huge benefits.

An extensive list of 92 drug repurposing examples of drugs that have been approved or orphan designated for a secondary use can be found at http://drugrepurposing.info/index.php.

1.3   ADVANTAGES OF DRUG REPURPOSING

The 505(b)(2) process is a regulatory pathway applicable in the United States for exactly this situation: it applies to previously approved drugs that have undergone small modifications, for instance, either as a new formulation or in terms of a new use. It stands in comparison to the 505(b)(1) pathway, which applies to new chemical entities. In recent years, the proportion of FDA approvals that are based on the 505(b)(2) regulatory pathway has been increasing markedly. In 2014, compared to the 41 FDA approvals *via* 505(b)(2), there were only 35 *via* the 505(b)(1) route (Camargo Pharmaceutical Services 2015). However, not all of the 505(b)(2) approvals relate to drug repurposing. Another, slightly earlier, statistic provides that 20% of the 84 new marketed drug products in 2013 derive specifically from repurposing (Graul et al. 2014). It has been estimated that repurposed drugs now reap $250 billion per year, constituting around 25% of the annual revenue of the pharmaceutical industry (Naylor et al. 2015b; Tobinick 2009). Some of the most prominent commercial examples are shown in Table 1.1. The last row in Table 1.1 represents a structural variant of the famous repurposing example, thalidomide, a strategy which is dealt with at the end of the chapter. These commercial successes have fuelled and ratified increased adoption of the repurposing strategy.

It would be a mistake to assume that repurposing overwhelmingly produces incremental advances. Beyond the commercial successes in Table 1.1, repurposing has also produced noteworthy advances in healthcare as a whole. Examples of drug repurposing products that have been effective in serious and intractable conditions are described in Table 1.2. In some cases, such as pirfenidone and espindolol, the products are first-in-class approaches to the new therapeutic indication.

Defined as a modern prospective strategy of R&D, as distinct from the historical approach reliant on clinical serendipity, drug repurposing started to be used widely in the first few years of the new millennium. The most notable reason for this is an attempt to solve the poor, and declining, drug R&D productivity in the pharmaceutical industry over the past 20–30 years. It has been estimated that the cost of developing a new drug *de novo* may be over $1800 million. In addition, due to stringent regulations regarding safety, efficacy, and quality, the time required has been estimated to be 10–17 years (Paul et al. 2010). The expected value of a drug discovery program at its inception has been estimated as less than zero for a small molecule drug discovery campaign, as a direct result of the time and cost of development and risk of failure; according to an analysis from 2009, the estimated net present value (NPV) for an average small molecule is −$65 million with an internal rate of return (IRR) of 7.5% (David et al. 2009).

1.3.1   ATTRITIONAL RISK

Around 10 drug candidates need to enter into human investigation in order to produce one new molecular entity product launch (DiMasi and Grabowski 2007) and, before that, many thousands of molecular library members may need to be screened, structurally optimized and tested for effects in animal toxicology studies in order for the preclinical candidates themselves to enter first-in-human studies. The risk of R&D failure is therefore reduced if one starts with a product that has already been

TABLE 1.1
Commercial Successes of Drug Repurposing

Generic Name	Brand Name	Original Indication	New Indication (Year)	Company	Peak Annual Sales ($Billion)
Gemcitabine	Gemzar	Antiviral	Cancers (breast, ovarian, lung and pancreatic) (various)	Lilly	1.72
Raloxifene	Evista	Osteoporosis	Breast cancer (2007)	Lilly	1.09[a]
Finasteride	Proscar	Hypertension	Benign prostatic hypertrophy (1992)	Merck	0.74
	Propecia	Hypertension	Male pattern baldness (1997)	Merck	0.43
Sildenafil	Viagra	Angina	Erectile dysfunction (1998)	Pfizer	2.05
	Revatio	Angina/erectile dysfunction	Pulmonary hypertension (2005)	Pfizer	0.53
Rituximab	Rituxan	Various cancers	Rheumatoid arthritis, Wegener granulomatosis, and microscopic polyangiitis (MPA) (2004)	Biogen & Roche	1.2
Dimethyl fumarate	Tecfidera	Psoriasis	Multiple sclerosis (2013)	Biogen	2.91
Thalidomide	Thalomid	Antinausea and insomnia	Leprosy (1998)	Celgene	0.54
			Multiple myeloma (2006)	Celgene	
Lenalidomide	Revlimid	Structural analogue of thalidomide	Multiple myeloma, myelodysplastic syndrome, and mantle cell lymphoma (2006)	Celgene	4.28

[a] Peak annual sales figure includes both osteoporosis and breast cancer numbers.

TABLE 1.2

Important Therapeutic Advances Based on Drug Repurposing

Generic Name	Original Indication	New Indication	Comments
Alemtuzumab	Chronic lymphocytic leukemia	Multiple sclerosis	Approved 2015
Ketamine	Anesthesia	Severe depression	Phase II
Pirfenidone	Anthelmintic	Idiopathic pulmonary fibrosis	Approved 2014
Espindolol	Hypertension	Cachexia	Phase II

through part of the developmental trajectory. When comparing the two, a report by Thompson Reuters suggests a success rate of around 25% for drug repurposing projects, higher than the 10% or so for conventional new drugs (Thayer 2012).

While most of the risk of failure is associated with uncertain efficacy, it is also worth mentioning that safety is a relative concept and cannot be assumed for the new indication, even for a well-known approved drug. Repurposed developments can (normally) take advantage of preexisting data; however, this does not mean that acceptable safety for the new product is unequivocally established. The safety threshold required for a chemotherapeutic agent is quite different from that of a new treatment for insomnia, as pointed out in the discussion around thalidomide. As a matter of fact, it is fairly hard to repurpose chemotherapeutic drugs, because there are scant examples within that class that would be acceptable for another use (Oprea and Mestres 2012). Furthermore, certain diseases have been highlighted by regulatory agencies for even more detailed examination of safety hazards, for example, new antidiabetic drugs are required to undergo extensive safety assessments because of the greater risk of adverse cardiovascular events among diabetic patients. There are also risks in situations where the route of administration is changed in the new development (for instance, from oral to inhaled). In other words, safety risks being a reason for discontinuing a repurposing development are reduced, but not eradicated.

1.3.2 Cost

The normal regulatory approval process for a pharmaceutical product based on a new chemical entity is discussed earlier (Paul et al. 2010). This headline figure ($1.8 billion) includes the costs of ultimately abandoned developments, as well as a component associated with the cost of the capital deployed while funds are allocated to these failures. Compared to the cost of developing an entirely new molecule, new product development costs for drug-repurposed products may comprise only the middle and later stages of clinical development. It has been estimated that these costs stack up to around $300 million, assuming that the candidate still has to undergo Phase II and Phase III clinical trials (Naylor et al. 2015a). This figure is predicated on the model proposed by Paul et al. (2010), but still represents an approximate saving of 84% against the cost of a *de novo* drug R&D program as referred to earlier.

Part of the reason for this striking difference in cost is attributable to the reduction in attrition resulting from better success rates for repurposing projects. Although the bulk of the development costs for a new chemical entity program are also encountered in repurposing projects, the reduced expenditure on a failed program, and the associated costs of capital, gives rise to this substantial cost saving for the repurposing strategy.

1.3.3 TIME

In comparison to *de novo* drug development, the evaluation of efficacy in a human setting is brought forward drastically when following a repurposing approach. In a new chemical entity project, there are years of research and discovery before the investigation into clinical utility can even begin, with substantial expenditure of time and money in this preclinical development phase. Aside from the economic impact, the benefit for the patient is dramatic, with reduced time frames for new product innovation and associated healthcare improvements. The improvement in time taken to proof-of-concept studies is especially important for repurposing relative to new chemical entity (NCE) discovery. As shown in Figure 1.1, the time required for preclinical activities can be shrunk to 1–3.5 years compared to the normal expected duration of 4–9 years (Ashburn and Thor 2004).

Apart from discovery, the formulation of new development plans and their undertaking, market analyses, IP, and regulatory diligence are still required (Ashburn and Thor 2004). The time requirements for later drug development and regulatory

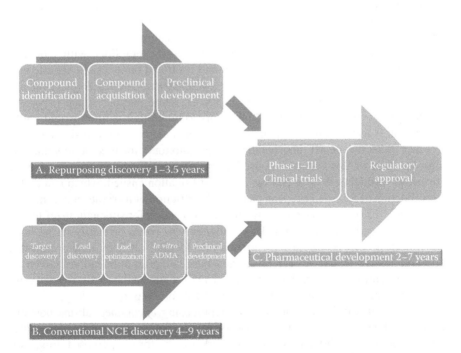

FIGURE 1.1 Comparison of the processes and times for preclinical development of drug repurposing projects (A) compared to conventional new chemical entity projects (B). The processes and times for clinical development of both types of projects are similar (C).

approval for a repurposed drug program are often comparable to those for traditional R&D (DiMasi 2013; Li et al. 2016) so that overall, the drug development cycle for a repositioned drug is predicted to be 3–12 years—a reduction on the usual 10–17 years.

1.4   DISADVANTAGES

A substantial advantage of drug repurposing is the ability to obtain so-called "method-of-use" patents, which are designed to promote the discovery of the secondary uses of compounds. Under international patent law, new uses of pharmaceuticals differ from new uses of other things in the way they are protected. These patents last for the same period as conventional "composition-of-matter" patents and are valid in most countries including the major pharmaceutical markets, although "the 'mere' identification of a new use" of a drug in India remains a significant exception. In some cases, they can be sufficient in themselves to protect a marketed repurposed product; however, they are seen as weaker and as more easily challenged.

The weakness is rooted in the greater difficulty for a patent holder in controlling drug use in parallel conditions. For instance, the patent that protected the use of sildenafil as a treatment for erectile dysfunction did not cover its use in pulmonary hypertension, which was the third purpose (taking the original anticipated use in angina as the first) for which the drug was used. This was not a problem for the marketing company, since Pfizer had both indications—erectile dysfunction and pulmonary hypertension—under its control but could have been a problem if two companies shared the franchise for sildenafil in two separate indications.

The second problem, that of legal challenge, is also exemplified with sildenafil, since Pfizer filed a patent both on the compound itself and on its use in erectile dysfunction. After Viagra was approved, other companies scrambled to discover and develop compounds of similar mechanism for the same purpose, since it became clear that profits from selective PDE5 inhibitors to treat this condition would not be insignificant. However, Pfizer's method-of-use patent claimed not just sildenafil, but also other "selective cGMP phosphodiesterase inhibitors" for erectile dysfunction; within the term of this "method-of-use" patent, it blocked the marketing of these competitors' products. A huge process of patent litigation ensued, which resulted in the invalidation of Pfizer's European patent and restriction of its rights under the U.S. Patent 6,469,012 (though the company retained protection for sildenafil itself for the treatment of erectile dysfunction). The challenge was based primarily on an issue of obviousness, that is, whether the effects of sildenafil (and, by extension, other PDE5 inhibitors) would have been predictable based on prior publications. Given the breadth of scientific literature that can be cited against any method-of-use patent, this remains an issue for any method-of-use patent filed for a repurposed drug.

The second major disadvantage of drug repurposing is commercial: the potential substitution of a repurposed product with a generic competitor. This substitution, which occurs off-label, can substantially limit the market niche for a repurposed drug, as is demonstrated with the use of aspirin referred to earlier. Sildenafil also clearly exemplifies this issue, because generic versions that are intended for the treatment of pulmonary hypertension can be used "off-label" in the United States for the treatment of erectile dysfunction (although multiple pills are required). Contrary to the fact that legally Pfizer has a monopoly on the sale of products containing sildenafil for erectile dysfunction under the method-of-use patent U.S. 6,469,012, it is difficult to police, and some generic substitution occurs. This is a result of the ability of a medical practitioner to prescribe a treatment which, according to their professional judgment, best suits the medical needs of the patient. This commercial disadvantage may limit the development of a significant proportion of drug-repurposing candidates, and pharmaceutical companies may be wary of drug-repurposing projects that might suffer reduced profits from generic substitution owing to insufficient market potential (Rastegar-Mojarad et al. 2015).

The international Repurposing Drugs in Oncology (ReDO) project is a collaboration initiated by a diverse group of researchers, clinicians, and patient advocates all working in the not-for-profit sector, seeking the repositioning of drugs to find new, effective cancer treatments (Pantziarka et al. 2014). From retrospective evidence, there are many examples of drugs currently used for other purposes that have been associated with reduced rates of cancer incidence. It would be beneficial to investigate these further, if only for their effects in cancer prevention or recurrence. However, the process of acquisition of further data and regulatory approval of a specific anticancer product based on these generic substances is very unlikely to be compensated by a preferentially priced product. In this sense, the very existence of generic products acts in opposition to repurposed drug innovation.

1.4.1 OFF-LABEL MEDICINE

When a new medicine is brought to market, the regulatory agencies judge its fitness for use by reference to a very specific set of conditions, the "product label," according to which a product may legally be promoted by the manufacturer. The safety and efficacy of the medicine when used under these restricted conditions are carefully assessed, and a judgment handed down. However, the remit of the regulatory agencies does not extend to the way the product is prescribed by the doctor; once the drug is approved, a medical practitioner may vary the use of the product outside the regulatory stipulations so long as, in their judgment, its prescription is in the interests of the patient. The regulatory stipulations may relate to the indication, the type of patient, the dose, or the length of time the medicine is administered; a variation on any of these would be classed as "off-label."

Commonly written for pediatric use and for certain therapeutic areas such as oncological and psychiatric cases, off-label prescriptions are estimated to amount for 20% of pharmaceutical prescriptions (Cavalla 2015). However, around 73%–79% of off-label uses lack any, or any good, scientific support (Eguale et al. 2012; Radley et al. 2006). An example of this is the use of antibiotics for viral infections; this category of off-label use has contributed to the increasing threat from bacterial resistance. In addition to the obvious efficacy issues, off-label use, the rate of adverse events, including serious adverse events, is doubled or tripled. In summary, there is less evidence for efficacy and good evidence for poorer safety. Acceptance of off-label use varies by country, with German practitioners only able to write such prescriptions for serious conditions where there is no approved treatment. In the

United Kingdom, the rules are only slightly less stringent, but as their enforcement is not rigidly policed, it is not clear how strictly they are followed. In the United States, doctors are comparatively much less bound, with statute blocking the FDA from becoming involved, and the American Medical Association unwilling to limit the freedom of their members to make their own prescription choices. Some proponents of these freedoms even argue against a requirement for doctors to tell their patients when a product is not labeled for their patient's condition.

Despite both the lack of evidential support and the greater risk of side effects, the freedom to prescribe off-label is a right guarded jealously by practitioners. Historically, the successful identification of ancillary medical uses often derives from this freedom; in rare diseases, for instance, where there are only approved treatments for around 5% of the known conditions, there is often no alternative. The efficacy of thalidomide in leprosy would certainly never have been discovered without this freedom; thus, its value lies in the saved lives and reduced suffering of many leprosy patients.

Notwithstanding this undeniable benefit, within the prescription of antipsychotics to dementia patients lies just one instance where there is a lack of benefit and a greater risk of stroke as a result of off-label prescription. In the United Kingdom, a report commissioned into the associated patient harm estimated 1800 excess deaths per year (Banerjee 2009). In another case, the drug NovoSeven (Factor VIIa) was approved for certain forms of hemophilia or congenital deficiencies in the coagulant protein, but the proportion of off-label prescriptions reached 98%, which is certainly an alarming proportion; ultimately, there was no clinical benefit in most of these uses, where again excess deaths arose (Yank et al. 2011). The erythropoietin-stimulating agent epoetin alfa was similarly originally approved for use in anemia in end-stage kidney disease and later with HIV. However, despite its narrow approbation in two orphan-designated uses, it was prescribed far more commonly: first to nearly all kidney dialysis patients, then to many more patients with anemia, disregarding the discrepancy in severity between the cases. With time, the doses rose incrementally, until in 2006, the sales of the product Epogen reached $5.6 billion. Thereafter, in the less serious cases, and at the higher doses, the safety of the product compared unfavorably with the benefit, and concerns about the way the drug was being used were duly reported (Mesgarpour et al. 2013). The FDA required new warnings to be added to the product label, as a result of which sales dropped off. The drug was also associated with athletic performance enhancement and provided part of the global "doping" scandal involving the cyclist Lance Armstrong. There are therefore multiple facets of erythropoietin: from the approved life-saving antianemic product in rare blood conditions, through to off-label use in the general improvement of "wellbeing and happiness" in much more common and less serious cases of anemia, and finally to a drug that can be used to cheat in sport.

1.5 VARIANTS

There are a number of different methods by which drug repurposing can come to fruition. The major source of variation is the starting point, or substrate, for the repurposing activity, which results in different commercial challenges, patent opportunities, and possible routes to regulatory approval.

1.5.1 Repurposing of Generic Drugs

In the simplest case, old, generic drugs without patent protection are used: there are 1527 such new molecular entities that have been approved by the FDA since its inception in 1938 to 2011 (FDA 2015). Repurposing opportunities that use exactly the same generic drug, formulated in the same way, delivered by the same route of administration are likely to be challenged by generic substitution as was demonstrated by the example of aspirin referred to earlier. As a result, a number of repurposing developments prioritize altering these variables. This may necessitate not only the alteration of the preclinical toxicology package, but also a reevaluation of the safety requirements (particularly if the original compound was approved in the more distant past). The process toward clinical testing therefore may not be as straightforward as can optimally occur with a repurposed product. This is particularly pertinent for inhaled drug repositioning, where the formulation of a respiratory product and its preclinical toxicology is relatively more expensive and time consuming than that of compounds with different administration methods.

1.5.2 Repurposing of Abandoned Assets

The repurposing of drugs that were never brought to market presents a solution to both the issue of the otherwise lost commercial value and time investment associated with these developments, and the issue of generic substitution. Here, the repurposed product, assuming success, would be the only product on the market with that particular active ingredient. Some years ago, the opportunity from stalled candidate drugs that had passed Phase I but not Phase II or III clinical trials was identified as a large resource of repositioning substrates (Tartaglia 2006). However, access to collections of failed drugs, and the necessary data to progress into clinical trials, is typically hard to obtain owing to intellectual property and confidentiality concerns (Li and Jones 2012).

This changed with the announcement of an imaginative collaborative partnership between public and private sectors in the United Kingdom. Specifically, it involved the Medical Research Council (MRC) and AstraZeneca. Named the "Mechanisms for Human Diseases Initiative," the approach was anchored in a library of 22 mid-stage stalled developmental compounds contributed by the industrial component. The model was replicated in the United States, after the NIH held a convention in 2011 for experts from academia, government, and private sector R&D to explore new uses for abandoned and approved therapeutics, at which pharmaceutical companies would create a pool of compounds worth pursuing. The NIH subsequently established NCATS (National Center for Advancing Translational Sciences) in 2012 and implemented the "Discovering New Therapeutic Uses for Existing Molecules" program. The strategy was the same as the UK prototype: the provision of partially developed compounds by the industry and public funds to spur their development into new indications. A third collaboration between the Broad Institute and Roche was begun in 2012, combining the former's screening technologies with some 300 compounds from the latter (GEN 2012).

From the governmental perspective, the advantage was the high-quality developmental compounds to which they were now privy. Translational grant schemes were established by each, for which academic investigators would compete. From the industry perspective, all they had to contribute were partly developed compounds and associated data, thereby making use of information would otherwise have been discarded. The value from working in a highly innovative area, together with engagement from an industrial partner, meant that the outcome from successful repositioning by the academic partner might be taken forward with alacrity, especially if there was significant patent life remaining.

An unpredicted boost to the campaigns that was discovered along the way was the use of crowdsourcing. In the NIH/NCATS scheme, examination of the submissions showed that 15 of the 16 compounds that received 5 or more applications had been proposed for at least 3 different indications. The diversity of the proposed developments emanating from one compound supports the potential of repurposing for new therapeutic outcomes (Colvis and Austin 2015).

Further examples of industry–academia partnerships forged with a drug repositioning objective include one between Pfizer and Washington University (St. Louis) (Washington University 2010) and one in Taiwan between AstraZeneca and the National Research Program for Biopharmaceuticals (NRPB) (AstraZeneca 2013).

1.6   OFF-TARGET *VERSUS* ON-TARGET

When considering drug-repurposing strategies, after considering the stage of development, it is important to consider the mechanism of action: whether it is the same as for the original use, or distinct. The difference may seem trivial but is far greater than semantic.

Most repurposing involves the identification of a new indication for an existing compound working through the original mechanism or target. This is called "on-target" repurposing (Li and Jones 2012). The principle is that new scientific understanding has connected the old mechanism to a new outcome *via* the same pathway or protein interaction as is responsible for the original indication.

By contrast, off-target repurposing involves a separate biological mechanism or pathway entirely. The use of doxycycline for the treatment of periodontitis is such an example. Repurposing of this type is rarer: the original research process will have involved a stage of optimizing the drug for that particular pathway, so it is unlikely to be an optimal compound for another mechanism. In the case of doxycycline, the original indication was antibacterial and functioned *via* preventing the attachment of aminoacyl-tRNA to the ribosomal acceptor (A) site, thereby blocking the synthesis of new protein strands (Chopra and Roberts 2001). Surprisingly, it was later discovered that doxycycline's more potent effects are against matrix metalloproteinases, and it was specifically *via* its effect on collagenase and gelatinase that it prevented the infiltration of polymorphonuclear cells into the gingival tissue of patients affected with periodontitis (Golub et al. 1995). Importantly, for this indication, the therapeutic benefit of doxycycline occurs at sub-antibacterial doses; this was crucial in the regulatory approval of the product, since overexposure of patients to an effective antibiotic increases the risk of resistant bacteria—which the FDA stipulated needed to be avoided.

A large proportion of the drugs on the market today have, besides the targets to which they are known to bind, far more targets which are unknown to us. The elucidation of these has been the purpose of recent studies evaluating the extent of polypharmacology by analyzing all known drug–target interactions. A human pharmacology interaction network that bridges proteins by the criteria of a shared interacting partner has been created (Paolini et al. 2006). In the database of 276,122 active compounds, 25% of the compounds bound targets from different protein families. In a second study, Mestres and coworkers combined the data from seven such interaction databases and found that, on average, each drug interacted with six different targets (Mestres et al. 2009). However, it is not sufficient that a drug merely binds to a target; it must do so at a therapeutically acceptable dose. It is, as a result, much rarer than on-target repurposing.

Tyrosine kinase inhibitors, given that they inhibit such a range of enzymes, are a highly promiscuous therapeutic class of marketed drugs. Generally, the mechanism is that they bind competitively to the ATP-active site, of which there are over 500 similar sites in the human protein kinome (Manning et al. 2002). Imatinib was first introduced for the treatment of chronic myelogenous leukemia (CML)—a condition that begins in the often asymptomatic chronic phase and over the course of several years progresses through an accelerated phase and ultimately to a blast crisis, by which time it is associated with a very high mortality rate. From diagnosis, the median survival period is just over 5 years. Imatinib was the first tyrosine kinase inhibitor of its kind to be developed: its primary target was bcr-abl, which is involved in a specific genetic abnormality associated with CML. However, there are two further tyrosine kinases through which it also acts, both with an applicable indication. The first is PDGF-R: inhibition of which accounts for its use in dermatofibrosarcoma. The second is c-kit: mutations in both c-kit and PDGF-R are central, albeit independent, events in gastrointestinal stromal tumor (GIST), thus inhibition of both these enzymes is proposed to underlie imatinib's benefit in this condition (Lasota and Miettinen 2006).

1.6.1 RARE DISEASES

Drug repurposing is a particularly attractive approach for rare diseases, for both scientific and commercial reasons; the term "rare-purposing" has even been trademarked by the company Healx Ltd., which is focused entirely on this strategy. Scientifically, these diseases are often poorly characterized in their pathophysiology, and scientists lack a clear understanding of the biological pathways or compounds with which they could be obstructed. *In silico* techniques for predictive repurposing accelerate the identification of testable hypotheses that may be clinically relevant. In addition, when considering the business potential, there are specific incentives designed to encourage research into rare diseases, which can provide commercial exclusivity in situations where repurposed products cannot be protected by patent, or where that patent is vulnerable.

When considering gaps in current healthcare, a huge opportunity exists: for the 8000 rare diseases which exist, over 95% lack an FDA-approved therapeutic agent (Global Genes 2014). Rare diseases are often equated with "orphan" disease status, given they lack an approved therapeutic treatment, and various regulatory legislation has been enacted in many countries to incentivize research into potential drugs for this type of condition.

In the United States, if a disease affects fewer than 200,000 people, it is classed as "rare" (Sardana et al. 2011). The Orphan Drug Act (ODA) was enacted because the economics for developing a drug for a rare disease, where the costs of R&D can only be amortized over a relatively small number of patients, were unfavorable. Legislating the allocation of additional market exclusivity in such situations, over 300 drugs and biological products for rare diseases have been approved by the U.S. FDA as a result of this act, which came into force in 1983 (compared with fewer than 10 such products 1973–1983). Orphan products represented 30% of products approved for the first time by the FDA in the 5-year period from 2004 to 2008 (Xu and Coté 2011). Today, the rare disease market is growing rapidly and is attractive for pharmaceutical R&D. Despite the small patient populations, high prices represent significant market sizes. Take for instance, the drug Soliris from Alexion Pharmaceuticals: priced at over $500k per patient per year, its sales in 2013 were over $1 billion.

The benefits enjoyed by pharmaceutical companies embarking upon the development of a drug for a rare disease include fast-track FDA approval, marketing protection, tax incentives, and funding for clinical research in rare diseases. As a boost to regular patent protection, which is also granted, once a drug is approved, a generic version in that indication cannot be approved for 7 years. The developer may also receive tax concessions, grants, and waivers of regulatory fees. As an indicator of the success of the ODA, similar legislation is now in place in Europe, Japan, Australia, and Singapore, with each jurisdiction having a slightly different definition of an orphan indication and applicable commercial incentive (Table 1.3). For instance, while an orphan designation in the United States may be handed down if the prevalence is less than 200,000 (approximately 6.25 in 10,000), in Europe the frequency must be slightly lower, at 5 in 10,000 (corresponding to approximately 250,000 patients in the EU28).

The regulatory protection in rare diseases is worthy of note when considering the relative weakness of patent protection for repurposed generic products, which is often restricted to "method-of-use" intellectual property. As would be expected,

TABLE 1.3
Legislation in Major Pharmaceutical Markets Offering Commercial Incentives for Orphan Drug Developments

	United States	Japan	Australia	European Union
Legislation date	1983	1993	1997/1998	2000
Prevalence	Fewer than 200,000 (6.25 per 10,000)	Fewer than 50,000 (4 per 10,000)	Fewer than 2,000 (1.1 per 10,000)	Fewer than 5 per 10,000
Market exclusivity	7 years	Reexamination period extended from 4 to 10 years	None	10 years
Fee waiver	Yes	No	Yes	At least partial

within this remit, repurposing activity is rife. A 2011 comparison of the FDA approvals database with those drugs that have received the coveted orphan drug designation revealed 236 tipped as "promising" for the treatment of a rare disease, though not yet granted permission for marketing (Xu and Coté 2011).

1.7 CONCLUSION

Drug repurposing is a growing purposeful strategy to identify new uses for pharmaceutical entities at any stage of their commercial or developmental life. It involves substantial improvements to the metrics of drug discovery, reducing the time to enter into clinical trials by 3–5.5 years, increasing the risk of developmental success by 150%, and reducing the cost of new product innovation by 83%. Drug repurposing projects may be protected by method-of-use patents, and sometimes formulation patents too. However, these patents are often more easily challenged and offer weaker exclusivity protection than composition-of-matter patents. In addition, repurposed products can suffer from off-label generic substitution which engenders pricing and reimbursement concerns and deters pharmaceutical company investment. Some or all of these disadvantages may be designed out by careful consideration of the project to be developed, and other exclusivity provisions can be brought to bear. Rare diseases are attractive areas for a repurposing strategy because of the regulatory exclusivity for products in this field, and, in addition, the commercial returns in such conditions may be more appropriate given the substantially reduced cost of new product introduction.

REFERENCES

Ashburn, T.T. and K.B. Thor. 2004. Drug repositioning: Identifying and developing new uses for existing drugs. *Nature Reviews Drug Discovery* 3 (8): 673–683.

AstraZeneca, 2013. AstraZeneca Announces 'Open Innovation' Research Collaboration with Taiwan's Research Program for Biopharmaceuticals.

Banerjee, S. 2009. The use of antipsychotic medication for people with dementia: Time for action. Department of Health. Available from http://www.bmj.com/content/342/bmj. d3514. Accessed August 8, 2014.

Camargo Pharmaceutical Services. 2015. 2014 505(b)(2) NDA approvals at the 505(b)(2). Blog accessed September 23. http://camargopharma.com/2015/01/2014-505b2-nda-approvals/.

Cavalla, D. 2015. *Off-Label Prescribing: Justifying Unapproved Medicine.* Wiley-Blackwell, New York, 216 pages.

Chopra, I. and M. Roberts. 2001. Tetracycline antibiotics: Mode of action, applications, molecular biology, and epidemiology of bacterial resistance. *Microbiology and Molecular Biology Reviews* 65 (2): 232–260.

Colvis, C.M. and C.P. Austin. 2015. The NIH-industry new therapeutic uses pilot program: Demonstrating the power of crowdsourcing. *ASSAY and Drug Development Technologies* 13 (6): 297–298.

David, E., T. Tramontin, and R. Zemmel. 2009. Pharmaceutical R&D: The road to positive returns. *Nature Reviews Drug Discovery* 8 (8): 609–610.

DiMasi, J.A. 2013. Innovating by developing new uses of already-approved drugs: Trends in the marketing approval of supplemental indications. *Clinical Therapeutics* 35 (6): 808–818.

DiMasi, J.A. and H.G. Grabowski. 2007. The cost of biopharmaceutical R&D: Is biotech different? *Managerial and Decision Economics* 28 (4–5): 469–479.

Eguale, T., D.L. Buckeridge, N.E. Winslade, A. Benedetti, J.A. Hanley, and R. Tamblyn. 2012. Drug, patient, and physician characteristics associated with off-label prescribing in primary care. *Archives of Internal Medicine* 172 (10): 781–788.

FDA. 2015. Summary of NDA approvals & receipts, 1938 to the present. WebContent.

GEN. 2012. Roche, Broad Institute Seek New Uses for Old Drugs, GEN News Highlights.

Global Genes. RARE diseases: Facts and statistics. Available from: https://globalgenes.org/rare-diseases-facts-statistics/. Accessed August 8, 2014.

Golub, L.M., T. Sorsa, H.M. Lee, S. Ciancio, D. Sorbi, N.S. Ramamurthy, B. Gruber, T. Salo, and Y.T. Konttinen. 1995. Doxycycline inhibits neutrophil (PMN)-type matrix metalloproteinases in human adult periodontitis gingiva. *Journal of Clinical Periodontology* 22 (2): 100–109.

Graul, A.I., E. Cruces, and M. Stringer. 2014. The year's new drugs & biologics, 2013: Part I. *Drugs of Today (Barcelona, Spain: 1998)* 50 (1): 51–100.

Lasota, J. and M. Miettinen. 2006. KIT and PDGFRA mutations in gastrointestinal stromal tumors (GISTs). *Seminars in Diagnostic Pathology* 23 (2): 91–102.

Li, Y.Y. and S.J.M. Jones. 2012. Drug repositioning for personalized medicine. *Genome Medicine* 4 (3): 27.

Li, J., S. Zheng, B. Chen, A.J. Butte, S.J. Swamidass, and Z. Lu. March 2016. A survey of current trends in computational drug repositioning. *Briefings in Bioinformatics* 17 (1): 2–12.

Manning, G., D.B. Whyte, R. Martinez, T. Hunter, and S. Sudarsanam. 2002. The protein kinase complement of the human genome. *Science (New York)* 298 (5600): 1912–1934.

Mesgarpour, B., B.H. Heidinger, M. Schwameis, C. Kienbacher, C. Walsh, S. Schmitz, and H. Herkner. 2013. Safety of off-label erythropoiesis stimulating agents in critically ill patients: A meta-analysis. *Intensive Care Medicine* 39 (11): 1896–1908.

Mestres, J., E. Gregori-Puigjané, S. Valverde, and R.V. Solé. 2009. The topology of drug-target interaction networks: Implicit dependence on drug properties and target families. *Molecular BioSystems* 5 (9): 1051–1057.

Mucke, H.A.M. 2014. A new journal for the drug repurposing community. *Drug Repurposing, Rescue, and Repositioning* 1 (1): 3–4.

Naylor, S., D.M. Kauppi, and J.M. Schonfeld. 2015a. Therapeutic drug repurposing, repositioning and rescue part II: Business review. *Drug Discovery World* (Spring 15). Available from: http://www.ddw-online.com/p-303325. Accessed August 8, 2014.

Naylor, S., D.M. Kauppi, and J.M. Schonfeld. 2015b. Therapeutic drug repurposing, repositioning and rescue part III market exclusivity using intellectual property and regulatory pathways. *Drug Discovery World* Summer: 62–69. Available from: http://www.ddw-online.com/p-303678. Accessed August 8, 2014.

Oprea, T.I. and J. Mestres. 2012. Drug repurposing: Far beyond new targets for old drugs. *The AAPS Journal* 14 (4): 759–763.

Pantziarka, P., G. Bouche, L. Meheus, V. Sukhatme, V.P. Sukhatme, and P. Vikas. 2014. The repurposing drugs in oncology (ReDO) project. *ecancer* 8 (442). doi:10.3332/ecancer.2014.442.

Paolini, G.V., R.H.B. Shapland, W.P. van Hoorn, J.S. Mason, and A.L. Hopkins. 2006. Global mapping of pharmacological space. *Nature Biotechnology* 24 (7): 805–815.

Paul, S.M., D.S. Mytelka, C.T. Dunwiddie, C.C. Persinger, B.H. Munos, S.R. Lindborg, and A.L. Schacht. 2010. How to improve R&D productivity: The pharmaceutical industry's grand challenge. *Nature Reviews Drug Discovery* 9 (3): 203–214.

Radley, D.C., S.N. Finkelstein, and R.S. Stafford. 2006. Off-label prescribing among office-based physicians. *Archives of Internal Medicine* 166 (9): 1021–1026.

Rastegar-Mojarad, M., Z. Ye, J.M. Kolesar, S.J. Hebbring, and S.M. Lin. 2015. Opportunities for drug repositioning from phenome-wide association studies. *Nature Biotechnology* 33 (4): 342–345.

Rothwell, P.M., F. Gerald, R. Fowkes, J.F.F. Belch, H. Ogawa, C.P. Warlow, and T.W. Meade. 2011. Effect of daily aspirin on long-term risk of death due to cancer: Analysis of individual patient data from randomised trials. *The Lancet* 377 (9759): 31–41.

Sardana, D., C. Zhu, M. Zhang, R.C. Gudivada, L. Yang, and A.G. Jegga. 2011. Drug repositioning for orphan diseases. *Briefings in Bioinformatics* 12 (4): 346–356.

Tartaglia, L.A. 2006. Complementary new approaches enable repositioning of failed drug candidates. *Expert Opinion on Investigational Drugs* 15 (11): 1295–1298.

Thayer, A. October 1, 2012. Drug repurposing. *Chemical & Engineering News* 90 (40): 15–25.

Tobinick, E.L. 2009. The value of drug repositioning in the current pharmaceutical market. *Drug News & Perspectives* 22 (2): 119–125.

Vane, J.R. 1971. Inhibition of prostaglandin synthesis as a mechanism of action for aspirin-like drugs. *Nature: New Biology* 231 (25): 232–235.

Washington University. 2010. Pfizer Announce Groundbreaking Research Collaboration, Newsroom, Washington University, St. Louis, MO.

Xu, K. and T.R. Coté. 2011. Database identifies FDA-approved drugs with potential to be repurposed for treatment of orphan diseases. *Briefings in Bioinformatics* 12 (4): 341–345.

Yank, V., C.V. Tuohy, A.C. Logan, D.M. Bravata, K. Staudenmayer, R. Eisenhut, V. Sundaram et al. 2011. Systematic review: Benefits and harms of in-hospital use of recombinant factor VIIa for off-label indications. *Annals of Internal Medicine* 154 (8): 529–540.

2 Repurposing for Alzheimer's and Parkinson's Diseases
The Ideas, the Pipeline, the Successes, and the Disappointments

Hermann Mucke

CONTENTS

2.1 INTRODUCTION

Only about 1 in 10,000 new chemical entities that enter active investigation as potential drugs for any medical condition ultimately make it to market, and even for drug candidates entering Phase II clinical trials the chance of success is less than 20%. Drug projects aiming for central nervous system (CNS) disorders traditionally have one of the longest development times and the highest pipeline attrition rates in the industry. According to the Tufts Center for the Study of Drug Development, mean clinical development time for CNS drugs approved for marketing in the United States from 1999 through 2013 was 12.8 months, or 18%, longer than for non-CNS compounds. In addition, the overall clinical approval success rate (i.e., the share of entities entering clinical testing that obtain marketing approval) for CNS compounds first tested in human subjects from 1995 to 2007 (and followed through 2013) was 6.2%, or less than half the 13.3% rate for non-CNS drugs. During 1999–2013, mean approval phase time for CNS compounds approved by the U.S. Food and Drug Administration (FDA) was 19.3 months, or 31% longer than the 14.7 months for non-CNS approvals (Tufts Center for the Study of Drug Development, 2014). The sector has also seen its share of postmarketing withdrawals of drugs, frequently because of cardiovascular side effects.

The reaction to this situation became apparent by 2010 when large pharmaceutical companies reduced their focus on neuropsychiatric drug development, closing several corporate neuroscience research facilities and discontinuing R&D programs (Kaitin and Milne, 2011). Another reactive measure by these companies was to step up existing efforts to accelerate drug development and to reduce the risk of failure. In the neuropsychiatry research community, where seeking new medical uses for known active pharmaceutical ingredients has featured prominently even before the term "drug repurposing" became popular, efforts in this segment were redoubled.

In terms of alternative uses for drugs and drug candidates, neurology and psychiatry have received from, and given to, each other much more than they have exchanged with medical fields not directly related to the human nervous system. An analysis of recently published international patent applications focused on drug repurposing has shown that neurology and psychiatry are not only "hotspots" of potentially patentable new-use findings, but also they frequently serve as sources of repurposing candidate compounds for each other (Mucke and Mucke, 2015). This is understandable given the intimate connection between these two fields. However, there are also plenty of examples showing that successful repurposing can happen with compounds that have never been used in neurology or psychiatry.

In this chapter, a wide variety of repurposing cases for Alzheimer's and Parkinson's disease that illustrate the diversity of ways a repurposing candidate can take toward ultimate success or failure will be discussed.

2.2 ALZHEIMER'S DISEASE

Primary degenerative dementia, or Alzheimer's disease as it is commonly called in the honor of Alois Alzheimer who first described it as a distinct psychiatric entity, still poses a huge challenge for science and medicine. While many of its downstream mechanisms have been uncovered and are now reasonably well understood, we still do not know the ultimate cause.

The first characteristic of Alzheimer's disease that provided relevant molecular targets for therapy was the deficit in cholinergic neurotransmission that results from the profound degeneration of cortically-projecting cholinergic neurons in the basal forebrain. If this loss of structure as well as function could be partially compensated by mechanisms that increase the capacity of the surviving elements of the central cholinergic system, it should be possible to restore some of its functions and hence improve impaired cognition, which depends on the neurotransmitter, acetylcholine.

While receptor-targeted approaches failed, the approach to increase intrasynaptic acetylcholine concentration by inhibiting the degrading enzymes—the cholinesterases—provided the first drugs to be approved for the treatment of Alzheimer's disease.

2.2.1 CHOLINERGIC ON-TARGET REPURPOSING: A SUCCESS STORY

2.2.1.1 Tacrine: Success and Side Effect Dilemma

The first drug to be specifically approved for Alzheimer's disease was a repurposed one: tacrine (1,2,3,4-tetrahydroacridin-9-amine) had been synthesized at Sydney University in the early 1940s in the course of a war-driven effort to develop antibiotics and antimalarials (Albert and Gledhill, 1945) but proved inactive in both respects. In 1949, tacrine was reported as an analeptic of unknown mechanism capable of causing rapid arousal in morphinized dogs and cats (Shaw and Bently, 1949), and the following years saw its use as a decurarizing agent (Romotal®) to restore muscle tone after anesthesia (Gerson and Shaw, 1958). In 1961, it was reported for the first time that tacrine inhibits acetyl- and butyrylcholinesterase (Heilbronn, 1961).

After the cholinergic hypothesis of geriatric memory dysfunction had been formally advanced in 1982, and oral controlled-release formulations of physostigmine (another cholinesterase inhibitor) had not been successful, the Warner Lambert Company (which later became part of Pfizer, Inc.) embarked on developing tacrine as the first specific drug for Alzheimer's disease. The development program had its rough spots, especially in 1988 and 1991 when the U.S. FDA's Peripheral and Central Nervous System Drugs Advisory Committee voted that the company's efficacy data did not yet warrant approval of the drug. A 6-week, double-blind study in patients who were selected for apparent responsiveness to tacrine revealed a reduction in the decline of cognitive function, which was not large enough to be detected by the study physicians' global assessments of the patients (Davis et al., 1992). A 30-week randomized, double-blind, placebo-controlled, parallel-group trial (Knapp et al., 1994) ended with only 263 patients (out of 663 who were enrolled) with evaluable data at endpoint, but it showed significant differences in favor of 160 mg/day of tacrine *versus* placebo on cognitive scales and quality-of-life assessments. The primary reasons for withdrawal of tacrine-treated patients were asymptomatic liver transaminase elevations (28%) and gastrointestinal complaints (16%).

Unfortunately, aminoacridines frequently possess poor side-effect profiles that may severely limit dosage and thus therapeutic benefits. As the mentioned data show, tacrine is no exception. The fact that only a small fraction of Alzheimer's patients could tolerate the 120 and 160 mg/day doses that provided the best cognitive benefit severely impaired the drug's overall therapeutic utility. A 1998 meta-analysis of 49 clinical trials published since 1981 showed that just over 20% of patients given tacrine at doses that they could tolerate experienced improvements in cognitive function (as evidenced by 3–4 points in Alzheimer's Disease Assessment Scale cognitive subscale and 2–3 points in Mini-Mental State Examination) and in functional ability at 3–6 months of treatment (Arrieta and Artalejo, 1998).

When tacrine was finally approved by the FDA in 1993 and marketed as Cognex®, the regulators imposed a strict dose titration scheme with strict liver transaminase parameter monitoring, which started with 4 weeks of treatment at 40 mg/day (in divided doses on a q.i.d. schedule) and could continue to 120 and 160 mg/day q.i.d. in 4-week intervals, provided that laboratory values allowed. For patients who develop ALT/SGPT elevations greater than two times the upper limit of normal, the dose and monitoring regimen should be modified according to yet another complicated schedule.

While tacrine had a pioneering role as the first drug specifically developed and approved for the treatment of Alzheimers' dementia, it clearly was a first-generation drug that was quickly superseded by donepezil (Aricept®; Eisai/Pfizer), a next-generation cholinesterase inhibitor that is much better tolerated, has significantly slower pharmacokinetics, and requires only a fraction of the dose. While tacrine has been discontinued in practically all larger markets, it provided the scaffold for many derivatives that continue to be reported to this day (Munawar et al., 2015).

2.2.1.2 Galantamine: A Tortuous Repurposing Success without Scientific Surprises

Like several other cholinesterase inhibitors, galantamine—first reported in 1947, chemically characterized by Soviet and Japanese researchers in the 1950s, and pharmacologically in 1960 (Irwin and Smith, 1960)—already had a track record of therapeutic use in myopathies, post-polio paralytic conditions, and in the reversal of neuromuscular blockade after anesthesia when the cholinergic hypothesis of geriatric memory dysfunction changed our understanding of Alzheimer's disease. Well before that—as early as 1977—galantamine had been shown to reverse the delirium-like condition that can be induced in cognitively normal humans by the muscarinic acetylcholine receptor blocker, scopolamine (Baraka and Harik, 1977).

This information could have suggested galantamine as the first-ever repurposing candidate for Alzheimer's disease. But in sharp contrast to tacrine, which is easy and cheap to synthesize on an industrial scale, galantamine is a complex compound that at that time was only available as an extremely expensive plant extract from Bulgarian, Turkish, and Soviet sources. Furthermore, the intellectual property situation was fragmented; several parties had to be brought to cooperate before the redevelopment of galantamine could commence in earnest.

These efforts ultimately succeeded. However, when galantamine was internationally launched by the Johnson & Johnson Group in 2000 (originally as Reminyl®, which was later changed to Razadyne® in the United States), it was already the third cholinergic drug for Alzheimer's disease on the market, and it never gained as much traction as its competitors—in part because it had only a few years of patent exclusivity left at this point. The redevelopment history of galantamine has recently been reviewed in great detail (Mucke, 2015), including the nonscientific and non-medical hurdles that a drug repurposing project can face even if clinical development for the new therapeutic indication goes smoothly.

With several industrially feasible pathways for its full synthesis available, the galantamine chemical scaffold also became a target for modifications aiming at better efficacy and tolerability in dementia treatment. The most progressed example is Memogain, a simple prodrug (Maelicke et al., 2010). More interesting in our context, the galantamine parent compound saw an impressive number of repurposing investigations that were undertaken to exploit its cholinergic action (which results from a combination of cholinesterase inhibition and allosteric modulation of nicotinic acetylcholine receptors) in other fields of neuropsychiatry. These include alcohol abuse, cigarette smoking (Diehl et al., 2006; Mann et al., 2006), autism (Ghaleiha et al., 2013; Niederhofer et al., 2002), and post-stroke aphasia (Hong et al., 2012). For a detailed discussion of these pilot studies (none of which has so far resulted in a full clinical development program), see Mucke (2015).

2.2.2 Nonsteroidal Anti-inflammatory Drugs: Not Entirely Out Yet?

The role of the cholinergic system in Alzheimer's disease is a central one, and not only because acetylcholine is the neurotransmitter that mediates cognition most directly. The central cholinergic tone also modulates many other aspects of brain physiology that are pathophysiologically relevant, for example, the deposition of β-amyloid plaques. In the absence of a comprehensive and generally accepted model of what actually causes primary degenerative dementia, choosing the cholinergic approach used to be the best rational starting point.

There are other factors that characterize Alzheimer's disease. One of these is chronic activation of astrocytes, which triggers a neuroinflammatory process. Inflammation is initially diffuse but becomes highly apparent around the amyloid deposits that form as the disease progresses. This reflects a brain immune reaction that might be beneficial initially, but subsequently it escapes physiologic control and contributes to the exacerbation of the neuronal damage.

Cyclooxygenase (COX) enzymes oxidize arachidonic acid to prostaglandins and thromboxanes, which are central mediators of the inflammatory response. In the CNS as well as in other organs, COX-1 is constitutively expressed while COX-2 is upregulated under pathophysiological conditions. Most nonsteroidal anti-inflammatory drugs (NSAIDs), broadly used as anti-inflammatory and mildly analgesic remedies, have reasonable brain penetrance. A logical conclusion—supported by cross-sectional epidemiological observations that elderly NSAID users are slightly less likely to develop Alzheimer's disease (Wang et al., 2015)—was to investigate NSAIDs, mostly in preventive settings.

Several large controlled clinical trials were initiated from the early 2000s onward, but none confirmed what seemed to be a reasonable assumption. One final "nail into the COX inhibitor coffin" was the Alzheimer's Disease Anti-inflammatory Prevention Trial Follow-up Study (ADAPT-FS), a 7-year cognitive follow-up of patients with a family history of dementia who had been treated with the nonselective COX inhibitor naproxen or the selective COX-2 inhibitor celecoxib for the primary prevention of Alzheimer's disease (Alzheimer's Disease Anti-inflammatory Prevention Trial Research Group, 2013). This confirmed the earlier results of the ADAPT core study, which had shown no cognitive benefit for either NSAID, and even had suggested a slight detrimental effect for naproxen (ADAPT Research Group, 2008).

The phenomenon that observational studies seem to support the long-term use of at least some NSAIDs for the prevention of Alzheimer's disease, while several huge randomized controlled trials have consistently refuted it, suggests that some crucial elements were not understood (or not considered) in these investigations. These might be related to our current concept of the disease process, to yet-unknown confounding factors, or to the designs of the studies. Indeed, an earlier analysis of the ADAPT data had suggested an NSAID treatment effect that differs at various stages of the disease—adverse effect in the later stages of pathogenesis, but protective for asymptomatic individuals if prophylactic NSAID intake is maintained for 2–3 years (Breitner et al., 2011).

The new concept of amnestic-type mild cognitive impairment (aMCI) that is believed to precede clinical Alzheimer's disease (while other types of MCI might even be reversible) allows to focus longitudinal studies, probably with adaptive designs, on groups at high genetic risk. Such clinical studies are ongoing.

Gastrointestinal side effects are frequent with long-term use of NSAIDs. They can be limited by concomitant administration of gastroprotectants, but in a purely preventive setting where there is no overt beneficial drug effect, this still limits patient compliance and might mask the drugs' full prophylactic potential. New brain-targeted drug presentations, such as liposomes or nanoparticles that facilitate blood–brain barrier penetration, might allow to reduce oral NSAID doses, and thereby gastric irritation. While the final verdict on repurposing some NSAIDs for preventing Alzheimer's disease in populations with a known genetic risk or in those diagnosed with aMCI is not yet out, all recent meta-analyses confirm that they are not an option for its treatment (Miguel-Álvarez et al., 2015).

Somewhat on the sidelines because it lacks COX-inhibitory effect is tarenflurbil, the R-stereoisomer of the racemic NSAID, flurbiprofen. This compound is a γ-secretase inhibitor and effectively reduces β-amyloid levels in the brains of transgenic mice. For several years, research and trials with the drug were conducted by Myriad Genetics, but the company announced discontinuation of development in June 2008 after a study had failed to demonstrate benefit in mild Alzheimer's disease (Green et al., 2009).

2.2.3 BEYOND THE NSAIDs

Recent research has revealed several additional networked pathways that are altered in Alzheimer's disease and might provide therapeutic targets that could be addressed for a more causative pharmacotherapy. To this end, drug repurposing can specifically target critical networks' nodes that are identified as relevant. However, it is also possible to use a parameter-free bioinformatics approach that combines differential gene expression signatures from healthy and diseased brain areas, their modulation upon drug exposure, and *in silico* repurposing algorithms.

One such integrative approach employed five disease-related microarray data sets of hippocampal origin, three different methods of evaluating differential gene expression, and four algorithmic drug repurposing tools. Inhibitors of protein kinase C, histone deacetylase, glycogen synthase kinase-3β, and arginase consistently appeared in the resultant drug list; they appear to act in a subpathway of Alzheimer's disease that is ultimately mediated by the epidermal growth factor receptor (Siavelis et al., 2015).

With a good recent review of the more remote possibilities available (Appleby et al., 2013), only the clinically most interesting therapeutic classes of repurposing candidates for Alzheimer's disease shall be briefly discussed.

2.2.3.1   Lithium

Lithium, the prototypical mood stabilizer, has been introduced for the prevention of manic episodes in bipolar disorder since the 1940s. In spite of its narrow therapeutic window and nephrotoxicity, it is still broadly used. Lithium is now known to exert this effect by inhibiting glycogen synthase kinase-3β, a "cellular nexus" that integrates several second messengers and a wide selection of pathway stimulants (Medina and Avila, 2014); this activity is believed to be responsible for the actions of lithium that modulate autophagy, oxidative stress, inflammation, and mitochondrial function and synergize into a neuroprotective effect (Forlenza et al., 2014). Only a few small studies evaluating its effect on cognition have been conducted, but these appear to show a positive effect (Matsunaga et al., 2015).

2.2.3.2   Antidiabetics

It has long been known that diabetes is a risk factor for dementia. Insulin is a signaling molecule that affects neuronal receptors in the central nervous system as well as peripheral ones; imbalances in brain insulin signaling can affect cognition. Therefore, it has been suggested that diabetes drugs, especially those that modulate the IRS-1/Akt pathway, could be useful in Alzheimer's disease by targeting insulin signaling in the brain, independent of their effects on glucose levels.

Peroxisome proliferator-activated receptor-gamma (PPAR-γ) agonists such as pioglitazone (Takeda Pharmaceuticals' Actos®) and rosiglitazone (GlaxoSmithKline's Avandia®) used to be among the best-selling drugs before cardiovascular problems, and also some oncological risks, have greatly compromised the hopes that these new antidiabetics used to carry. However, they might be repurposed for Alzheimer's disease: in rodent models, PPAR-γ activation degrades β-amyloid deposits (Mandrekar-Colucci et al., 2012), prevents oxidative damage, normalizes levels of brain-derived neurotrophic factor, and improves memory impairment (Prakash and Kumar, 2014).

After pilot studies had shown cognitive and functional response in early Alzheimer's disease with pioglitazone that was sustained for 4 years (Read et al., 2014), Takeda and Zinfandel Pharmaceuticals commenced a Phase III trial (TOMMORROW; ClinicalTrials.gov code: NCT01931566) to investigate a low-dose formulation of pioglitazone (designated AD-4833) in August 2013. TOMMORROW is designed to investigate the prevention of progression of mild cognitive impairment to Alzheimer's disease within 5 years; the study uses a biomarker-based risk assignment algorithm for predicting risk of progression.

Multiyear observation periods in a preventive setting might indeed be essential with this type of intervention. Years earlier, a 693-patient, 6-month Phase III study had found no evidence of efficacy of 2 or 8 mg rosiglitazone monotherapy in cognition or global function in patients with mild-to-moderate Alzheimer's disease (Gold et al., 2010). Two other Phase III studies that investigated the same rosiglitazone dosing scheme adjunctive to donepezil and to any approved cholinesterase inhibitor, respectively, also detected no clinically significant efficacy in cognition or global function (Harrington et al., 2011).

A recently published meta-analysis has concluded that the clinical evidence for cognitive benefits from PPAR-γ agonists in Alzheimer's disease remains insufficient, but that hopes remain especially for pioglitazone, provided that larger trials are conducted (Cheng et al., 2015).

2.2.3.3   Latrepirdine and Bexarotene: Two Controversial Cases

Many years ago, latrepirdine had been marketed as a nonselective antihistamine (Dimebon®) in Russia. It modulates several targets involved in Alzheimer's disease pathology, including lipid peroxidation, mitochondrial permeability, voltage-gated calcium ion channels, and neurotransmitter receptor activity (Bharadwaj et al., 2013).

Following up on a Russian Phase II/III 18-month study that had promising results in 2008, Pfizer and Medivation conducted an international clinical program for Alzheimer's disease, emphasizing the compound's potential to address mitochondrial imbalances as a new complex target in dementia. However, in January 2012, the Phase III CONCERT trial, which had evaluated latrepirdine as an add-on to donepezil in patients with mild-to-moderate Alzheimer's disease, failed to achieve significant results for either of the two coprimary endpoints. A Cochrane Group meta-analysis of six clinical trials confirmed the lack of effect on cognition and function (Chau et al., 2015). The totally disparate results precipitated disputes between the Russian investigators and the other participants in the program.

One clinical study (DIMOND) indicated some cognitive benefit of latrepirdine in Huntington' disease (Kieburtz et al., 2010) while another one (HORIZON) did not (HORIZON, 2013). However, it might improve some aspects of clinical and cognitive status in schizophrenic patients (Morozova et al., 2012).

The retinoid bexarotene (Targretin®; Eisai and Valeant Pharmaceuticals), a retinoid X receptor (RXR) agonist, is indicated for skin manifestations associated with cutaneous T-cell lymphoma and has been investigated (and is sometimes used off-label) for other types of cancer. Because the RXR transcriptionally regulates the expression of apolipoprotein E, in conjunction with PPAR-γ and the liver X receptor, it has been suggested that the drug could assist β-amyloid clearance. An amyloid-imaging-driven study (BEAT-AD; NCT01782742) had a negative result on primary outcome. The preclinical data from rodent models have caused some discord, with some authors claiming an emerging role for bexarotene (Tousi, 2015) while others have disputed the significance and the interpretations of these findings (O'Hare et al., 2016).

2.2.3.4   VX-745—A Discontinued MAPK Inhibitor

EIP Pharma, LLC is redeveloping this orally bioavailable, highly selective, and potent inhibitor of the alpha isoform of p38 mitogen-activated protein kinase

(Duffy et al., 2011). VX-745 had previously completed a full chronic toxicology program and had demonstrated significant clinical and anti-inflammatory activity in a 12-week Phase IIa treatment study in rheumatoid arthritis patients. The original developer, Vertex Pharmaceuticals, had discontinued development of VX-745 in 2001 because high doses had produced adverse neurological side effects in one of two animal species.

In June 2015, EIP Pharma commenced two Phase IIa clinical studies—the first (NCT02423200) for clinical pharmacology, the second one (NCT02423122) focusing on PET imaging of amyloid plaque load—in patients with mild Alzheimer's disease or mild cognitive impairment presumably due to incipient Alzheimer's disease. By October 2015, these studies were presumed to be recruiting patients.

2.3 PARKINSON'S DISEASE

Levodopa, now given in fixed combinations with decarboxylase inhibitors to extend its bioavailability, has provided symptomatic relief from the motor symptoms of Parkinson's disease since the 1960s. Because levodopa actually accelerates the underlying neurodegenerative process in the substantia nigra (the part of the basal ganglia where the degeneration of dopaminergic neurons relevant to the motor symptoms occurs), selective dopamine receptor agonists are nowadays preferred in the early stages of Parkinson's disease. A disease-modifying treatment is still lacking.

While judgment is still out whether the nature of the neurodegenerative process in Parkinson's disease fundamentally differs from that seen in Alzheimer's disease, the affected brain areas are different ones and have very different functions. Anti-excitatory approaches might be more important in Parkinson's disease, and therefore many anti-convulsant drugs (such as topiramate) and epilepsy drug candidates have undergone clinical studies. As with Alzheimer's disease, longer-term use of NSAIDs appears to exert some protection against the development of Parkinson's disease. More interesting are the off-target drug repurposing cases and ideas, which are discussed here.

2.3.1 AMANTADINE: FROM INFLUENZA TO NEUROLOGY

A very early example of drug repurposing is amantadine (1-aminoadamantane) that entered the scientific literature in 1964 and was initially marketed as an antiviral compound (Symmetrel®, Viregyt®) primarily as a prophylactic for influenza, but also for Rubella virus infections (Davies et al., 1964). It was soon found that amantadine is not virucidal at physiological concentrations but had to somehow interfere with viral binding, entry, or replication (Hoffmann et al., 1965). The World Health Organization discouraged the use of amantadine for influenza prevention in 1969 (Couch, 1969), but in the same year reports of its utility in Parkinson's disease began to appear (Schwab et al., 1969). It was a classic case of repurposing by serendipity: a patient with Parkinson's disease had noticed relief in her motor symptoms after taking amantadine for influenza. Two years later, a controlled study confirmed these findings (Barbeau et al., 1971).

It took another 20 years to unravel the mechanism of amantadine in this new indication, in which it was soon used as an adjunct to the standard drug, levodopa.

Evidence for dopaminergic actions on various levels had been found, but these were far too weak to contribute substantially to amantadine's motor actions in Parkinson's disease. Only in 1992 it was reported that amantadine is a noncompetitive NMDA receptor antagonist that inhibits striatal acetylcholine release and acts as an anti-excitatory neuroprotectant (Stoof et al., 1992). A year later, the molecular basis of its anti-influenza action was also revealed: amantadine blocks the proton channel function of the viral M2 integral membrane protein (Wang et al., 1993). The only common factor to these two activities is that the cage-like adamantan moiety enters and sterically blocks the membrane protein channels and does so only in their open configuration. This remarkable case of adventitious drug repurposing, which could not have been expected from available data, has recently been reviewed (Hubsher et al., 2012).

Today amantadine still holds drug approvals for both indications, but it no longer finds much use in either. However, attempts at further repurposing are ongoing. In June 2015, U.S.-based Adamas Pharmaceuticals commenced a Phase II clinical trial with an amantadine-based extended-release formulation (ADS-5102) to treat gait impairment in multiple sclerosis patients. Around 60 individuals will be included in this multicenter, randomized, double-blind, placebo-controlled, two-arm, parallel-group Phase II trial. Dosing is 340 mg once daily at bedtime, well within the dose range used in both influenza and Parkinson's indications. ADS-5102 has already completed a Phase II/III study for levodopa-induced dyskinesia in Parkinson's patients; the data suggest additional utility in Huntington's disease.

It should be mentioned at this point that memantine, the only noncholinesterase inhibitor that is broadly approved for the symptomatic treatment of Alzheimer's disease (marketed as Namenda®, Axura®, and Ebixa®), is a simple bis-methyl derivative of amantadine, with a longer residence time in the NMDA receptor channel.

2.3.2 PIOGLITAZONE—ALSO FOR PARKINSON'S DISEASE?

The neuroprotective mechanisms that suggest utility of PPAR-γ agonists in Alzheimer's disease are sufficiently general to warrant investigation of these "glitazones" as potentially disease-modifying drugs for Parkinson's disease. The fact that PPAR-γ, a key regulator of enzymes involved in mitochondrial respiration, is highly expressed in neurons of the substantia nigra further underscores this. However, clinical safety data are not yet fully conclusive and until a very short while ago studies had been limited to diabetic Parkinson's patients (Carta and Simuni, 2015).

In a recently published Phase II multicenter trial (NINDS Exploratory Trials in Parkinson Disease (NET-PD) FS-ZONE Investigators, 2015) funded by the U.S. National Institute of Neurological Disorders and Stroke, 210 early-stage Parkinson's disease patients who were on a stable dopaminergic agonist drug regimen were randomized to additionally receive pioglitazone (15 or 45 mg/day) or placebo for 44 weeks. The results, based on the Unified Parkinson's Disease Rating Scale (UPDRS) score, clearly refuted the hypothesis that pioglitazone, at least at these doses (which correspond to those used for diabetes), could modify progression in early Parkinson's disease.

Motor symptoms of Parkinson's disease become apparent only when about half of the substantia nigra neurons are already nonfunctional. This suggests that beginning treatment with PPAR-γ agonists at this point probably is already too late for clinically significant effects; it might be necessary to commence pharmacotherapy in the prodromal phase. However, the known safety problems of PPAR-γ agonists would make studies anywhere but in the very small population at high genetic risk for Parkinson's disease difficult. Higher oral doses might work better in manifest disease, but the safety risks from systemic exposure to thiazolinediones would be even higher. Unless ways are found to deliver glitazones selectively to the brain, their utility as neuroprotectants remains a theoretical one.

2.3.3 Exenatide—Another Antidiabetic

Animal experiments had suggested that stimulating the glucagon-like peptide-1 receptor with exenatide could be neuroprotective in Parkinson's disease (and perhaps also in ischemic stroke) (Li et al., 2009). Exenatide (exendin-4; an endogenous insulinotropic peptide) is approved for the treatment of type 2 diabetes and is marketed by AstraZeneca and Bristol-Myers-Squibb as Byetta® and Bydureon®.

A single-blind proof of concept study assigned 45 patients with moderate Parkinson's disease to receive subcutaneous exenatide injections (10 μg b.i.d.) for 12 months or to act as controls. Exenatide patients showed a mean improvement of 2.7 points on the UPDRS score, while controls had a mean decline of 2.2 points—a statistically and clinically significant benefit (Aviles-Olmos et al., 2013). Motor and cognitive advantages persisted 12 months after exenatide exposure (Aviles-Olmos et al., 2014). The open-label design of this study (necessitated because no placebo-filled injector pens outwardly similar to Byetta pens were available) significantly limited its power; strong placebo effects are quite common in Parkinson's disease patients, even in a double-blind setting. A larger, well-controlled clinical study would be required to corroborate the disease-modifying effect of exenatide (Simuni and Brundin, 2014).

2.3.4 Minocycline—An Antibiotic That Disappointed

The semisynthetic tetracycline minocycline, has many effects that are totally unrelated to its antibiotic effects. It ameliorates neuroinflammation and neurodegeneration in many animal models of human conditions, including some for Parkinson's disease. Minocycline counteracts the changes in the nigrostriatal mitochondrial proteome induced by maneb and paraquat (Dixit et al., 2013) (agricultural chemicals that target dopaminergic neurons and cause Parkinsonian syndromes in rodents and man). This effect is deemed particularly important given the recent insights into mitochondrial dysfunction in Parkinson's disease.

The Neuroprotective Exploratory Trials in Parkinson's Disease, Futility Study 1 (NET-PD FS-1), a small double-blind 12-month Phase II clinical pilot trial in previously untreated early-stage Parkinson's patients with a 6-month follow-up that was completed in 2005, did not demonstrate safety concerns for minocycline (200 mg/day) that would preclude a large Phase III trial (NINDS NET-PD Investigators, 2008). However, the discontinuation rate was much higher with minocycline (23%) than with creatine (10 g/day; evaluated in parallel because of its beneficial effect on mitochondrial function) or placebo (9% and 6%, resp.), and there is ample evidence of systemic minocycline toxicity in the scientific literature.

Shortly afterward it was shown that attenuation of chronic microglial activation, another proven effect of minocycline that suggests its potential utility in Parkinson's disease, is insufficient to modulate striatal dopaminergic neurotoxicity in animal models (Sriram et al., 2006); transient activation of microglia may suffice to initiate neurodegeneration in these experimental paradigms. This is consistent with the results from the prospective multinational MEMSA trial, which randomized 63 patients with multiple-system atrophy of the Parkinson's-type to minocycline (200 mg/day) or placebo for 48 weeks in a double-blind setting. It revealed some benefit of minocycline in terms of microglial activation but failed to improve symptom severity as assessed by clinical motor function (Dodel et al., 2010). Today minocycline is no longer considered a repurposing candidate for Parkinson's disease, although tailored tetracycline derivatives are still being investigated as new chemical entities.

2.3.5 NILOTINIB—A MODERN LEUKEMIA DRUG

The proto-oncogenic tyrosine kinase, c-Abl, is chronically activated in the brain of Parkinson's disease patients. In this state, it causes neuronal cell death through interference with components of the ubiquitin-proteasome system, which leads to accumulation of the parkin and α-synuclein proteins. Nilotinib, a c-Abl inhibitor that is approved for the treatment of chronic myelogenous leukemia (Novartis' Tasigna®), shows reasonable brain penetration and prevents dopaminergic neuron loss and behavioral deficits in the animal models of Parkinson's disease (Karuppagounder et al., 2014), causing enhanced autophagic clearance of relevant proteins through lysosomes (Hebron et al., 2013). It would seem to be a candidate for drug repurposing, especially if the treatment of Parkinson's disease requires lower doses than those needed for leukemia.

In 2015, the Parkinson's Disease Center of Excellence at Georgetown University completed a 6-month open-label study (ClinicalTrials.gov: NCT02281474) to test the ability of nilotinib to alter the abnormal protein buildup in patients with Parkinson's disease or diffuse Lewy body disease. The investigated doses were 150 and 300 mg/day; up to 800 mg/day is used for the treatment of myeloid leukemia. Results have not been reported at the time of writing.

2.3.6 CANDESARTAN—A GENERIC ANTIHYPERTENSIVE

The angiotensin-II receptor (AT1) antagonists of the "sartan" class, widely marketed for hypertension and now generically available for the most part, can inhibit the microglial inflammatory response and dopaminergic cell loss in animal models of Parkinson's disease, independent of their vascular actions. For telmisartan, this effect has been shown to be mediated by PPAR-γ activation (Garrido-Gil et al., 2012). That might apply to a lesser degree to candesartan, marketed as a cilexetil ester prodrug, Atacand® (by AstraZeneca and Takeda) and also available in generic versions. Candesartan also has good brain penetration (although its peripheral bioavailability is low). Its anti-Parkinsonian effects in animal models of Parkinson's disease might be more attributable to antioxidant and anticytokine actions (Mertens et al., 2011; Muñoz et al., 2014; Wu et al., 2013). Some of these data show that candesartan also reduces levodopa-induced dyskinesia.

At the time of writing (October 2015), there were no registered clinical trials ongoing to investigate candesartan cilexetil, specifically for Parkinson's disease. Uncontrolled hypertension is associated with mild cognitive and motor impairments even in the absence of clinical neurodegeneration; obviously, candesartan (as well as other antihypertensive drugs) could have benefits in such patients. This is currently examined by the CAndesartan vs LIsinopril Effects on the BRain (CALIBREX) Phase III study, sponsored by the U.S. National Institutes of Aging.

2.3.7 COMPUTATIONAL APPROACHES

The drug screening company In Silico Biosciences, Inc., used a humanized quantitative systems pharmacology platform containing 30 CNS targets, to screen pharmacological profiles of serotonergic drugs in the Prestwick compound library. The approach is designed to simulate neuronal network interactions between supplemental motor cortex and striatum based on preclinical neurophysiology and human electrophysiology data. The identified hits included the old antidepressant, trazodone, which in a previously reported study improved clinical scores of Parkinson's disease symptoms when given as part of a drug augmentation strategy (Spiros et al., 2013).

Another computational approach exploits the topology of drugs in a tripartite indication-drug-target network, as well as the significance of their targets in the Parkinson-specific protein–protein interaction network. The researchers, based at the University of L'Aquila in Italy and the Bose Institute in India, identified nine non-Parkinson's drugs as significant candidates: melatonin; the analgesic sodium channel blocker lidocaine; the calcium channel blockers nicardipine and nifedipine (all of these drugs are known neuroprotectants); the hormones testosterone (also neuroprotective) and diethylstilbestrol; and the tyrosine kinase inhibitors erlotinib, dasatinib, and sorafenib, all of which are approved and marketed for cancer by major pharmaceutical companies (Rakshit et al., 2015).

2.4 OUTLOOK

At first sight, what has been presented here might suggest that drug repurposing for Alzheimer's and Parkinson's disease has been successful only in the past, and mostly where the "therapeutic leap" had not been too wide, while the newer attempts at repurposing had either failed in the clinic or are at the preclinical stage. In other words, it could seem as if drug repurposing strategies were no more effective than classical drug development working with new chemical entities.

That would be a fallacy of temporal perspective: we are looking at an early snapshot. Full development of a drug takes many years, and regulatory authorities' strict guidelines all but guarantee that the process for repurposed agents cannot be assumed to be quicker (or cheaper, on a single-project basis) than for a new chemical entity. Systematic and/or rational drug repurposing (which are the modalities that we view as the most relevant ones today, as opposed to the serendipity-driven early successes exemplified by amantadine) simply has not been around for a sufficiently long period to allow a valid efficacy judgment.

Furthermore, as has been mentioned in the introduction to this discussion, the great majority of new chemical entities fail at some stage of their development process, and even proof-of-principle in clinical Phase II is far from guaranteeing a place on the pharmacy shelves. While drug repurposing does not automatically accelerate drug development, it should certainly de-risk it—and this is true for on-target repurposing as exemplified by the cholinesterase inhibitors that were repositioned for Alzheimer's disease based on their known mechanism of action.

In contrast, we must never assume that a drug or candidate compound that has revealed promising off-target effects, and has been developed for an entirely different therapeutic indication, would be any less prone to clinical safety or efficacy failures. How it will behave in the new patient population—how its well-established effects (which are now likely be side effects) will combine with the newly discovered ones—is much more difficult to predict than with on-target repurposing. This caveat will always hold for compounds that represent the most unexpected (and hence, most innovative) embodiments of drug repurposing—unless of course the required dose is a fraction of what has been tried earlier, along with highly convincing results in highly representative animal models of the human condition. This is known to happen, but only rarely.

A consortium of authors that included scientists from The Alzheimer's Drug Discovery Foundation, The Michael J. Fox Foundation for Parkinson's Research, Faster Cures, AstraZeneca, and various consultants, and from various universities have recently outlined the difficulties that drug repurposing for chronic neurodegenerative diseases faces (Shineman et al., 2014). The main obstacles which they describe include the following: limited option for patent protection because only new use or formulation patents can be secured; only 3 years of new use exclusivity granted by the U.S. FDA; challenges with drug pricing, reimbursement, and off-label use if the compound is available as a generic drug. These are common to all drug repurposing projects. This is also true for the authors' actionable proposals, such as forming consortia and collaborations between nonprofit organizations, industry, and government entities, and to implement innovative policies to incentivize industry investment in repurposing drugs. However, they apply smoothly to projects addressing Alzheimer's and Parkinson's disease.

Chronically progressive neurodegenerative diseases could very well be the "door opener" for creating a regulatory, commercial, and intellectual property environment that takes the specific challenges of drug repurposing into account and allows developers to fully exploit its advantages. If it was possible to develop dimethyl fumarate—a chemical that was originally used as a preservative for leather furniture—into an approved drug for relapse prevention in multiple sclerosis (Biogen Idec's Tecfidera®), realistic repurposing options should exist not only for Alzheimer's and Parkinson's disease, but also for the much rarer Huntington's disease and amyotrophic lateral sclerosis—all of which lack the type of truly effective disease-modifying therapies that had been firmly established for multiple sclerosis when dimethyl fumarate was launched.

REFERENCES

ADAPT Research Group. Cognitive function over time in the Alzheimer's Disease Anti-inflammatory Prevention Trial (ADAPT): Results of a randomized, controlled trial of naproxen and celecoxib. *Arch Neurol.* 2008; 65(7):896–905.

Albert A, Gledhill W. Improved synthesis of aminoacridines V. Substituted 5-aminoacridines. *J Soc Chem Ind.* 1945; 64:169–172.

Alzheimer's Disease Anti-inflammatory Prevention Trial Research Group. Results of a follow-up study to the randomized Alzheimer's Disease Anti-inflammatory Prevention Trial (ADAPT). *Alzheimers Dement.* 2013; 9(6):714–723.

Appleby BS, Nacopoulos D, Milano N et al. A review: Treatment of Alzheimer's disease discovered in repurposed agents. *Dement Geriatr Cogn Disord.* 2013; 35:1–22.

Arrieta JL, Artalejo FR. Methodology, results and quality of clinical trials of tacrine in the treatment of Alzheimer's disease: A systematic review of the literature. *Age Ageing.* 1998; 27(2):161–179.

Aviles-Olmos I, Dickson J, Kefalopoulou Z et al. Exenatide and the treatment of patients with Parkinson's disease. *J Clin Invest.* 2013; 123(6):2730–2736.

Aviles-Olmos I, Dickson J, Kefalopoulou Z et al. Motor and cognitive advantages persist 12 months after exenatide exposure in Parkinson's disease. *J Parkinsons Dis.* 2014; 4(3):337–344.

Baraka A, Harik S. Reversal of central anticholinergic syndrome by galanthamine. *J Am Med Assoc.* 1977; 238(21):2293–2294.

Barbeau A, Mars H, Botez MI, Joubert M. Amantadine-HCl (Symmetrel) in the management of Parkinson's disease: A double-blind cross-over study. *Can Med Assoc J.* 1971; 105(1):42–46 passim.

Bharadwaj PR, Bates KA, Porter T et al. Latrepirdine: Molecular mechanisms underlying potential therapeutic roles in Alzheimer's and other neurodegenerative diseases. *Transl Psychiatry.* 2013; 3:e332.

Breitner JC, Baker LD, Montine TJ et al. Extended results of the Alzheimer's disease anti-inflammatory prevention trial. *Alzheimers Dement.* 2011; 7(4):402–411.

Carta AR, Simuni T. Thiazolidinediones under preclinical and early clinical development for the treatment of Parkinson's disease. *Expert Opin Invest Drugs.* 2015; 24(2):219–227.

Chau S, Herrmann N, Ruthirakuhan MT et al. Latrepirdine for Alzheimer's disease. *Cochrane Database Syst Rev.* 2015; 4:CD009524.

Cheng H, Shang Y, Jiang L et al. The peroxisome proliferators activated receptor-gamma agonists as therapeutics for the treatment of Alzheimer's disease and mild-to-moderate Alzheimer's disease: A meta-analysis. *Int J Neurosci.* 2016; 126(4):299–307.

Couch RB. Use of amantadine in the therapy and prophylaxis of A2 influenza. *Bull World Health Organ.* 1969; 41(3):695–696.

Davies WL, Grunert RR, Haff RF et al. Antiviral activity of 1-adamantanamine (amantadine). *Science.* 1964; 144(3620):862–863.

Davis KL, Thal LJ, Gamzu ER et al. A double-blind, placebo-controlled multicenter study of tacrine for Alzheimer's disease. The Tacrine Collaborative Study Group. *N Engl J Med.* 1992; 327(18):1253–1259.

Diehl A, Nakovics H, Croissant B et al. Galantamine reduces smoking in alcohol-dependent patients: A randomized, placebo-controlled trial. *Int J Clin Pharmacol Ther.* 2006; 44(12):614–622.

Dixit A, Srivastava G, Verma D et al. Minocycline, levodopa and MnTMPyP induced changes in the mitochondrial proteome profile of MPTP and maneb and paraquat mice models of Parkinson's disease. *Biochim Biophys Acta.* 2013; 1832(8):1227–1240.

Dodel R, Spottke A, Gerhard A et al. Minocycline 1-year therapy in multiple-system-atrophy: Effect on clinical symptoms and [(11)C] (R)-PK11195 PET (MEMSA-trial). *Mov Disord.* 2010; 25(1):97–107.

Duffy JP, Harrington EM, Salituro FG et al. The discovery of VX-745: A novel and selective p38? Kinase inhibitor. *ACS Med Chem Lett.* 2011; 2(10):758–763.

Forlenza OV, De-Paula VJ, Diniz BS. Neuroprotective effects of lithium: Implications for the treatment of Alzheimer's disease and related neurodegenerative disorders. *ACS Chem Neurosci.* 2014; 5(6):443–450.

Garrido-Gil P, Joglar B, Rodriguez-Perez AI et al. Involvement of PPAR-γ in the neuroprotective and anti-inflammatory effects of angiotensin type 1 receptor inhibition: Effects of the receptor antagonist telmisartan and receptor deletion in a mouse MPTP model of Parkinson's disease. *J Neuroinflammation.* 2012; 9:38.

Gerson S, Shaw FH. Tetrahydroaminacrin as a decurarizing agent. *J Pharm Pharmacol.* 1958; 10:638–641.

Ghaleiha A, Ghyasvand M, Mohammadi MR et al. Galantamine efficacy and tolerability as an augmentative therapy in autistic children: A randomized, double-blind, placebo-controlled trial. *J Psychopharmacol.* 2013; 28(7):677–685.

Gold M, Alderton C, Zvartau-Hind M et al. Rosiglitazone monotherapy in mild-to-moderate Alzheimer's disease: Results from a randomized, double-blind, placebo-controlled phase III study. *Dement Geriatr Cogn Disord.* 2010; 30(2):131–146.

Green RC, Schneider LS, Amato DA et al. Effect of tarenflurbil on cognitive decline and activities of daily living in patients with mild Alzheimer disease: A randomized controlled trial. *J Am Med Assoc.* 2009; 302(23):2557–2564.

Harrington C, Sawchak S, Chiang C et al. Rosiglitazone does not improve cognition or global function when used as adjunctive therapy to AChE inhibitors in mild-to-moderate Alzheimer's disease: Two phase 3 studies. *Curr Alzheimer Res.* 2011; 8(5):592–606.

Hebron ML, Lonskaya I, Moussa CE. Tyrosine kinase inhibition facilitates autophagic SNCA/α-synuclein clearance. *Autophagy.* 2013; 9(8):1249–1250.

Heilbronn E. Inhibition of cholinesterase by tetrahydroaminoacridine. *Acta Chem Scand.* 1961; 15:1386–1390.

Hoffmann CE, Neumayer EM, Haff RF, Goldsby RA. Mode of action of the antiviral activity of amantadine in tissue culture. *J Bacteriol.* 1965; 90(3):623–628.

Hong JM, Shin DH, Lim TS, Lee JS, Huh K. Galantamine administration in chronic post-stroke aphasia. *J Neurol Neurosurg Psychiatry.* 2012; 83(7):675–680.

HORIZON Investigators of the Huntington Study Group and European Huntington's Disease Network. A randomized, double-blind, placebo-controlled study of latrepirdine in patients with mild to moderate Huntington disease. *JAMA Neurol.* 2013; 70(1):25–33.

Hubsher G, Haider M, Okun MS. Amantadine: The journey from fighting flu to treating Parkinson disease. *Neurology.* 2012; 78(14):1096–1099.

Irwin RL, Smith HJ 3rd. Cholinesterase inhibition by galanthamine and lycoramine. *Biochem Pharmacol.* 1960; 3:147–148.

Kaitin KI, Milne CP. A dearth of new meds: Drugs to treat neuropsychiatric disorders have become too risky for big pharma. *Sci Am.* 2011; 305(2):16.

Karuppagounder SS, Brahmachari S, Lee Y et al. The c-Abl inhibitor, nilotinib, protects dopaminergic neurons in a preclinical animal model of Parkinson's disease. *Sci Rep.* 2014; 4:4874.

Kieburtz K, McDermott MP, Voss TS et al.; Huntington Disease Study Group DIMOND Investigators. A randomized, placebo-controlled trial of latrepirdine in Huntington disease. *Arch Neurol.* 2010; 67(2):154–160.

Knapp MJ, Knopman DS, Solomon PR et al. A 30-week randomized controlled trial of high-dose tacrine in patients with Alzheimer's disease. The Tacrine Study Group. *J Am Med Assoc.* 1994; 271(13):985–991.

Li Y, Perry T, Kindy MS et al. GLP-1 receptor stimulation preserves primary cortical and dopaminergic neurons in cellular and rodent models of stroke and Parkinsonism. *Proc Natl Acad Sci USA*. 2009; 106(4):1285–1290.

Maelicke A, Hoeffle-Maas A, Ludwig J, Maus A, Samochocki M, Jordis U, Koepke AK. Memogain is a galantamine pro-drug having dramatically reduced adverse effects and enhanced efficacy. *J Mol Neurosci*. 2010; 40(1–2):135–137.

Mandrekar-Colucci S, Karlo JC, Landreth GE. Mechanisms underlying the rapid peroxisome proliferator-activated receptor-γ-mediated amyloid clearance and reversal of cognitive deficits in a murine model of Alzheimer's disease. *J Neurosci*. 2012; 32(30):10117–10128.

Mann K, Ackermann K, Diehl A et al. Galantamine: A cholinergic patch in the treatment of alcoholism: A randomized, placebo-controlled trial. *Psychopharmacology (Berl)*. 2006; 184(1):115–121.

Matsunaga S, Kishi T, Annas P et al. Lithium as a treatment for Alzheimer's disease: A systematic review and meta-analysis. *J Alzheimers Dis*. 2015; 48(2):403–410.

Medina M, Avila J. New insights into the role of glycogen synthase kinase-3 in Alzheimer's disease. *Expert Opin Ther Targets*. 2014; 18(1):69–77.

Mertens B, Varcin M, Michotte Y, Sarre S. The neuroprotective action of candesartan is related to interference with the early stages of 6-hydroxydopamine-induced dopaminergic cell death. *Eur J Neurosci*. 2011; 34(7):1141–1148.

Miguel-Álvarez M, Santos-Lozano A, Sanchis-Gomar F. Non-steroidal anti-inflammatory drugs as a treatment for Alzheimer's disease: A systematic review and meta-analysis of treatment effect. *Drugs Aging*. 2015; 32(2):139–147.

Morozova MA, Beniashvili AG, Lepilkina TA, Rupchev GE. Double-blind placebo-controlled randomized efficacy and safety trial of add-on treatment of dimebon plus risperidone in schizophrenic patients during transition from acute psychotic episode to remission. *Psychiatr Danub*. 2012; 24(2):159–166.

Mucke HA, Mucke E. Sources and targets for drug repurposing: Landscaping transitions in therapeutic space. *Assay Drug Dev Technol*. 2015; 13(6):319–324.

Mucke HAM. The case of galantamine: Repurposing and late blooming of a cholinergic drug. *Future Sci OA*. 2015; 1(4):FSO73.

Munawar R, Mushtaq N, Arif S, Ahmed A, Akhtar S, Ansari S, Meer S, Saify ZS, Arif M. Synthesis of 9-aminoacridine derivatives as anti-Alzheimer agents. *Am J Alzheimers Dis Other Dement*. 2016; 31(3):263–269.

Muñoz A, Garrido-Gil P, Dominguez-Meijide A, Labandeira-Garcia JL. Angiotensin type 1 receptor blockage reduces l-dopa-induced dyskinesia in the 6-OHDA model of Parkinson's disease. Involvement of vascular endothelial growth factor and interleukin-1β. *Exp Neurol*. 2014; 261:720–732.

Niederhofer H, Staffen W, Mair A. Galantamine may be effective in treating autistic disorder. *Br Med J*. 2002; 325(7377):1422.

NINDS Exploratory Trials in Parkinson Disease (NET-PD) FS-ZONE Investigators. Pioglitazone in early Parkinson's disease: A phase 2, multicentre, double-blind, randomised trial. *Lancet Neurol*. 2015; 14(8):795–803.

NINDS NET-PD Investigators. A pilot clinical trial of creatine and minocycline in early Parkinson disease: 18-month results. *Clin Neuropharmacol*. 2008; 31(3):141–150.

O'Hare E, Jeggo R, Kim EM et al. Lack of support for bexarotene as a treatment for Alzheimer's disease. *Neuropharmacology*. 2016; 100:124–130.

Prakash A, Kumar A. Role of nuclear receptor on regulation of BDNF and neuroinflammation in hippocampus of β-amyloid animal model of Alzheimer's disease. *Neurotox Res*. 2014; 25(4):335–347.

Rakshit H, Chatterjee P, Roy D. A bidirectional drug repositioning approach for Parkinson's disease through network-based inference. *Biochem Biophys Res Commun*. 2015; 457(3):280–287.

Read S, Wu P, Biscow M. Sustained 4-year cognitive and functional response in early Alzheimer's disease with pioglitazone. *J Am Geriatr Soc*. 2014; 62(3):584–586.

Schwab RS, England AC Jr., Poskanzer DC, Young RR. Amantadine in the treatment of Parkinson's disease. *J Am Med Assoc*. 1969; 208(7):1168–1170.

Shaw FH, Bently G. Some aspects of the pharmacology of morphine with special reference to its antagonism by 5-amino-acridine and other chemically related compounds. *Med J Aust*. 1949; 2(25):868–875.

Shineman DW, Alam J, Anderson M et al. Overcoming obstacles to repurposing for neurodegenerative disease. *Ann Clin Transl Neurol*. 2014; 1(7):512–518.

Siavelis JC, Bourdakou MM, Athanasiadis EI et al. Bioinformatics methods in drug repurposing for Alzheimer's disease. *Brief Bioinform*. 2016; 17(2):322–335.

Simuni T, Brundin P. Is exenatide the next big thing in Parkinson's disease? *J Parkinsons Dis*. 2014; 4(3):345–357.

Spiros A, Roberts P, Geerts H. Phenotypic screening of the Prestwick library for treatment of Parkinson's tremor symptoms using a humanized quantitative systems pharmacology platform. *J Parkinsons Dis*. 2013;3(4):569–580.

Sriram K, Miller DB, O'Callaghan JP. Minocycline attenuates microglial activation but fails to mitigate striatal dopaminergic neurotoxicity: Role of tumor necrosis factor-alpha. *J Neurochem*. 2006; 96(3):706–718.

Stoof JC, Booij J, Drukarch B. Amantadine as *N*-methyl-D-aspartic acid receptor antagonist: New possibilities for therapeutic applications? *Clin Neurol Neurosurg*. 1992; 94(Suppl):S4–S6.

Tousi B. The emerging role of bexarotene in the treatment of Alzheimer's disease: Current evidence. *Neuropsychiatr Dis Treat*. 2015; 11:311–315.

Tufts Center for the Study of Drug Development. Pace of CNS drug development and FDA approvals lags other drug classes. Impact Report, Vol. 16(6), November/December 2014. Abstract available at http://csdd.tufts.edu/news/complete_story/pr_ir_nov_dec_ir. Accessed March 14, 2017.

Wang C, Takeuchi K, Pinto LH, Lamb RA. Ion channel activity of influenza A virus M2 protein: Characterization of the amantadine block. *J Virol*. 1993; 67(9):5585–5594.

Wang J, Tan L, Wang HF et al. Anti-inflammatory drugs and risk of Alzheimer's disease: An updated systematic review and meta-analysis. *J Alzheimers Dis*. 2015; 44(2):385–396.

Wu L, Tian YY, Shi JP et al. Inhibition of endoplasmic reticulum stress is involved in the neuroprotective effects of candesartan cilexitil in the rotenone rat model of Parkinson's disease. *Neurosci Lett*. 2013; 548:50–55.

3 Contribution of Not-for-Profit Organizations to Drug Repurposing

Bruce Bloom

CONTENTS

3.1 WHY NOT-FOR-PROFIT ORGANIZATIONS (NPOs) GET INVOLVED IN DRUG REPURPOSING

3.1.1 INDUSTRY AND GOVERNMENT DRUG REPURPOSING

There are currently over 7000 unsolved diseases (National Organization of Rare Diseases, 2015). The pharmaceutical/biotech industries have been engaged with drug reprofiling, repositioning, and repurposing that can provide a profit, such as the repurposing of thalidomide for the blood cancer multiple myeloma (National Cancer Institute, 2015) and the repurposing of sildenafil for both erectile dysfunction (Pfizer, 2015) and pulmonary arterial hypertension (Pulmonary Hypertension Association, 2015). While industry drug repurposing is growing

(Gaffney, 2015), even at its most productive, the pharmaceutical/biotech industries will not be able to fund *de novo* or repurposed solutions for all of the thousands of unsolved diseases, and even if they could, many of the potential repurposed solutions will not create sufficient profit to suit the pharmaceutical/ biotech industries model.

This lack of economic incentive for repurposing is true for generic drugs that are inexpensive and widely available from a variety of manufacturers, and especially true when the drug can be taken in the approved dosage and formulation in the new indication. In these situations, it is difficult to establish intellectual property (IP) or marketing protection that would allow one company to create sufficient market share to recoup the costs of completing the safety, manufacturing, and clinical work necessary to gain regulatory marketing approval. Sometimes a company can secure short-term market exclusivity, but the time frame is often insufficient to offset the development marketing costs. This is exacerbated when the patient population is small or is very poor, when the disease is acute so there is no long-term use of the treatment, when the disease is less serious and many patients will choose no treatment, or when there is already a partial therapy on the market. Because of all of these issues, the pharmaceutical/biotech industry does not have any economic reason to fund these generic drug repurposing projects. Philanthropists and not-for-profit organizations (NPOs) have started funding these types of repurposing projects, but they often run into the same issues of cost *versus* return on investment.

Occasionally, government programs fund the repurposing of generic drugs, but as in the case of the NCATS program of the U.S. National Institutes of Health (NCATS, 2015), the funding went to reposition shelved pharmaceutical compounds rather than generic drugs. The reason the government chose repositioning of shelved compounds *versus* repurposing of generic drugs is not specified, but one rationale is that the shelved compounds have sufficient IP protection that, if a new indication were to be discovered, there would be economic incentive for the company holding the composition of matter patent to invest the additional capital to bring the product to market. This would be unlikely with a generic drug with less or no IP protection.

In a further example of repositioning shelved compounds, on December 2014, 68 deprioritized pharmaceutical compounds were made available to academic researchers through a partnership between the UK's Medical Research Council (MRC) and 7 global drug companies (Medical Research Council, 2014a). UK scientists were encouraged to apply for MRC funding to use any of the compounds in medical research studies to investigate the underlying mechanics of disease, which may lead to the development of more effective treatments for a range of conditions.

The same year, the MRC engaged a number of charity partners to fund a repositioning initiative directed at brain diseases, called The Neurodegeneration Medicines Acceleration Programme (Neuro-MAP) (Medical Research Council, 2014b). In both of these repositioning initiatives, all compounds still maintain their patent protection, so there is profit potential if a new indication is found, assuming that the patent life is still robust, the patient population is large enough, the disease is chronic, and the use is efficacious and safe.

3.1.2   NOT-FOR-PROFIT DRUG REPURPOSING

There are two significant reasons that NPOs have started to participate in drug repurposing. First, there is a huge funding gap in the generic drug repurposing market. The pharmaceutical/biotech industries cannot afford to be in the generic drug repurposing business, and most governments have not yet created generic drug repurposing funding programs. One of the chief reasons that NPOs exist in all areas of social need is to fill funding gaps when industry and government are absent. This is true in generic drug repurposing.

Second, NPOs have decades of frustrating experience funding *de novo* drug development research, because it is very expensive, time-consuming, and risky (Paul et al., 2010), with costs of $1.5B+ and a timeline of 12–19 years. NPOs have poured billions of dollars into *de novo* medical research for the last 50+ years, and yet the FDA approves an average of 20–30 new drugs per year, and many of them are simply improvements over drugs that already work. For every 10,000 new compounds discovered and tested, only one makes it to market, and it is getting more difficult to get approvals each year (Michigan Bio, 2015). NPOs fund a significant portion of that *de novo* discovery and have very little progress to show for it in most diseases.

Many NPOs have surveyed the market, discussed drug development with their science advisors and academic research partners, and come to the conclusion that adding drug repurposing to their *de novo* drug development research portfolios makes sense, especially since their mission is to improve patient healthcare outcomes as quickly as possible. A recent study determined that the average cost to bring a repurposed drug to market was $8.4M–$41.3M (Persidis, 2015), which is significantly less than *de novo* drug development's cost of over $1B. Couple this much lower cost with the much shorter 3–7 year timelines to approval (Phelps, 2012) and the 10%–30% success rate of drug repurposing (Roundtable on Translating Genomic-Based Research for Health, Board on Health Sciences Policy, Institute of Medicine, 2014), and the case for adding drug repurposing to an NPO research portfolio is strong. These NPOs also realize that generic drug repurposing can create inexpensive "new" therapies, which can potentially save healthcare dollars for patients and payers.

3.1.3   ADDITIONAL ADVANTAGES TO GENERIC DRUG REPURPOSING

There are additional advantages for an NPO to engage in a generic drug repurposing strategy:

1. There are more than a thousand generic drugs (Clinical Leader, 2015) available to be tested for repurposing to impact patients, and an equal or greater number of nutriceuticals, which are bioactive compounds, such as vitamins, minerals, and other supplements, that have never been through a drug regulatory approval process but have been shown to be safe for human use.

2. These drugs and nutriceuticals have been marketed for many years, so much is known about their side effects, drug interactions, and overall safety, which can reduce the time and cost required to secure FDA, NDA, and IRB approvals for clinical research, and it may reduce the difficulty of accruing patients and receiving their informed consent to participate in a clinical trial.

3. There is usually an ample supply of drug for testing from many manufacturers, and the drug is often inexpensive to purchase for test purposes.

4. Starting with an approved drug significantly reduces the time and costs involved in securing marketing approval (Roundtable on Translating Genomic-Based Research for Health, Board on Health Sciences Policy, Institute of Medicine, 2014), because the development process can skip most of the steps between compound discovery and completion of Phase II clinical trials (Thayer, 2012). When market approval is required or optimal, repurposed drugs can often be approved through an abbreviated FDA 505(b)(2) pathway in the United States (Camargo Pharmaceutical Services, 2015), saving as much as $100M–$600M+ and reducing approval time by as much as 10 years.

5. Proof-of-concept generic drug repurposing clinical trial projects often cost less than $500,000, and often the time from the initiation of the project to the conclusion of a robust clinical trial is 36 months. In many cases, this is all the time and cost that is necessary, as the drug may be used off-label once the data of the clinical trial are published (Ventola, 2009).

6. Since the drug is already available, a government or other payer could simply choose to allow it to be prescribed off-label once proof-of-concept success is verified and published, reducing the cost to the patient and increasing the likelihood of physician usage. The resulting repurposed treatment will often be inexpensive for the patients and payers, significantly improving patient outcomes and reducing the healthcare costs (AKU Society, 2015).

7. Physicians will be more familiar with the repurposed drug than with a newly discovered drug, and potentially more likely try it in a patient who does not have a currently successful therapy.

8. Since many of these repurposed drugs have been used for more than a decade by millions of patients, the chances of finding a significant long-term side effect are reduced.

9. If a repurposed drug helps a patient population in one country, it is likely to help in any other country where the drug is available. If a repurposed drug reduces the cost for one payer, it is likely to reduce costs for all payers. And many drugs repurposed for one new indication can be repurposed for additional indications (Teachey, 2016).

3.1.4   MEETING THE CHALLENGES OF DRUG REPURPOSING

While some NPOs, such as the Alzheimer's Drug Discovery Foundation (ADDF) and the Michael J. Fox Foundation (The Michael J. Fox Foundation, 2015) for Parkinson's Research (MJFF), have started to participate in generic and proprietary drug repurposing (Shineman et al., 2014), many other NPOs have been reluctant to participate. There are many reasons for this.

First, it may be initially difficult to raise funds for drug repurposing when the NPO has traditionally been involved in funding *de novo* drug development. The reason for this may include the following:

1. NPO science advisors who have been reviewing and selecting *de novo* research for many years may have a bias against drug repurposing as not innovative, or they may not have the experience to review repurposing projects with the same confidence as they review *de novo* research. In some cases, it takes years for scientific and clinical advisors to change their stance on drug repurposing; sometimes it takes a change of advisors before this can happen. This is true even when the lay leadership of an NPO is eager to support drug repurposing. However, drug repurposing is becoming more of a mainstay of medical research, both for profit and nonprofit (Persidis, 2015), and more scientists and clinicians are becoming involved in repurposing efforts.

2. Funders of the NPO, much like the science advisors, often initially perceive that drug repurposing is not innovative, and it can take quite a while for them to begin to see the value of drug repurposing, either as an alternative to or an addition to *de novo* drug discovery. In the same way that scientists and clinicians are beginning to see repurposing as a critical part of medical research, funders are starting to embrace repurposing (Allarakhia, 2013).

3. Even when there is wide support at an NPO for drug repurposing, the best result that can be achieved is often a robust clinical trial leading to the possibility of off-label use, instead of regulatory and marketing approval, because there is not enough NPO funding for the many millions of dollars it takes to secure regulatory approval. Many lay and scientific advisors to NPOs worry, rightly so, that the chance of widespread adoption of a new standard of care using a repurposed off-label therapy is lower than the chance of adoption of a new regulatory approved therapy. However, the publication of clinical research results in searchable online journals, coupled with the use of social media by NPOs and patient advocacy groups, allows the off-label use of repurposed drugs to become more widespread more quickly (Wittich et al., 2012).

4. When a repurposed off-label therapy is adopted by physicians, NPO supporters are often concerned, again rightly so, that payers will not cover the expense of the repurposed therapy, since there is no regulatory approval (Butcher, 2009). Many patient advocacy groups and social science organizations have proposed ideas to help with this situation (Ewing Marion Kauffman Foundation, 2015).

Second, researchers who typically apply for research grants may be disinclined to participate in drug repurposing, because drug repurposing results may not get published in highly ranked journals, drug repurposing research does not generate large and long-term grants, and patient populations are often small, so enrolling enough patients to get meaningful data can take a long time.

These challenges are being overcome on several levels. Many peer-reviewed journals have become more inclined to review and accept drug repurposing submissions. In 2014, Mary Ann Liebert Publishing launched the first journal dedicated to drug repurposing, called the *Journal of Drug Repurposing, Rescue, and Repositioning*, of which this author was an inaugural editor. The *Journal of Drug Repurposing, Rescue, and Repositioning* later became an integral portion of the *Journal ASSAY and Drug Development Technologies* (ASSAY and Drug Development Technologies, 2016).

The grants for drug repurposing are becoming larger and the timelines longer, as some NPOs start to think of drug repurposing in terms of market approval instead of off-label use. The AKU Society raised more than $13M for the repurposing of nitisinone for the rare disease alkaptonuria (AKU Society, 2015), and the Leukemia and Lymphoma Society raised over $1M for a clinical trial of auranofin for CLL (Leukemia and Lymphoma Society, 2015). In the world of neurodegenerative diseases, Cures Within Reach, the ADDF, and the Alzheimer's Society of Canada are cofunding a project repurposing the cannaboid nabilone for agitation in Alzheimer's disease (Alzheimer's Society of Ontario, 2015).

Accruing patients is always a challenge for clinical trials. There are several online resources that are working to solve this issue. Researchmatch.org is a collaborative project, led by the Vanderbilt Institute for Clinical & Translational Research, and involves a number of other not-for-profit U.S. institutions. As of April 2015, the site has enlisted over 75,000 patients willing to participate in clinical trials, 2,500 researchers, and has engaged them to undertake almost 400 research projects at over 100 institutions (ResearchMatch. org, 2015). The U.S. National Cancer Institute has a web platform called AccrualNet to help researchers improve patient accrual (National Cancer Institute, 2015). The European Organisation for Research and Treatment of Cancer (EORTC) has many resources across Europe to help with patient accrual (EORTC, 2015). The DevelopAKUre trial to test nitisinone for the ultra-rare disease alkaptonuria has brought together researchers in 13 countries to help with patient accrual (AKU Society, 2015).

3.2   HOW DO IDEAS FOR DRUG REPURPOSING COME TO NPOs?

Typically, drug repurposing ideas come to NPOs through Requests for Proposals (RFPs). A few NPOs specifically solicit drug repurposing proposals through their RFPs (AntiCancer Fund, 2015), and this is becoming more popular. However, the most common mechanism is that some repurposing projects come through a general therapy discovery RFP published by an NPO. The NPO Cures Within Reach funded over 200 medical research projects over a 12-year period from 1998 to 2009, never specifically requesting drug repurposing projects in any of its RFPs. Over that time, however, 10 of the 200 projects selected for funding happened to involve drug or device repurposing. When the Cures Within Reach staff reviewed the 190 *de novo* projects funded, they found little direct patient impact. When they reviewed the 10 repurposing projects, they found that 4 of the projects created therapies that were actually being used clinically to help patients. In two such cases, the projects had developed treatments that had created solutions to once-deadly pediatric diseases (Anderson and Rubin, 2005; Teachey, 2009). The fact that the *de novo* research wasn't reaching patients and cost more money per project prompted Cures Within Reach to begin to focus exclusively on funding repurposing medical research starting in 2010.

Where do researchers get the repurposing ideas to submit to these RFPs? Sometimes the ideas are developed from new information about a disease. Sometimes this new disease information points directly to a drug repurposing opportunity. In the disease autoimmune lymphoproliferative syndrome, researchers discovered a specific genetic defect that led to the disruption of normal apoptosis in lymphocytes (Rieux-Laucat et al., 2003). This genetic discovery led to further elucidation of the pathways involved in the disease (Teachey, 2009), and it became clear that mTOR inhibitors, such as sirolimus, could be a possible drug repurposing choice. The research team applied for a grant from the NPO Goldman Philanthropic Partnerships to complete the mouse model research (Teachey et al., 2006), and later a human clinical trial (Teachey, 2009) proved sirolimus was effective in refractory autoimmune lymphoproliferative syndrome. Sirolimus is now being tested as a repurposed therapy for five additional pediatric autoimmune diseases by the same researcher (ClinicalTrials.Gov, 2016).

On other occasions, new discoveries about a disease don't point to obvious drug repurposing candidates. In these cases, researchers often create an assay representative of the disease and then screen a library of drugs and drug-like compounds against the assay to determine if any of these appear to be potentially effective. An example in neurotherapeutics is in the rare disease Friedreich's ataxia (FA), for which there is no successful treatment. Researchers identified a new redox deficiency in FA cells and used this to model the disease (Sahdeo, 2014). The researchers then screened a 1600-compound library to identify existing drugs that could be of therapeutic benefit and identified the topical anesthetic dyclonine as protective. The researchers applied for a grant from the NPO Friedreich's Ataxia Research Alliance and completed a human clinical proof-of-concept study in eight FA patients dosed twice daily using a 1% dyclonine rinse for 1 week. Results showed that dyclonine represented a novel therapeutic strategy that can potentially be repurposed for the treatment of FA. This use of a repurposed drug would be a "known compound-new target" (Grau, 2015) use.

Sometimes ideas for drug repurposing come from new scientific knowledge about a drug, usually about a newly discovered mechanism of action (Iorio, 2010). This would be thought of as a "known target-new indication" (Grau, 2015) drug repurposing, since the new information is not about the disease, but about the drug. A neurotherapeutic example of this is the discovery of additional mechanisms of action for the antibiotic minocycline, which provide evidence that it might be effective in the treatment of ischemic stroke (Hess, 2010). In addition to showing that minocycline was effective as a single agent in animal models of stroke, Murata and colleagues reported that minocycline was effective when used in combination with tPA (Murata et al., 2008).

One very common way that an NPO will find out about a drug repurposing idea is from clinical observations of clinicians treating patients who are associated with the NPO. This often happens when patients have more than one disease condition. The patient takes a drug for condition A and ends up having an impact on condition B. An example of this off-label use informing physicians, and eventually NPOs about a repurposing opportunity, involves the drug thalidomide. This drug was given to a leprosy patient in 1964 to help with the comorbidities of sleep deprivation and pain, but it was found to actually help the leprosy patient. Eventually, it was discovered that thalidomide impacted blood vessel growth in a way that would help not only leprosy, but potentially the blood cancer multiple myeloma. A number of NPOs, including Goldman Philanthropic Partnerships, funded thalidomide repurposing research in the late 1990s and early 2000s that eventually led to the approval of thalidomide as a validated treatment for multiple myeloma (National Cancer Institute, 2013).

Another method for researchers and clinicians to propose repurposing research to NPOs is through evaluation of scientific information already available in the literature. Some NPOs actually have researchers and clinicians on staff whose responsibility is to sift through the scientific and clinical literature to uncover repurposing opportunities that already exist. GlobalCures (Global Cures, 2017), an NPO headquartered in the United States, and the AntiCancer Fund (The AntiCancer Fund, 2017) in Belgium, both scan the literature for drug repurposing opportunities in cancer. Both groups have a mission to expand evidence-based cancer treatment options. The Anticancer Fund is also selectively funding the development of promising therapies.

There are also many for-profit companies and academic centers using computational biology, alone or in combination with other research methods, to find both obvious and nonobvious drug repurposing opportunities. Some computational biology methods comb through published and unpublished literature to find information that already points to a drug repurposing opportunity. Other computational biology methods use molecular docking modeling to determine what shapes of molecules might fit a certain disease receptor and then scan existing drugs to determine which drugs, if any, might meet those spatial requirements. Other methods, such as pharmacophore modeling and mapping, molecular similarity calculation, and sequence-based virtual screening, are also used (Ou-Yang, 2012).

Many of these computational biology companies and academic centers are working with NPOs, either at the behest of the NPO (Accelerating Paediatric Orphan Drug Development, 2016; Persidis, 2013), or as applicants through RFPs by NPOs or the government, once the computational biology company finds drug repurposing opportunities. As computational biology continues to expand, and perhaps decreases in cost, it will create significant additional value to NPOs.

3.3 SPECIFIC EXAMPLES OF NPOs ENGAGED IN DRUG REPURPOSING

3.3.1 GLOBALCURES

GlobalCures is a Massachusetts-based 501(c)3 nonprofit medical research organization (www.global-cures.org) dedicated to helping patients with diseases by promoting clinical research on scientifically promising, readily available, and cost-effective treatments that are not being pursued due to lack of financial incentives, which they term "financial orphans." GlobalCures focuses on these financial orphans through a three-step process:

1. *FIND*: Identify promising treatment options that lack financial incentive and prioritize candidate treatments for clinical development.
2. *FUND*: Facilitate clinical development by recruiting clinical researchers, writing protocols, writing grants, and finding other sources of funding for sponsoring clinical studies.
3. *SHARE*: Educate physicians and patients regarding these novel ideas.

3.3.1.1   FIND

GlobalCures staff uses several methods to identify promising therapies. The most time intensive of these involves PubMed searches of published small clinical trials, case reports, or preclinical papers that collectively point to potential drugs for repurposing or to other treatment options that fall under the financial orphan category. In addition to PubMed searches, GlobalCures reviews websites from patient advocacy groups; alternative or integrative medicine clinics; and suggestions from scientists, clinicians, and patients.

The prioritization of drugs and treatments is based on the following criteria:

- Availability of the drug/treatment
- Cost
- Feasibility (e.g., shorter treatments are better)
- Manageable toxicity (nonoverlapping with standard treatment)
- Human data (clinical trials or case reports, even if limited, trump animal data)
- Understanding of the mechanism of action, especially if it might be synergistic with current approaches
- Applicability across diverse tumor types
- Availability of biomarkers (for patient stratification, prognosis, or monitoring)
- Potential for significant improvement over the current standard of care

3.3.1.2   FUND

GlobalCures facilitates clinical development by recruiting clinical researchers, writing protocols, writing grants, and finding other sources of funding in order to sponsor the clinical trials. GlobalCures' FUND program is to conceive, design, and write clinical protocols and then raise funds to enact them. The goal is to ultimately publish in peer-reviewed journals the results of these studies so that they become part of the standard of care through off-label or on-label use and be reimbursed. The goal is not to formally gain a new labeling indication, though a drug company that has commercial interest in the treatment may be motivated to conduct a confirmatory trial for this purpose.

3.3.1.3   SHARE

GlobalCures educates physicians and patients regarding these novel ideas. It is designing a web-based patient-reported outcomes database to catalog the clinical course of patients who may have undergone some of the treatment options in FIND or other treatments altogether. Though not as robust as a formal clinical study, GlobalCures believes that by sharing such data with others like them, patients and physicians might note an early signal of efficacy and safety, which could inform and reprioritize future clinical studies. GlobalCures will disseminate the results of studies it sponsors in the widest possible way, much as pharma markets the drugs it develops and spreads the word on publications that involve them.

3.3.2   Cures Within Reach

Cures Within Reach is the world's only disease agnostic NPO focused solely on Repurposing Research. Cures Within Reach facilitates the sourcing, evaluation, funding, commencement, and completion of proof-of-concept repurposing clinical trials. The group also supports final preclinical work that can lead directly to a clinical trial. Cures Within Reach has been funding Repurposing Research since 2010 and has funded 50 clinical trials that have led to 12 repurposed therapies that are being used clinically or have advanced to phase III trials, in autoimmune lymphoproliferative syndrome, familial dysautonomia, types 1 and 2 diabetes, lung cancer, multiple sclerosis, prostate cancer, myelodysplastic syndrome, and five additional pediatric autoimmune diseases. In 2015, Cures Within Reach was supporting research in autism, pediatric delirium, multiple sclerosis, Parkinson's disease, Alzheimer's disease, stroke, lung cancer, Batten disease, diabetes, myelodysplastic syndrome, melanoma, pancreatic cancer, breast cancer, bladder cancer, multiple myeloma, neuroblastoma, leukemia, nocturia, prematurity, neutropenia, ascites, lupus, autoimmune lymphoproliferative disease, Evans disease, autoimmune hemolytic anemia, idiopathic thrombocytopenic purpura, common variable immune deficiency, and polycystic kidney disease.

Cures Within Reach processes all of its research through the web portal CureAccelerator™, which Cures Within Reach built with a grant from the Robert Wood Johnson Foundation. CureAccelerator provides all Repurposing Research stakeholders (researchers, clinicians, funders, patients, industry, academia, and government) a collaboration hub and Repurposing Research marketplace so that they can work together to drive more Repurposing Research to patients as quickly as possible. In the first 4 months of CureAccelerator use, 7 Repurposing Research projects were posted, evaluated, and funded.

Cures Within Reach funds 20–30 Repurposing Research projects each year. Among the projects funded in 2014 and 2015, those that involve the central or peripheral nervous systems are the following:

1. *Autism Spectrum Disorder (ASD)*: Cures Within Reach is funding several repurposing projects in ASD. The first involves evaluating a particular over-the-counter omega-3, -6, and -9 fatty acids (FAs) combination in reducing symptoms of ASD. The central hypothesis is that children 18–42 months of age in a high-risk group and already displaying ASD symptoms will demonstrate reductions in ASD and ADHD symptoms. The scope of the proposed effort includes ongoing follow-up data collection beyond the 90-day intervention to evaluate symptoms over time. The project will follow children (n = 40) with ASD symptoms, aged 18–42 months and born extremely or very preterm (≤29 weeks' gestation), who completed a 90-day double-blind, randomized early intervention trial with an Omega 3–6–9 dietary supplement compared to placebo, to evaluate efficacy in improving symptoms of ASD and ADHD (primary endpoints). If this trial is successful, it will likely lead to a larger phase III trial as well as correlative studies to determine the mechanism of action. In addition, since the nutriceuticals are publicly available, it is conceivable that parents of ASD patients could decide to use this off-label, hopefully under the supervision of the patient's physician.

A second project involves using the drug ketamine with autistic patients. To address the significant need for the effective treatment of core symptoms of ASD, the researchers propose a double-blind, placebo-controlled parallel-group pilot study of intranasal ketamine with open-label extension in 24 adults with ASD aged 18–50 years using novel quantitative outcome measures of social and communication impairment.

The two main aims of this project are: *Aim #1*, to determine if intranasal ketamine shows initial evidence of efficacy, safety, and tolerability in adults with ASD. Hypothesis: Ketamine will be safe and show initial signs of efficacy targeting core impairment in ASD. *Aim #2*, to further the development of quantitative and objective measurement of social and communication change in ASD utilizing an eye-tracking paradigm and expressive language sampling. If this trial is successful, it will likely lead to a larger confirmatory trial to determine long-term impacts. It is possible that one or more companies working on various intranasal and other formulations of ketamine would take this project on a commercialization path to marketing approval.

2. *Multiple Sclerosis, Parkinson's Disease, Stroke, and Traumatic Brain Injury*: Cures Within Reach is supporting the repurposing of a tongue stimulation device for these neurological disorders. Cures Within Reach funded the proof-of-concept clinical trial at the University of Wisconsin in 2009 (University of Wisconsin, 2009) and has continued to support the research. Over the last 6 years, over $3M in philanthropy and government grants supported this research. In 2014, a company, Helius Medical, was formed to raise the capital required to support the research trials (Helius Medical Technologies, 2015) and regulatory work required to secure FDA and other marketing approval.

3. *Alzheimer's disease*: Cures Within Reach has partnered with a number of other philanthropies to support Alzheimer's repurposing research. In 2015, Cures Within Reach and its cofunders supported a trial repurposing the cannaboid nabilone to combat the severe agitation that often arises in the course of Alzheimer's disease. Nabilone is a synthetic version of tetrahydrocannabinol, an active ingredient in marijuana plants. Cancer patients use it to treat chemotherapy-induced nausea, and those with anorexia use it to help stimulate appetite and regain weight. But it has never been tested in people with Alzheimer's. The research team's study will assess 40 participants with Alzheimer's disease. Half will be randomly assigned to take nabilone for 6 weeks and then a placebo for 6 weeks, with a week in between when they take neither. The other half will begin with 6 weeks of placebo, followed by a week off, and then 6 weeks of nabilone. The group will measure changes in agitation, reported pain, and weight gain. Because such assessments are carefully vetted, the results won't be released until 2018, and if results are positive it will progress to a larger clinical trial.

3.3.3 AKU Society

The experience of the AKU Society, a patient group based in the United Kingdom, provides a strong case study of how a nonprofit can be heavily involved in repurposing a drug that otherwise would never be developed for that particular indication.

AKU is short for alkaptonuria, a rare genetic disease. AKU is a monogenic disease that causes an enzyme—homogentisate 1,2 dioxygenase—to malfunction, leading to a 2000-fold increase in a molecule called homogentisic acid. This acid binds to cartilage and bone and turns black—a process called ochronosis—which explains AKU's common name of black bone disease.

There is no cure, but there is a promising treatment. It's a drug called nitisinone. Originally developed as a weedkiller, nitisinone was already once repurposed to treat an ultra-rare disease called Hereditary Tyrosinemia Type 1 (HT1) (Sobi, 2015), which kills children often by age three by causing liver cancer. Nitisinone works on the tyrosine metabolic pathway in plants and in humans, hence its success in treating HT1 patients.

AKU is caused by a defect in the same metabolic pathway. In theory, nitisinone should also be able to treat it, since it stops the accumulation of the homogentisic acid that causes all the damage in AKU. In practice, this is much more difficult to prove. While nitisinone prevents HT1 patients from dying—a clear endpoint for any clinical study—in AKU, it would take years to show any clinical benefit using this endpoint.

The company that owned the license to nitisinone in 2010—Swedish Orphan Biovitrum International (Sobi)—was not interested in AKU; the company thought it was too difficult and too costly to study AKU in a clinical trial. Furthermore, the patent for nitisinone was to run out within a few years, making it difficult to protect against generic production, even if orphan drug protection was granted.

The situation was a classic drug repositioning conundrum for an NPO: a promising drug in limbo because of a series of market failures caused by inadequate intellectual property protection and a capitalist system focused on low risk and shorter-term investments. Without the drive of the AKU Society, it is unlikely the drug would ever have been given a second chance for AKU.

The AKU Society gathered a consortium to develop the drug through a robust clinical development program. Having just completed a natural history study—funded by the AKU Society and a foundation called the Childwick Trust—lead PI Professor Lakshminarayan Ranganath from the Royal Liverpool University Hospital had enough data to put together a severity score index. This would allow clinicians to evaluate the severity of a patient's disease by scoring all its different aspects—such as joint deterioration, the spine, and ochronosis in the ears and eyes. Over time, the evolution of the score would give an indication of how the disease was progressing and would provide a strong endpoint for a clinical trial.

Meanwhile, the AKU Society had also been working closely with Professor Jim Gallather and Professor Jonathan Jarvis from the University of Liverpool to develop a mouse model of AKU, thanks to funding it secured from the UK's Big Lottery Fund. AKU mice, given nitisinone shortly after birth, did not show any symptom of AKU, even at the microscopic level. Those given the drug halfway through their lives showed a complete halt in the progress of their disease.

Armed with robust preclinical data, a severity score index, a draft clinical development plan, and the agreement of three clinical trial sites, the AKU Society approached Sobi to encourage it to reconsider its plans and join the consortium for a last attempt at developing nitisinone for AKU.

After much internal deliberations, Sobi accepted (Sobi, 2014). Even though nitisinone would probably not offer much financial return, the company felt that the project was in line with its values of helping patients with rare diseases and that it offered a new model of collaboration that was worth exploring.

The final hurdle was to secure the funding necessary for the clinical development program. For this, the group turned to the European Commission, which had an upcoming call for proposals for rare diseases as part of its Seventh Framework Programme. The group put together a 140-page proposal, which scored 15/15 and was therefore funded to the tune of €6M.

The proposal was for a three-stage clinical development program. The first stage, in 2013, was a phase 2 dose-response study in which 40 patients were randomized to five groups: no treatment, 1, 2, 4, and 8 mg/day. The trial lasted 4 weeks, with comparison of homogentisic acid levels at the start and the end of the time period.

The results were astonishing, with one member of the consortium saying they had never seen such a "beautiful" dose-response curve. Indeed, it showed a very clear relationship between the dose of the drug and the reduction in homogentisic acid.

In the end, based on the curve, a dose of 10 mg/day was chosen for the second stage of the clinical development program, a phase 3 trial on 140 patients over 4 years. This was for simplicity, since a 10 mg dose was a single pill, as opposed to four pills for an 8 mg dose. The phase 3 trial started officially in spring 2014. By January 2015, the target of 140 subjects enrolled was achieved (AKU Society, 2016).

3.3.4 THE MICHAEL J. FOX FOUNDATION

The Michael J. Fox Foundation is one of the NPOs in the neurodegenerative disease area that has embraced repurposing and repositioning. The MJFF launched a repositioning specific Request for Applications in 2010, funding nine research teams for awards totaling $3.4M. These teams were investigating repurposing of a tuberculosis vaccine, the high blood pressure drug isradipine, and the antidepressant duloxetine, and the repositioning of a drug-like compound originally used to treat ADHD, to test whether these drugs might have a disease-modifying impact on PD. MJFF is also supporting a Phase 4 clinical trial of naltrexone, a drug marketed for alcohol abuse, to treat compulsive disorders associated with PD. This NPO is also supporting research to determine if alpha-galactosidase A is a therapeutic target for Parkinson's Disease. Drugs that increase alpha-Gal A are already in use for treating Fabry disease, a rare lysosomal storage disorder. If alpha-Gal A is found to be decreased in Parkinson's disease brain, this would suggest a novel drug target and drugs that are already approved for clinical use may be "repurposed" for treating Parkinson's disease. Such treatments may actually delay progression of Parkinson's disease rather than just treat the symptoms as only current therapies can offer (The Michael J. Fox Foundation, 2014a).

The MJFF has done some repurposing in the past that has generated enough data to draw some conclusions. One such project proposed to repurpose the tricyclic antidepressant medications amitriptyline, nortriptyline, and venlafaxine, for effects on protecting dopamine nerve cells from degeneration in a toxin model of parkinsonism. No protective effect of any of the drugs or doses tested was found. This is in contrast to the MJFF's previous supported work showing that amitriptyline can protect dopamine cells, albeit in a less severe degeneration model. The tentative conclusion is that while antidepressant medications may have a modest protective effect, this positive influence can be overwhelmed in a model of severe, sudden damage to the dopamine system (The Michael J. Fox Foundation, 2014b).

The MJFF is also testing some compound repositioning, in addition to drug repurposing. One such project is testing AVE8112, a PDE4 inhibitor that was in development by the pharmaceutical company Sanofi for the treatment of cognitive impairment in Alzheimer's disease patients and had shown promise in several preclinical models. The goal of this study is to determine the safety and tolerability of AVE8112 in subjects with PD at two sites (Glendale, CA and Baltimore, MD) to establish the safety and tolerability of various doses of AVE8112 in patients with PD to enable wider exploration in a subsequent Phase II trial for the symptomatic treatment of cognitive impairment in PD patients (The Michael J. Fox Foundation, 2012).

3.3.5 TIRCON

Drug repurposing is the focus of therapeutic testing from NPOs that support rare neurodegenerative diseases. One example is the group TIRCON, which is an acronym for Treat Iron-Related Childhood-Onset Neurodegeneration. TIRCON focuses on neurodegeneration with brain iron accumulation (NBIA), a clinically and genetically heterogeneous group of rare, hereditary, neurodegenerative disorders characterized by high levels of brain iron. Many NBIA cases are characterized by early childhood onset and rapid progression to disability and death. The most frequent form of NBIA is pantothenate kinase–associated neurodegeneration (PKAN). Currently, there is no proven therapy to halt or reverse PKAN or any other form of NBIA. This is especially unfortunate as both the iron accumulation in NBIA and the biochemical defect in PKAN are predicted to be amenable to drug-based treatment. Thus, the absence of adequately powered randomized clinical trials is not due to a lack of therapeutic options but due to the rarity of the disease, the lack of patient registries, and the fragmentation of therapeutic research worldwide.

In TIRCON, for the first time, an international group of scientists and clinicians have elaborated a collaborative project with patient representatives and innovative companies committed to orphan products.

TIRCON's goals are the following:

1. Set up an international NBIA patient registry
2. Establish a biobank
3. Develop biomarkers for the disease
4. Conduct randomized clinical trial of the iron-chelating drug deferiprone in PKAN (TIRCON, 2016)

There is a further work at TIRCON to make sure that any clinical research is disseminated to clinicians and patients (TIRCON, 2017).

In the world of neurodegeneration research, companies are supporting NPOs for creating tools for drug repurposing. In the United Kingdom, a small biotech called Parkure has developed a system using fruit flies as a proxy for Parkinson's Disease. There is strong evidence to show that the fruit fly *Drosophila melanogaster* develops symptoms of Parkinson's disease when challenged by the expression of certain human proteins—known to cause Parkinson's disease in man—in the fruit fly brain. Fruit flies and man share many molecular neurological mechanisms. Genes that cause hereditary Parkinson's disease in man were discovered in the fruit fly, where they act in an identical fashion (Christmann, 2014). Discoveries made in the fruit fly will in almost all cases be adaptable to the human system. Parkure has refined the use of fruit flies as a platform for drug screening against Parkinson's disease. Candidate drug compounds are administered to flies with Parkinson's disease. Candidate "hits" that reverse the symptoms of Parkinson's disease are then followed up with histological and biochemical studies. Pilot experiments suggest that 0.1%–0.4% of compounds have an effect. Parkure will screen as many chemicals as possible to maximize the probability of discovering and providing new efficacious drugs and reduce the cost of drug development.

There are many more examples of NPOs involved with drug repurposing in neurotherapeutics and other therapies. Every year, these NPOs learn more about and invent new ways to overcome some of the issues involved with drug repurposing, including sourcing the drug, raising the capital, creating or overcoming government involvement, the value and challenge of creating public private partnerships, getting physicians and patients involved, and handling regulatory issues. NPOs fill market gaps that naturally occur between the profit incentives of industry and the public policy issues of government. Often, these NPOs develop systems that can be leveraged by other NPOs, as well as by government and industry. With 7000 unsolved diseases to be tackled, the drug repurposing efforts of NPOs are critical to creating many affordable, safe, and effective treatments, as quickly as possible.

REFERENCES

Accelerating Paediatric Orphan Drug Development. Crowdfunding campaign officially launched! http://www.apoddfoundation.org/news/press-releases/38/, May 25, 2016.

AKU Society, About DevelopAKUre, http://www.developakure.eu/aku-clinical-trials.html, July 24, 2016.

AKU Society. AKU Society compilation of sample patient costs, http://webcache.google-usercontent.com/search?q=cache:TdJwYPYfoRkJ:https://fundamentaldiseases.files.wordpress.com/2011/05/aku-michael-craig-study-cost-of-patient-final.ppt+&cd=4&hl=en&ct=clnk&gl=us&client=safari, April 7, 2015.

AKU Society. DevelopAKUre, http://www.developakure.eu, April 12, 2015.

Allarakhia M. Open-source approaches for the repurposing of existing or failed candidate drugs: Learning from and applying the lessons across diseases. *Drug Des Develop Ther* August 2013;2013(7):753–766.

Alzheimer's Society of Ontario. New hope for soothing agitation in people living with Alzheimer's, http://www.alzheimer.ca/en/on/Research/Meet-the-researchers/Krista-Lanctot, January 6, 2015.

Anderson SL, Rubin BY. Tocotrienols reverse IKAP and monoamine oxidase deficiencies in familial dysautonomia. *Biochem Biophys Res Commun* October 14, 2005;336(1):150–156.

AntiCancer Fund. Call for project: Launch of the second call, http://www.anticancerfund.org/news/call-for-project-launch-of-the-second-call, April 12, 2015.

ASSAY and Drug Development Technologies, Mary Ann Liebert Publishers, New Rochelle, NY, 2016. http://www.liebertpub.com/overview/assay-and-drug-development-technologies/118/, March 2017.

Bird PD. The treatment of autism with low-dose phenytoin: A case report. *J Med Case Rep* 2015;9:8.

Butcher L. When should insurers cover off-label drug usage?, http://managedcaremag.com/archives/0905/0905.offlabel.html, May 1, 2009.

Camargo Pharmaceutical Services. Understanding the 505(b)(2) pathway, http://www.camargopharma.com/Userfiles/white-paper/Cmrgo_WhitePaperApprovalPthwy_VFb.pdf, April 11, 2015.

Christmann B. Translational findings: How fruit flies are helping us understand Parkinson's disease, http://blogs.brandeis.edu/flyonthewall/translational-findings-how-fruit-fliesare-helping-us-understand-parkinsons-disease/, August 22, 2014.

Clinical Leader. New report reveals FDA-approved new molecular entities: 1827–2013, http://www.clinicalleader.com/doc/new-report-reveals-fda-approved-new-molecular-entities-0001, April 7, 2015.

ClinicalTrials.Gov. Sirolimus for autoimmune disease of blood cells, https://www.clinicaltrials.gov/ct2/show/NCT00392951?term=teachey+sirolimus&rank=1, April 22, 2016.

EORTC. Clinical studies patient accrual, http://www.eortc.org/clinical-trials/clinical-studies-patient-accrual/, April 12, 2015.

Ewing Marion Kauffman Foundation. A new market access path for repurposed drugs, http://www.kauffman.org/~/media/kauffman_org/research%20reports%20and%20covers/2014/05/new_market_access_path_for_repurposed_drugs.pdf, April 12, 2015, pp. 3–4.

Gaffney A. An increasing number of companies are using a once-obscure FDA drug approval pathway, http://www.raps.org/Regulatory-Focus/News/2015/04/08/21933/An-Increasing-Numberof-Companies-Are-Using-a-Once-Obscure-FDA-Drug-Approval-Pathway/, April 8, 2015.

Global Cures. Promising therapies to enhance your treatments, https://www.global-cures.org, March 10, 2017.

Grau D. Innovative strategies for drug repurposing. *Drug Discov Develop*, http://www.dddmag.com/articles/2007/09/innovative-strategies-drug-repurposing, April 12, 2015.

Helius Medical Technologies. Clinical trials, http://www.heliusmedical.com/our-research/clinical-trials, March 2015.

Hess D. Repurposing an old drug to improve the safety and use of tissue plasminogen activator for acute ischemic stroke: Minocycline. *Pharmacotherapy* July 2010;30(702):55S–61S.

Iorio F. Discovery of drug mode of action and drug repositioning from transcriptional responses. *PNAS* 2010;107(33):14621–14626.

Leukemia and Lymphoma Society. Groundbreaking clinical trial for leukemia patients begins in Kansas, http://www.lls.org/aboutlls/news/newsreleases/11111_groundbreakingclinicaltrial, April 12, 2015.

Medical Research Council. Charities pledge to jump start drug development for brain diseases, http://www.mrctechnology.org/charities-pledge-jump-start-drug-development-brain-diseases/, November 18, 2014b.

Medical Research Council. Seven pharma companies offer up compounds to UK researchers, http://www.mrc.ac.uk/news-events/news/seven-pharma-companies-offer-up-compounds-to-uk-researchers/, July 21, 2014a.

Michigan Bio. Mapping a path to market: Creating a comprehensive drug development strategy, http://c.ymcdn.com/sites/www.michbio.org/resource/resmgr/BioToolBox_-_BioResearch/Final_Path_to_Market_Infogra.pdf, April 11, 2015.

Murata Y et al. Extension of the thrombolytic time window with minocycline in experimental stroke. *Stroke* 2008;39:3372–3377.

National Cancer Institute. AccrualNet, https://accrualnet.cancer.gov/protocol_accrual_life-cycle/recruitingandcommunicatingwithparticipants#.VSq8Prqsk6Q, April 12, 2015.

National Cancer Institute. FDA approval for thalidomide, http://www.cancer.gov/cancertopics/druginfo/fda-thalidomide, April 7, 2015.

National Cancer Institute. FDA approval of thalidomide, http://www.cancer.gov/about-cancer/treatment/drugs/fda-thalidomide, July 3, 2013.

National Organization of Rare Diseases. Rare disease database, https://www.rarediseases.org/rare-disease-information/rare-diseases, April 10, 2015.

NCATS. Discovering new therapeutic uses for existing molecules, http://www.ncats.nih.gov/research/reengineering/rescue-repurpose/therapeutic-uses/therapeutic-uses.html, April 11, 2015.

Ou-Yang SS. Computational drug discovery. *Acta Pharmacol Sin* September 2012;33(9):1131–1140.

Paul SM et al. How to improve R&D productivity: The pharmaceutical industry's grand challenge. *Nat Rev Drug Discov* 2010;9:203–214.

Persidis A. *Biovista and the CFIDS Association of America Achieve Drug Repositioning Milestone with Pre-IND Meeting*, http://www.biovista.com/biovista-and-the-cfids-association-of-america-achieve-drug-repositioning-milestonewith-pre-ind-meeting/, May 30, 2013.

Persidis A. The benefits of drug repositioning, DDW Online, http://www.ddw-online.com/business/p142737-the-benefits-of-drug-repositioning-spring-11.html, April 11, 2015.

Pfizer. Viagra, https://www.viagra.com, April 7, 2015.

Phelps K. Taking the 505(b)(2) route, http://www.dddmag.com/articles/2012/08/taking-505b2-route, August 9, 2012.

Pulmonary Hypertension Association. Treatments for pulmonary hypertension, http://www.phassociation.org/Patients/Treatment/Sildenafil, April 7, 2015.

ResearchMatch.org. RM metrics as of 04/11/2015, https://www.researchmatch.org/index_pubsitemetrics.php, April 12, 2015.

Rieux-Laucat F, Le Deist F, Fischer A. Autoimmune lymphoproliferative syndromes: Genetic defects of apoptosis pathways. *Cell Death Differ* 2003;10:124–133.

Roundtable on Translating Genomic-Based Research for Health, Board on Health Sciences Policy, Institute of Medicine. *Drug Repurposing and Repositioning: Workshop Summary*. Washington, DC: National Academies Press, p. 1. http://www.ncbi.nlm.nih.gov/books/NBK235874/, August 8, 2014. Accessed April 11, 2015.

Sahdeo S. Dyclonine rescues frataxin deficiency in animal models and buccal cells of patients with Friedreich's ataxia. *Hum Mol Genet* December 20, 2014;23(25):6848–6862.

Shineman DW et al. Overcoming obstacles to repurposing for neurodegenerative disease. *Ann Clin Transl Neurol* July 2014;1(7):512–518.

Sobi. Development pipeline, http://www.sobi.com/Pipeline/Development-pipeline/, June 5, 2014.

Sobi. Orfadin® (nitisinone), http://www.sobi.com/en/Pipeline/Development-pipeline/OrfadinR/, November 2, 2016.

Teachey D. Treatment with sirolimus results in complete responses in patients with autoimmune lymphoproliferative syndrome. *Br J Haematol* April 2009;145(1):101.

Teachey D. Sirolimus is effective in relapsed/refractory autoimmune cytopenias: Results of a prospective multi-institutional trial. *Blood* January 2016;127(1):17–28.

Teachey DT et al. Rapamycin improves lymphoproliferative disease in murine autoimmune lymphoproliferative syndrome (ALPS). *Blood* 2006;108:1965–1971.

Thayer A. Drug repurposing. *Chem Eng News* 2012;90:15–25.

The AntiCancer Fund. Promising therapies to enhance your treatments, http://www.anti-cancerfund.org, March 10, 2017.

The Michael J. Fox Foundation. Phase I-B clinical trial of AVE8112 for cognitive impairment in Parkinson's disease, https://www.michaeljfox.org/foundation/grant-detail.php?grant_id=1107, January 2012.

The Michael J. Fox Foundation. Alpha-galactosidase A as a therapeutic target for Parkinson's disease, https://www.michaeljfox.org/foundation/grant-detail.php?grant_id=1428, January 2014a.

The Michael J. Fox Foundation. Amitriptyline as a dopamine neuroprotective therapy, https://www.michaeljfox.org/foundation/grant-detail.php?grant_id=1034, January 2014b.

The Michael J. Fox Foundation. Repositioning existing therapies to get novel Parkinson's drug leads 'Off the Ground', https://www.michaeljfox.org/foundation/news-detail.php?repositioning-existing-therapies-to-get-novel-parkinson-drug-leads-off-the-ground, April 12, 2015.

TIRCON. A randomized, double-blind, placebo-controlled trial of deferiprone in patients with PKAN, http://tircon.eu/tircon-workpackages/wp4, September 29, 2016.

TIRCON. Dissemination, http://tircon.eu/tircon-workpackages/wp7, March 2017.

University of Wisconsin. Reducing symptoms of multiple sclerosis using NINM, https://tcnl.bme.wisc.edu/projects/completed/ms-ninm, June 2009.

Ventola CL. Off-label drug information regulation, distribution, evaluation, and related controversies. *Pharm Ther* August 2009;34(8):428–440.

Wittich CM, Burkle CM, Lanier WL. Ten common questions (and their answers) about off-label drug use. *Mayo Clin Proc* October 2012;87(10):982–990.

Section II

Repositioning Approaches and Technologies

From Serendipity to Systematic and Rational Strategies

4 Systematic Drug Repositioning

Spyros N. Deftereos, Aris Persidis,
Andreas Persidis, Eftychia Lekka,
Christos Andronis, and Vassilis Virvillis

CONTENTS

4.1 INTRODUCTION

The idea of linking disparate scientific disciplines through intermediate (or shared) concepts was first described by Swanson in 1986, in what has come to be known as the ABC model (Swanson, 1991). In that model, A, B, and C denote separate scientific concepts, where A is reported to be related to B in one set of scientific publications and B is reported to be related to C in another set, while A is not reported to be directly related to C. The two known relations of A-to-B and B-to-C allow one to infer that A may be indirectly related to C, through B. The unknown A–B–C relation, from which the name of the model is derived, might thus constitute a novel scientific discovery.

Discovery under the ABC model can be pursued as either a closed or an open process. A closed discovery process is used to determine whether a potential link between two prespecified concepts exists. It begins with known starting and target concepts, A and C. The task in this case is to identify and evaluate all relevant intermediate (B) concepts that support the relation of A and C. An open discovery process, on the other hand, aims to explore all potential correlates of a concept of interest. It begins with a known starting concept A, while the relevant target concepts C are not known beforehand and should be identified by the algorithm. Several implementations of the original ABC algorithm have been described in the literature, differing in the types of intermediate concepts used and the algorithms by which target concepts are ranked (for a review, see Deftereos et al., 2011).

In the late 1980s, Swanson used this methodology to discover that fish oil might be beneficial in Raynaud's syndrome (Swanson, 1986). This was the first published medical discovery that was based on the computational linking of disparate concepts, which paved the way to the scientific discipline that is nowadays called *Drug Repositioning* or *Drug Repurposing*. A second hypothesis, that magnesium deficiency might be implicated in the pathophysiology of migraine (Swanson, 1988), was not entirely novel, as the topic had been discussed in earlier publications (Altura, 1985; Altura and Altura, 1985; Vosgerau, 1973). However, papers predating Swanson's publication were admittedly scarce. It was later shown, in a double-blind placebo-controlled clinical trial, that fish oil improves tolerance to cold exposure and delays the onset of vasospasm in patients with primary Raynaud's syndrome (DiGiacomo et al., 1989), while the role of magnesium in the prophylaxis of migraine, particularly pediatric and premenstrual, is now established (Schiapparelli et al., 2010). Swanson's group later proposed novel mechanistic hypotheses for the known-at-the-time effects of estrogens (Smalheizer and Swanson, 1996a) and indomethacin (Smalheizer and Swanson, 1996b) in the treatment of Alzheimer's disease.

Building on Swanson's original work, early efforts in literature-based discovery have explored the potential of the methodology and the effect of several proposed improvements on the final output. Although they were not specifically focused on drug repositioning, some have identified new indications for existing drugs that proved correct in subsequent clinical trials. Unquestionably, these efforts have paved the way to contemporary drug repositioning (Deftereos et al., 2011).

Serendipitous drug repositioning, on the other hand (i.e., based on clinical observations), has been part of the regular drug discovery process since the very first days of the pharmaceutical industry. Important new uses of existing drugs have been discovered by chance; in most cases, a drug that was administered for a certain indication also proved to be beneficial in a comorbid condition, or what was known to be an adverse event of a drug ameliorated an entirely different disease. Thalidomide is an important example of serendipitous drug repositioning. In 1965, it was given to leprosy patients as a sedative. It was immediately discovered that it also inhibited erythema nodosum leprosum, a complication of lepromatous (multibacillary) leprosy (Sheskin, 1965). Similarly, sildenafil was originally evaluated as the treatment of angina pectoris, where it did not show much success. It was observed, however, that patients presented penile erection as an adverse effect, which led to its further development in its current indication, erectile dysfunction (Andronis et al., 2012).

These and other success stories of repositioned drugs, together with the dearth of approval of new chemical entities (NCEs) by the U.S. Food and Drug Administration (FDA) and the "patentcliff" facing many of the currently approved drugs, are pushing many groups in the pharmaceutical industry to pursue directed strategies toward drug repositioning. The recent interest in drug repositioning also stems from the fact that the sequencing of the human genome has not yet resulted, as many had initially expected, in an associated proliferation of the druggable target space (Ashburn and Thor, 2004; Fleming and Ma, 2002), limiting our ability to generate novel biological hypotheses based on newly characterized drug targets. From this, it becomes clear that a new way of looking at the existing drug target space is needed, based on the fact that many drug targets are ultimately shared by more than one drug (Andronis et al., 2011).

The serendipitous repositioning that was practiced in the early years has now been replaced by systematic approaches, which look for repositioning opportunities from the very early days of a drug's lifecycle, that is, from its infancy as an investigative compound, until after it has matured in the market and its patent coverage has expired. The scientific discipline that looks systematically for new uses of existing drugs, marketed or investigative, is called *Systematic Drug Repositioning* (SDR). The methods employed, ranging from *in silico* (i.e., computer based) to testing in cell cultures or animal models of diseases, are elaborated on in other chapters of this section. Here we will focus on the basic elements of computational drug repositioning and we will overview the various *in silico*, *in vivo*, and *in vitro* approaches (Section 4.2). We will also touch upon the use of SDR technologies to predict hitherto unknown adverse drug reactions (ADRs), or to derive mechanistic explanations of known ADRs (Section 4.3). In Section 4.4, we will use recent examples of discovery of new indications for existing drugs and for the prediction of ADRs to illustrate these concepts. In Section 4.5, we will discuss the intellectual property (IP) and commercial aspects of SDR, which are among the major factors that will determine its success.

4.2 TECHNOLOGICAL OVERVIEW

4.2.1 COMPUTATIONAL DRUG REPOSITIONING

4.2.1.1 Literature-Based Methods

Virtually all computational approaches to drug repositioning are based on the high-throughput processing of large biomedical *corpora*. Depending on the specifics of each algorithm, said corpora may be PubMed, the database of biomedical scientific publications,* patent repositories such as the U.S. Patent and

---

* PubMed. Available at: http://www.ncbi.nlm.nih.gov/pubmed/ [Accessed April 7, 2017].

Nortriptyline i-MoA		
Genes	472	
Pathways		66
Diseases		808
Adverse events	78	
Drugs	629	
Compounds		70

Cytokine release syndrome		
Genes	85	
Pathways		25
Diseases		70
Adverse drug reactions		79
Drugs	39	
Compounds		11

FIGURE 4.1 An example multidimensional profile of a drug (nortriptyline) and a pathological condition (cytokine release syndrome). Each profile can be visualized as the barcode of the drug or outcome of interest and is used by the drug repositioning algorithm to determine strength of association. Here, nortriptyline is associated in the biomedical literature with 465 genes, 66 pathways, etc. Similarly, cytokine release syndrome is associated with 85 genes, 25 pathways, etc. The common elements among these two entities (circle) determine their strength of association.

Trademark Office (USPTO),* and the European Patent Organization databases (EPO).† Here, the computer processes the corpora looking for text matches of biological entities of interest, such as the names of genes, proteins, biological pathways, posttranslational modifications, diseases, ADRs, drugs, and classes of organisms. It then creates pairwise associations (also called *links*) between the various entities found and expresses these associations in terms of their *strength*. Each *corpus* yields many millions of entity associations. The step of discovering a new use for a drug or a novel ADR comprises an exhaustive search of the database for indirect links between seemingly disparate entities. Typically, both the starting drug and all outcomes of interest (new indications, novel ADRs, etc.) are represented in terms of their *profiles*. A profile usually includes all the entities covered by each particular algorithm to which the starting drug and all outcomes of interest are associated in the corpus. Figure 4.1 shows the example profiles of a drug (nortriptyline) and a pathological condition (cytokine release syndrome), extracted from the biomedical literature. The common elements of these profiles determine the strength of association between the two entities.

Finally, the outcomes of interest (new indications, novel ADRs, etc.) are ranked according to the strength of their association of their profiles with that of the starting drug. This process is illustrated in Figure 4.2.

---

* United States Patent and Trademark Office. Available at: http://www.uspto.gov/ [Accessed April 7, 2017].
† European Patent Office. Available at: http://www.epo.org/ [Accessed April 7, 2017].

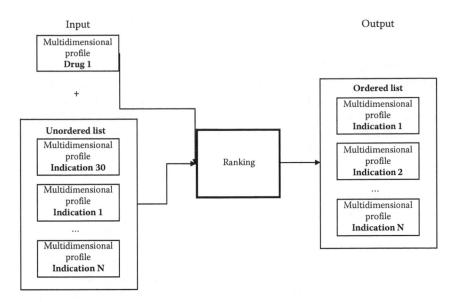

FIGURE 4.2 An example ranking and prediction workflow, for drug repositioning. The algorithm compares the multidimensional profile of the drug of interest (drug 1) with the profiles of all its possible indications (indication 1, ..., N). It then ranks the indications according to the similarity (strength of association) of their profiles with that of drug 1. The contents and details of the profiles depend on the algorithm used.

Swanson implemented his original algorithm in a tool called Arrowsmith (Smalheiser et al., 2009), currently available online.* This work has been expanded by Gordon and Lindsay, who used lexical statistics over titles and abstracts to recreate Swanson's discoveries (Lindsay and Gordon, 1999), while Weeber pursued the same goal using the Unified Medical Language System (UMLS) and lexical tools to map natural language text to UMLS concepts (Bodenreider, 2004). The tool they built, called DAD, was based on the MetaMap program (Aronson, 2001) to map words in the abstract to UMLS concepts. Weeber's work resulted in four proposed novel therapeutic applications for thalidomide: myasthenia gravis, chronic hepatitis C, *Helicobacter pylori*–induced gastritis, and acute pancreatitis (Weeber et al., 2003). LitLinker (Yetisgen-Yildiz and Pratt, 2006), Telemakus (Fuller et al., 2004), the Associative Concept Space (van der Eijk et al., 2004), and TransMiner (Wren et al., 2004) are other tools and algorithms that employ literature-based methods for SDR.

In more recent approaches, the biomedical literature has been enriched with protein expression data to arrive at mechanism-based disease similarity classifications (Liu et al., 2014) or with drug–disease associations from the U.S. database of clinical trials (ClinicalTrials.gov) to produce a corpus of extraction of accurate drug–disease treatment pairs (Xu and Wang, 2013). In the Clinical Outcome Search Space, platform (COSS™) PubMed has also been enriched with information extracted from the

---

* Arrowsmith. Available at: http://arrowsmith.psych.uic.edu/arrowsmith_uic/index.html [Accessed October 19, 2015].

biomedical patent literature, in an effort to increase the novelty of the predicted indications and to cover the IP issues related to such predictions.

4.2.1.2 Information Extraction

Central to all literature-based SDR efforts are the algorithms for detecting the entities of interest (genes, proteins, biological pathways, drugs, diseases, etc.) in the respective corpora. These algorithms comprise an analytical step that is intuitively called *Information Extraction* (*IE*) and produces the basic building blocks upon which all further discovery steps build. IE usually begins with Named Entity Recognition (NER), which attempts to detect biomedical terms in free text (Andronis et al., 2011). Terms might be identified using controlled vocabularies, such as UMLS (Bodenreider, 2004), MeSH for diseases,* Uniprot (UniProt Consortium, 2010) and NCBI Entrez Gene† for genes, and Reactome (Matthews et al., 2009a) for pathways, or through Natural Language Processing (NLP) and machine learning techniques (Cohen and Hunter, 2008; Zweigenbaum et al., 2007). NER is an important first step in IE. The quality of NER determines the quality of the output of IE and as a result the quality of the final output of the drug repositioning algorithm. However, the lack of standardization of names and the issues of synonymy and polysemy make it a difficult, sometimes impossible, task (Cohen and Hersh, 2005; Jensen et al., 2006). These problems are most evident with genes/proteins. Genes/proteins are described with a variety of descriptors, such as the gene symbol, the gene name, the gene product name, and various synonyms. However, in most cases, there is more than one gene symbol per gene and scientists tend to refer to a gene with different names in the literature, many times not using the "official" symbol. A good example is the p38 MAP kinase (Entrez Gene ID: 1432). The official gene symbol for that gene is MAPK14 and the official full name is mitogen-activated protein kinase 14. However, very few scientists refer to that gene with the name MAPK14—at least in the abstract of the article. To alleviate this issue, the Entrez Gene database contains a variety of synonyms for MAPK14, including p38. However, the issue of synonymy and lack of standardization are not the only challenges for determining the identity of a gene in free text. p38 is a synonym for over 20 different genes from a variety of organisms, including humans, flies, and viruses. This phenomenon is called polysemy and refers to the capacity of a name to have multiple meanings (i.e., functions). Recent methods have addressed the issue of gene disambiguation by integrating various features of the gene names, as their context and their linguistic, character, semantic, and case pattern characteristics (Andronis et al., 2011; Wei et al., 2015).

The early implementations of IE were mainly based on the identification of concepts that are coreported (i.e., they co-occur) within the same abstract or sentence (Ding et al., 2002). Despite its simplicity, co-occurrence-based IE continues to yield good results, even in recent publications (Kastrin et al., 2014). Because, however, it only detects coreported terms and does not attempt to determine whether the terms are semantically linked within the text, several groups have used NLP techniques

---

* Medical Subject Heading. Available at: http://www.ncbi.nlm.nih.gov/mesh/68046650 [Accessed April 7, 2017].
† NCBI Entrez Gene. Available at: http://www.ncbi.nlm.nih.gov/sites/entrez?db=gene [Accessed April 7, 2017].

that attempt to follow the semantics of a body of text and to identify concepts of interest therein. While NLP methodologies have improved in recent years (Chen et al., 2008; Cohen and Hunter, 2008), the task remains challenging (Shivade et al., 2015). Recent research, however, has produced combined approaches, where NLP-based NER is augmented by machine learning. Such methods have been applied to selected clinical problems with good results (López Pineda et al., 2015).

Finally, significant research efforts have been devoted to graph-based methods, where the semantics of words is sought in their interconnections within a body of text that is treated as a graph (Jiang et al., 2016), and to unsupervised or supervised text mining, which exploits such algorithms as the k-Nearest Neighbor and support vector machines (Quan et al., 2014).

4.2.1.3 Exploiting Ontologies and Other Databases

Similar to literature-based drug repositioning, where biomedical concepts are extracted from biomedical texts and combinatorially analyzed to produce novel results, various groups have exploited known associations between genes, proteins, diseases, drugs, biological pathways, ADRs, etc., which are already stored in public biomedical ontologies and in other databases. UMLS, Reactome, and Entrez Gene are among the most widely used ontologies for this purpose and have been referred to in the earlier section. Other well-known ontologies are *The Kyoto Encyclopedia of Genes and Genomes* (KEGG),* which includes genes, biological pathways, diseases, and drugs; the *Gene Ontology*,† which includes genes, proteins, and biological processes; and *OMIM*,‡ an online catalog of human genes and genetic disorders.

In DrugBank, the focus is shifted to drugs, their mechanism of action, and their chemical properties. This comprehensive resource combines detailed drug (i.e., chemical, pharmacological, and pharmaceutical) data with drug target (i.e., sequence, structure, and pathway) information. It contains FDA-approved small molecule drugs and biotechnology products, as well as neutraceuticals and experimental drugs.§ The FDA also maintains a database of approved products,¶ while the FDA Adverse Event Reporting System (FAERS)** systematically collects drug adverse events, which are voluntarily entered by the observing physicians. SIDER Side Effects†† is a resource similar to FAERS that contains information on marketed medicines and their known ADRs, extracted from public documents and package inserts. Finally, umbrella projects such as STITCH‡‡ and the disease–drug correlation ontology (DDCO) (Qu et al., 2009) attempt to bring more than one resource together.

---

*   Kyoto Encyclopedia of Genes and Genomes (KEGG). Available at: http://www.genome.jp/kegg/ [Accessed April 7, 2017].
†   Gene Ontology. Available at: http://geneontology.org/ [Accessed April 7, 2017].
‡   Online Mendelian Inheritance in Man. Available at: http://www.omim.org/ [Accessed April 7, 2017].
§   DrugBank. Available at: http://www.drugbank.ca/ [Accessed April 7, 2017].
¶   FDA Drug Approved Products. Available at: http://www.accessdata.fda.gov/scripts/cder/drugsatfda/ [Accessed April 7, 2017].
** FDA Adverse Event Reporting System. Available at: http://www.fda.gov/Drugs/GuidanceCompliance RegulatoryInformation/Surveillance/AdverseDrugEffects/ [Accessed April 7, 2017].
†† SIDER Side Effect Resource. Available at: http://sideeffects.embl.de/ [Accessed April 7, 2017].
‡‡ STITCH 4.0. Available at: http://stitch.embl.de/cgi/show_network_section.pl?identifier=-4594&input_ query_species=9606 [Accessed April 7, 2017].

In the same way that several reswearch groups have used concept associations in the literature, in ontologies, or in drug and ADR databases to identify novel drug indications, others have employed transcriptional data produced by microarray experiments for the same purpose (Huang et al., 2014; Iorio et al., 2013; Jahchan et al., 2013; Zerbini et al., 2014). Here the transcriptional profile of a disease or a disease stage (e.g., renal cancer or advanced nonsmall cell lung cancer) is obtained through microarray experiments and consists of hundreds of genes that are up- or downregulated. This profile is then matched against a list of drugs, in pursue of novel candidates that have the potential to target the specific pathways involved in the disease of interest and to reverse, if possible, the molecular changes indicated by the microarray results.

The use of the resources discussed offers an approach to SDR that is alternative to the biomedical literature. While some of this information can also be found in published papers, such as the final interpretation of a microarray experiment or the results of a study based on reported ADRs, raw data are not usually published. Processing of said raw data by the SDR software can lead to novel results, which may have been missed by those who have originally collected and processed the data. Thus, SDR based on the literature and on the resources discussed in this section can be used in tandem, one approach expanding and validating the results of the other. Many groups that offer drug repositioning services in fact advertise this multidimensional approach to SDR (Biovista, 2017; Melior Discovery, 2017).

Chapter 5 elaborates on the details of computational drug repositioning algorithms and discusses their advantages and disadvantages in depth.

4.2.2   CHEMINFORMATICS, *IN VITRO* AND PHENOTYPIC SCREENING

4.2.2.1   Cheminformatics

When repositioning an existing drug to a new indication, in most cases we make one of the following two hypotheses: either we find the known molecular target of the drug to be (unexpectedly) involved in the proposed indication, or we believe that the drug has additional, hitherto unknown, molecular targets that are involved in the pathogenesis of the indication. Duloxetine, for example, is approved for the treatment of depression, where it works by inhibiting the reuptake of serotonin and norepinephrine at the synaptic cleft (CYMBALTA Package Insert, 2017). It has also been found that duloxetine ameliorates an overactive bladder, through the same mechanism of action: augmentation of norepinephrine levels, albeit at a different neural site than that responsible for its antidepressive effect (Wang et al., 2015). Pirlindol, on the other hand, is an inhibitor of monoamine oxidase type A, also approved in certain countries as an antidepressant. It has been shown in animal models that pirlindol is also effective against progressive multiple sclerosis (MS) due to an entirely different mechanism of action, namely due to its antioxidant effects (Deftereos et al., 2012). The former is an example of an "on-target" effect, while the latter is an example of an "off-target" effect. Most drugs, particularly small-molecule drugs as opposed to monoclonal antibodies, have off-target effects and are called *promiscuous*. Another example of a promiscuous drug is mirtazapine, primarily an antagonist of the α_2-adrenergic and 5-HT1a serotoninergic receptors, currently approved as an antidepressant (Remeron Package Insert, 2015). The drug, however, is also a potent antagonist of H1 histamine receptors, a property that makes it an excellent sleep inducer. In fact, mirtazapine is often used for the treatment of insomnia (Kamphuis et al., 2015).

Cheminformatics, also known as chemical informatics, is the scientific discipline that uses shape, conformation, and other descriptors of chemical compounds to predict their binding affinity to molecular, usually protein, targets (Sukumar et al., 2008). In a typical scenario, *in silico* algorithms are used to predict the binding affinity of one or more drugs of interest against a wide range of potential target receptors. Both the drugs and the receptors are represented in three-dimensional structures with additional properties, such as the electrical charge of their constituent atoms; the algorithm tries to find the best conformation of a drug that can fit the active sites of the receptors and to predict the corresponding binding affinities. Three-dimensional representation of many compounds can be found in databases such as ZINC,* while the Protein Data Bank (PDB) is a widely used resource for three-dimensional models of proteins (Rose et al., 2015).

Computational approaches to assess the binding affinity of such receptor/drug complexes are very helpful during the initial screening of candidate compounds, or when experimental measurements are brought to their limits, and aim to have a predictive function to assess such binding affinities. They can be roughly divided into two classes, docking and free energy calculations from molecular dynamics (MD) simulations. In the docking approach, a binding site is defined in a, typically, rigid protein. A drug is then fitted into this binding site by flexible rotation of functional groups within the drug and rotation of the drug itself. Electrostatic and van der Waals interactions are calculated for different conformations and a scoring function evaluates the drug conformations, which energetically fit best. Several applications with different scoring functions, such as Autodock (Morris et al., 1998), FlexX (Rarey et al., 1996), or Gold (Jones et al., 1997), have been developed over the past years. It is common to these applications that an implicit solvent environment is used and the flexibility of the protein is often neglected. Despite the advantage to screen large libraries of drugs in a short amount of time, the different scoring functions often lead to different, inconsistent, results, especially in cases where water molecules in the binding pocket are important (de Graaf et al., 2006).

MD is predominantly used to refine and rescore already decently docked ligand poses, that is, poses in which the drug is already in the correct local minimum at the beginning of the simulations and thus equilibrium MD can be applied (Alonso et al., 2006). Typically, several nanoseconds need to be sampled in order for a binding affinity to be estimated with reasonable accuracy and many such simulations of different ligands, or of different poses of the same ligand, need to be run within the context of a single project. Cheminformatics methods for SDR are covered in more detail in Chapter 5.

4.2.2.2 *In Vitro* and *In Vivo* Drug Screening

While computational SDR uses the mechanism of action of the drugs under study to identify novel indications, *in vitro/in vivo* screening uses the opposite approach. A library of compounds is tested against an experimental end point, namely efficacy

* ZINC Database. Available at: http://zinc.docking.org/ [Accessed April 7, 2017].

in an *in vitro* or *in vivo* model of the disease of interest. An end point in an *in vitro* model may be the chemically measured binding affinity of all compounds in the test library against a target receptor. Here the receptor is usually available in a soluble form. An *in vitro* or *in vivo* disease model, on the other hand, comprises either a cell culture or an animal population that has been modified genetically or by other means to mimic the disease of interest. Cellular models are very often used in cancer, as well as in other diseases (Briand, 1970). Animal models exist for a multitude of diseases as well; the experimental allergic encephalomyelitis mouse model of MS, the kainic acid mouse model of epilepsy, and the ischemia-reperfusion mouse model of macular degeneration are some examples.

During *in vivo* screening, the drugs of interest are tested in the culture or in animals. Those drugs that show efficacy are candidate treatments for the respective indication. Obviously, the quality of the predictions depends on how accurately the model used represents the corresponding human disease. Because *in vivo* screening mainly comprises the evaluation of disease phenotypes as represented in cells or animals, this process is also called *phenotypic screening* (Vincent et al., 2015). Phenotypic screening has been used for the discovery of antituberculosis agents (Hervé et al., 2015), in neurodegenerative diseases (Khurana et al., 2015), and in cancer (Boone et al., 2015) among many examples.

In phenotypic screening, the mechanism of action of novel drug candidates in the target diseases is not necessarily known. A drug previously approved for a different indication may act in the new disease through its known (on-target) or hitherto unknown (off-target) mechanisms. The study of proteins' expression levels and of other biochemical parameters in the culture or in the experimental animals used may be attempted in order to elucidate the observed drug effects.

Phenotypic SDR is discussed in detail in Chapter 7.

4.3   DRUG REPOSITIONING *VERSUS* BENEFIT/ RISK ASSESSMENT: TWO SIDES OF THE SAME COIN

Drug repositioning and ADR prediction are two sides of the same coin; the same methods that are used for computational SDR have also been applied to the prediction of unknown drug ADRs, or to elucidate the mechanistic cause of an ADR and to predict at-risk patient subpopulations (Gronich et al., 2015). This makes sense: an ADR is just another clinical outcome, as is a novel indication. In the former case, the outcome is unwanted, while in the latter case it is desirable. And vice-versa: many cases of repositioned drugs, especially in the early days, sprung from the observation of an ADR that was undesirable under certain circumstances, but desirable under different circumstances. The most well-known example is that of sildenafil, which was originally evaluated for the treatment of angina pectoris. It was observed, however, that patients presented penile erection as an adverse effect, which led to its further development in its current indication, erectile dysfunction (Andronis et al., 2012).

Further to their obvious importance with regard to patient safety, ADRs can also limit the market size of a drug or may lead to its withdrawal. Thus, the better the understanding of the safety profile of a drug candidate, the more likely its commercial success. Some believe that extensive use of ADR prediction technologies could reduce the cost of drug development by 50% (PricewaterhouseCoopers, 2002/2003). Contemporary approaches focus on the use of cellular assays and animal models for the characterization of the pharmacokinetic and safety profiles of drugs. They comprise a separate discipline in their own right and are collectively referred to as ADME/tox (Kerns and Di, 2008). *In silico* ADME/Tox models are increasingly used, especially when screening large compound libraries, in an effort to reduce the cost of massive *in vitro/in vivo* screening. The latter is reserved for the most promising candidates (Wang et al., 2015). *In silico* ADME models range from plain filtering of candidate compounds by the well-known Lipinski rule of five (Amat-Ur-Rasool and Ahmed, 2015; Leeson, 2012) to computer-simulated measurements of oral bioavailability (The HP et al., 2011; Zhu et al., 2011).

Despite the medical and fiscal importance of adverse event prediction, the number of publications on the issue is surprisingly small, especially when the methodology of prediction is required to include the computation drug repositioning methods discussed in Section 4.2.2 (Deftereos et al., 2011). In a recent report, the Center for Food Safety and Applied Nutrition of the FDA combined multiple sources of data in an attempt to predict cardiac ADRs in humans (Matthews and Frid, 2010). The authors created a database of cardiac ADRs and used it to (1) construct quantitative structure–activity relationship (QSAR) models that could predict cardiac ADRs of untested chemicals, (2) identify different properties of pharmaceutical molecules that correlate with rare and unexpected cardiac ADRs observed in patients, and (3) identify plausible mechanisms by which the drugs might have caused the ADRs, on the basis of these *in silico* data. In this approach, drugs were classified according to (1) the clinical indications for which they were prescribed, (2) their primary target, (3) their mechanism of action, and (4) their structural similarity to other drugs, known to bind to specific receptors. Drug-related ADRs were derived from FDA's Spontaneous Reporting System (SRS) and Adverse Event Reporting System (FAERS), postmarket surveillance databases, and a supplement of adverse event data from published medical literature while drug mechanisms of action and target affinities were compiled for 2124 FDA-approved drugs through quantitative structure–activity relationship (QSAR) modeling. It was found that cardiac ADRs correlate with a small number of mechanisms of action, namely those affecting cardiovascular functions (such as alpha-adrenoceptor, beta-adrenoceptor, and calcium channel blocker) and cardioneurological functions (5-hydroxytryptomine receptor, dopamine receptor, and acetylcholinesterase) (Matthews and Frid, 2010). The authors suggested that screening of new chemical entities for the presence of these mechanisms of action could predict a major portion of cardiac ADRs that might occur in patients, and that this technology might be used proactively for the early detection of ADRs in clinical trials and for the investigation of rare, unexpected, and idiosyncratic ADRs that are identified by pharmacovigilance and postmarket surveillance. A similar approach had been used by the same authors to predict hepatobiliary and urinary tract ADRs (Matthews et al., 2009b; Ursem et al., 2009). Association rule mining, a well-established data mining method, has been employed to detect in FAERS associations between multiple drugs and potential ADRs (Harpaz et al., 2010).

More recently, we have used our (COSS) platform, which integrates multiple sources of biomedical data, including the biomedical and patent literature, to elucidate the mechanism through which statins may cause *de novo* diabetes mellitus, an ADR that has been detected in the last few years and has raised concerns about the current widespread use of these drugs (Gronich et al., 2015). We have also identified risk factors that may increase an individual's susceptibility to this ADR. We discuss this example more extensively in Section 4.4.1.

An important step in predicting hitherto unknown ADRs is the identification of novel off-target effects of the drugs under study. One way to achieve this is to find indirect connections between a drug and a new target in a corpus, such as the biomedical literature, potentially integrated with other disparate data sets, such as FAERS, and genomic information on target proteins contained, for example, in KEGG (Takarabe et al., 2012). Another way is to exploit chemical structures, either by estimating the binding affinity of a drug to a set of potential targets that are known to mediate certain ADRs through docking and the related technologies (Section 4.2.1) (Ehrlich et al., 2015; Yang et al., 2011), or by using said structures as an additional data set in the predictive approaches discussed here (Vilar et al., 2014).

The efficacy of *in silico* methods for ADR prediction is expectedly limited by the weaknesses that are inherent is all algorithms, discussed in Section 4.2. Despite its shortcomings, *in silico* ADR prediction can be very useful in the early stages of drug development, where it can inform the design of clinical trials. If an ADR is expected, it is fairly easy to study it in an upcoming clinical trial. If not, it may be reported in the postmarketing period, where understanding and handling it is more difficult and can require more resources significantly. Even if used post marketing, however, *in silico* ADR prediction can assist in understanding the mechanism of action underlying an ADR, in identifying patient subpopulations that are susceptible to it, and in developing appropriate biomarkers. This can reduce the impact of an ADR on patients, on healthcare expenditures, and on the companies that market the responsible drugs.

4.4   CASE STUDIES

In this section, we will walk through examples of successful drug repositioning and ADR prediction that will, hopefully, illustrate some of the concepts discussed earlier.

4.4.1   IDENTIFYING RISK FACTORS FOR THE DEVELOPMENT OF STATIN-INDUCED DIABETES MELLITUS

It was around 2010 when a meta-analysis of clinical trials involving statins, the widely used class of cholesterol-lowering drugs, confirmed that they increase the risk for development of new-onset diabetes mellitus, by approximately 9% (Sattar et al., 2010). This discovery fuelled intense discussions in the scientific community, regarding the risk *versus* benefit of the use of statins in various subpopulations, especially in younger patients; it became less obvious that we should start statin treatment in, for example, a 40-year old male with high blood cholesterol, but without other risk factors for atherosclerosis such as hypertension or smoking (Bleakley et al., 2015).

One way to work around this problem would be to identify risk factors for the development of statin-induced diabetes and to avoid their use in patients having said risk factors. However, epidemiological studies have only identified very broad risk groups, such as the elderly, women, and people of Asian origin, which are not of practical importance to clinicians (Goldstein and Mascitelli, 2013). It is not useful, for example, to know that the elderly are at increasing risk for developing statin-induced diabetes, since it is exactly this age group that also benefits the most from the reduction of cholesterol. Furthermore, these factors cannot be modified.

We, therefore, at Biovista, in cooperation with the Office of Clinical Pharmacology of the FDA, decided to use our COSS platform for drug repositioning and ADR prediction to elucidate the mechanism of action through which statins cause diabetes and to identify risk factors that could assist clinicians in making more educated decisions on whom to treat. Using a literature-based approach, as described in Section 4.2.1.1, we discovered that hypothyroidism, either overt or subclinical, can increase the risk for development of diabetes, for a number of reasons (Gronich et al., 2015). First, insulin resistance is found in patients with hypothyroidism (Dimitriadis et al., 2006). Second, impaired translocation of GLUT4 glucose transporters on the plasma membrane in patients with hypothyroidism and subclinical hypothyroidism (Teixeira et al., 2012) as well as downregulation of the hepatic glucose transporter GLUT2 (Maratou et al., 2009) have been demonstrated. Even low-normal free thyroxin levels in patients with euthyroidism have been associated with insulin resistance (Gronich et al., 2015; Roos et al., 2007; Weinstein et al., 1994). Third, thyroid disease induces mitochondrial dysfunction (Kvetny et al., 2010), and statins reduce levels of coenzyme Q10, a component of the electron transport chain involved in the process of ATP generation (Giudetti et al., 2006), also leading to mitochondrial dysfunction, as well as causing reduced insulin release and pancreatic β-cell failure, and contributing to the development of diabetes. Thus, COSS pinpointed these common mechanistic links between statins, diabetes, and thyroid disease and suggested that the latter may be a modifiable risk factor. We confirmed this hypothesis in a large epidemiological study, where we found that in highly compliant patients taking at least 80% of their prescribed statin doses in a follow-up of 5–7 years, the risk for developing diabetes was 2.5-fold (Gronich et al., 2015).

This case study illustrates a process through which computational drug repositioning methods can suggest mechanistic hypotheses that can help explain an ADR and propose modifiable risk factors. Epidemiological analyses can then be used to validate the hypotheses. Such mechanisms of action-guided epidemiological studies have higher changes of success, compared to nonguided studies that attempt to find statistically significant associations by randomly correlating patient data.

What we found means that if we request a thyroid stimulating hormone (TSH) measurement, an inexpensive and reliable test for thyroid dysfunction, before initiating statin treatment, we can detect a significant proportion of those patients that will develop diabetes mellitus. Correcting the dysfunction by thyroid hormone supplementation takes the risk back to normal, thus the development of diabetes can be prevented.

4.4.2 Elucidating the Biological Mechanism Underlying an ADR: The Case of Telithromycin

Telithromycin was the first ketolide antibiotic to enter clinical use. It is used to treat community-acquired pneumonia of mild-to-moderate severity (KETEK summary of product characteristics, 2017). Telithromycin has been associated with a spurious foursome of ADRs: liver failure, loss of consciousness, temporary vision loss, and exacerbation of myasthenia gravis (KETEK summary of product characteristics, 2017; Prescrire Editorial Staff, 2014). We were asked to identify mechanistic underpinnings for these ADRs.

Here we used drug–ADR associations for all FDA-approved drugs, which are extracted from the biomedical literature and are stored in COSS. We also used ADR reports in FAERS and those referred to in the drug SPCs. We looked for other drugs with similar adverse event profiles to telithromycin and we found that the drugs that can cause the same foursome of ADRs at a frequency comparable to that of telithromycin are not too many (in the order of a few tens). We then studied the mechanisms of action of these drugs and concluded that the specific set of ADRs can be justified by an anticholinergic mechanism of action (data not published). This work was carried out in 2008. Later, in 2010, a published report showed that a pyridine moiety that is part of the telithromycin molecule acts as an antagonist on cholinergic receptors located in the neuromuscular junction, the ciliary ganglion of the eye, and the vagus nerve innervating the liver (Bertrand et al., 2010). Other macrolides, such as azithromycin and clarithromycin, and the fluoroketolide, solithromycin, do not contain the pyridine moiety and do not antagonize these cholinergic receptors significantly. This elucidated the mechanism underlying the ADRs observed with telithromycin and allowed other drugs of the same or similar classes to be confidently differentiated, in terms of the specific ADRs.

4.4.3 From Psoriasis to Multiple Sclerosis: The Case of Dimethyl Fumarate

Dimethyl fumarate (DMF) is the methyl ester of fumaric acid. DMF was initially recognized as a very effective hypoxic cell radiosensitizer (Held et al., 1988). Later, DMF combined with three other fumaric acid esters was licensed in Germany as oral therapy for psoriasis (Mrowietz et al., 2007). Research in psoriasis leads to the realization that DMF has a number of immunomodulatory properties that were thought to be relevant to MS as well. DMF, in particular, has been shown to induce interleukin (IL)-10, IL-4, and IL-5 expression in peripheral blood mononuclear cells *in vitro* without changing interferon (IFN)-γ, IL-12, and IL-2 levels and to increase the production of IL-4 and IL-5 in T cells *in vitro* (Schilling et al., 2006). Tumor necrosis factor (TNF)-α levels are affected by DMF, while other *in vitro* studies have shown that DMF can inhibit the transcription of many proinflammatory cytokines and this inhibition appears to correlate with a blockade of the TNF-induced nuclear translocation of an NF-κB p65 (Schilling et al., 2006).

The relevance of these mechanistic effects of DMF to MS motivated Schimrigk et al. to conduct a small open-label clinical trial of DMF in MS patients

(Schimrigk et al., 2006). This initial trial yielded positive results; it was followed up by further Phase II and Phase III trials, which finally led to the approval of the drug for the treatment of relapsing–remitting MS.

The case of DMF illustrates the full life cycle of a drug, from marketing in one disease to repositioning and approval for another disease, based on the research conducted in the context of the former. What is also important with DMF is that the owner, Biogen Idec Inc., funded a full clinical development program in MS, despite the fact that this is an older drug with limited patent coverage. While the drug was launched in the United States in April 2013, its patent coverage is reported to expire in 2019 (DrugPatentWatch, 2017). This 6-year IP protection was adequate to justify development expenditures. Thus, SDR is a commercially viable strategy.

4.4.4 COPING WITH MORE THAN ONE PATIENT'S NEEDS AT THE SAME TIME: USING PIRLINDOL, AN ANTIDEPRESSANT, FOR MULTIPLE SCLEROSIS

The MS market is crowded with treatments for the relapsing–remitting forms. Where there is a real medical need, however, are the progressive forms, in which the currently available treatments have not proved effective. Using the COSS platform discussed in Section 4.2.1, we discovered that Pirlindol, an antidepressant developed and marketed at the former Russian Federation, has additional properties that can be beneficial to progressive MS.

In particular, Pirlindol reduces oxidative stress and inhibits lipid peroxidation, a process that is involved both in relapsing–remitting and in the progressive forms of MS. In relapsing–remitting MS, Pirlindol protects neural cells and myelin during the inflammatory attack and prevents the axonal damage that is evident even from the onset of the disease. Prevention of axonal damage is even more important in primary and secondary progressive MS, where this pathological abnormality is more prominent (Deftereos et al., 2012; Kvetny et al., 2010).

We showed that Pirlindol is neuroprotective in MOG-induced experimental allergic encephalomyelitis, a mouse model of progressive MS. The drug exerted a sizable, statistically significant reduction in disease severity and axonal damage, which was independent of any anti-inflammatory effects (Lekka et al., 2011). Furthermore, in its original use as an antidepressant, Pirlindol proved to have similar efficacy to tricyclic antidepressants, selective serotonin reuptake inhibitors, and nonselective, irreversible MAO inhibitors. Thus, it could offer additional benefits to MS patients with concomitant depression and could help treat both conditions without having to resort to additional medications.

4.5 INTELLECTUAL PROPERTIES AND BUSINESS CONSIDERATIONS

There are three key issues relating to IP of repositioned drugs and their business development potential. First, can one actually obtain IP protection for a repositioned drug. Second, do IP-protected new uses have real-world value that can be monetized. Finally, is it possible to promote off-label new uses in the marketplace. We now examine each of these issues in turn.

4.5.1 Obtaining IP Protection

Prevailing perception is that in the case of drugs, composition of matter IP is all that matters. This usually means that if an entity owns the IP on the actual chemical composition and structure of a drug, then any other IP is irrelevant and not as powerful; in other words, it is a "second class" patent. In some cases, especially with biologics that already exist in nature, the original IP covers methods of synthesis, since composition of matter protection for such drugs is not typically possible. Here, too, the perception is that such IP is more powerful than new use IP. These perceptions lack key understanding of what a patent actually is and what it allows its owners to accomplish.

First, there is no distinction in the eyes of the patent office between "first class" patents and others. If a patent is issued, then it is equal to any other issued patent. All things being equal, and any two patents surviving enforceability challenges, they will be equally powerful in terms of what they allow their owner to do.

Second, a patent is a blocking tool. The only real use of a patent is to block others from using the invention for monetary gain. This blocking element is key, since it allows the owner of a new use patent to block the original owner of a composition of matter to actually use their drug in the new use. If the original developer of a drug who owns its composition of matter does not obtain a new use patent on diseases beyond the original uses, then they cannot develop or sell their own drug in the new use, because they are blocked by the new use patent that somebody else owns. This is particularly important since follow-on drugs are also affected, if they are similar to the original drug. Here, it is important to understand that the blocking function of a patent works both ways: the new use IP owner blocks the composition of matter owner from the new use, but the composition of matter owner also blocks the new use IP owner from monetizing, until such time as the drug becomes generic. At that point, the new use IP owner is the only entity that can promote and sell the drug in the new use—assuming it is FDA approved in the new use following appropriate studies, including 505(b)2 type studies.

As an example, the 2014 lawsuit between Gilead and AbbVie serves to illustrate how this may play out in the courts (Terry, 2014). In addition to this specific lawsuit, where the originator is attempting to invalidate the follower's repositioning IP, there are multiple examples of new use patents being published for drugs owned by others, often involving major pharmaceutical companies and drugs. For example, Sanofi's May 23, 2013, patent WO/2013/072328 claims Vertex's Hepatitis C drug telaprevir in atherosclerosis and other uses; Bionor Immuno AS December 12, 2013, patent WO/2013/182660 claims Celgene's anticancer Istodax in HIV; Mass Eye & Ear Infirmary's March 13, 2014, patent WO/2014/039781 claims Lilly's semagacestat drug, which failed in advanced clinical trials for Alzheimer's disease in hearing loss. The latter case is especially important, since failed drugs are often shelved. They can find use in other indications, however, and if the originator does not pursue a strategy of systematic protection of all their assets, even shelved ones, then this opens up the door to third parties to realize significant new value.

4.5.2 MONETIZING A NEW USE

Prevailing perception is that it is not possible for a new use to generate any real-world monetary value. This is not the case.

As an example, Novartis acquired the rights to MS of GSK's cancer drug ofatumumab in August 2015 for $1 billion (Bushey, 2015). This shows how an originator can parse new uses and monetize them separately.

To be clear, significant studies still have to be done to validate a new use clinically and obtain approval, but the time and cost savings are significant, and the monetization potential is as relevant as for any new drug for which there is a medical and market need.

4.5.3 OFF-LABEL USES

The FDA does not allow the promotion of an existing drug for off-label uses by anyone, unless the drug has been FDA-approved for these new uses. In August 2015, the FDA lost a case in a U.S. federal court brought by Amarin. If there are credible and valid data supporting an off-label use, then the sponsor can promote said uses (Burton, 2015).

This is particularly important because it is now possible to reach the market earlier than previously possible, improving even further the return on investment (ROI) associated with developing the new use. The key challenge is passing the test of relevant and sufficient data, which will ultimately be judged as such by individual doctors, considering the off-label use of the drug. This is, however, appropriate, since it means that patients may benefit sooner from particular drugs.

In summary, the IP and business landscape for repositioned drugs is highly favorable. IP can be obtained, it is as powerful as composition of matter and has the same benefits, it can be monetized to the same extent, and it even has the added benefit of reaching the market even sooner as an off-label option than the acceleration it already enjoyed.

REFERENCES

Alonso H, Bliznyuk AA, Gready JE. Combining docking and molecular dynamic simulations in drug design. *Med Res Rev* 2006;26:531–568.

Altura BM. Calcium antagonist properties of magnesium: Implications for antimigraine actions. *Magnesium* 1985;4(4):169–175.

Altura BM, Altura BT. New perspectives on the role of magnesium in the pathophysiology of the cardiovascular system. I. Clinical aspects. *Magnesium* 1985;4(5–6):226–244.

Amat-Ur-Rasool H, Ahmed M. Designing second generation anti-Alzheimer compounds as inhibitors of human acetylcholinesterase: Computational screening of synthetic molecules and dietary phytochemicals. *PLoS One* 2015;10(9):e0136509.

Andronis C, Sharma A, Deftereos S, Virvilis V, Konstanti O, Persidis A, Persidis A. Mining scientific and clinical databases to identify novel uses for existing drugs. In: Barratt M, Frail D (eds.), *Drug Repositioning-Bringing New Life to Shelved Assets and Existing Drugs*. John Wiley & Sons, Hoboken, NJ, 2012, pp. 137–162.

Andronis C, Sharma A, Virvilis V, Deftereos S, Persidis A. Literature mining, ontologies and information visualization for drug repurposing. *Brief Bioinform* 2011;12(4):357–368.

Aronson AR. Effective mapping of biomedical text to the UMLS Metathesaurus: The MetaMap program. *Proceedings of the AMIA Symposium*, 2001, pp. 17–21.

Ashburn TT, Thor KB. Drug repositioning: Identifying and developing new uses for existing drugs. *Nat Rev Drug Discov* 2004;3(8):673–683.

Bertrand D, Bertrand S, Neveu E, Fernandes P. Molecular characterization of off-target activities of telithromycin: A potential role for nicotinic acetylcholine receptors. *Antimicrob Agents Chemother* 2010;54:5399–5402.

Biovista. Available at: http://www.Biovista.com/ [Accessed April 7, 2017].

Bleakley C, Pumb R, Harbinson M, McVeigh GE. A reappraisal of the safety and cost-effectiveness of statin therapy in primary prevention. *Can J Cardiol* 2015;31(12):1411–1414.

Bodenreider O. The Unified Medical Language System (UMLS): Integrating biomedical terminology. *Nucleic Acids Res* 2004;32(Database issue):D267–D270.

Boone JD, Dobbin ZC, Straughn JM Jr., Buchsbaum DJ. Ovarian and cervical cancer patient derived xenografts: The past, present, and future. *Gynecol Oncol* 2015;138(2):486–491.

Briand P. Malignant transformation of cells cultivated in vitro. A model of carcinogenesis with special reference to "spontaneous" malignant transformation. *Dan Med Bull* 1970;17(8):217–225.

Burton TM. Amarin wins off-label ruling against FDA. *The Wall Street Journal*, 2015. Available at: http://www.wsj.com/articles/amarin-wins-off-label-case-against-fda-1438961747 [Accessed April 7, 2017].

Bushey R. Novartis to spend $1B on rights to GSK's MS drug. *Drug Discovery & Development*, 2015. Available at: http://www.dddmag.com/news/2015/08/novartis-spend-1b-rights-gsks-ms-drug?et_cid=4763542&et_rid=230067327&type=cta [Accessed April 7, 2017].

Chen ES, Hripcsak G, Xu H, Markatou M, Friedman C. Automated acquisition of disease drug knowledge from biomedical and clinical documents: An initial study. *J Am Med Inform Assoc* 2008;15(1):87–98.

Cohen AM, Hersh WR. A survey of current work in biomedical text mining. *Brief Bioinform* 2005;6(1):57–71.

Cohen KB, Hunter L. Getting started in text mining. *PLoS Comput Biol* 2008;4(1):e20.

CYMBALTA Package Insert. Available at: http://www.accessdata.fda.gov/drugsatfda_docs/label/2010/022516lbl.pdf [Accessed April 7, 2017].

de Graaf C et al. Catalytic site prediction and virtual screening of cytochrome P450 2D6 substrates by consideration of water and rescoring in automated docking. *J Med Chem* 2006;49:2417–2430.

Deftereos SN, Andronis C, Friedla EJ, Persidis A, Persidis A. Drug repurposing and adverse event prediction using high-throughput literature analysis. *Wiley Interdiscip Rev Syst Biol Med* 2011;3(3):323–334.

Deftereos SN, Dodou E, Andronis C, Persidis A. From depression to neurodegeneration and heart failure: Re-examining the potential of MAO inhibitors. *Expert Rev Clin Pharmacol* 2012;5(4):413–425.

DiGiacomo RA, Kremer JM, Shah DM. Fish-oil dietary supplementation in patients with Raynaud's phenomenon: A double-blind, controlled, prospective study. *Am J Med* 1989;86(2):158–164.

Dimitriadis G, Mitrou P, Lambadiari V, Boutati E, Maratou E, Panagiotakos DB, Koukkou E, Tzanela M, Thalassinos N, Raptis SA. Insulin action in adipose tissue and muscle in hypothyroidism. *J Clin Endocrinol Metab* 2006;91:4930–4937.

Ding J, Berleant D, Nettleton D, Wurtele E. Mining MEDLINE: Abstracts, sentences, or phrases? *Pacific Symposium on Biocomputing*, 2002, pp. 326–337.

DrugPatentWatch. Report on Tecfidera. Available at: http://www.drugpatentwatch.com/ultimate/tradename/TECFIDERA [Accessed April 7, 2017].

Ehrlich VA, Dellafiora L, Mollergues J, Dall'Asta C, Serrant P, Marin-Kuan M, Lo Piparo E, Schilter B, Cozzini P. Hazard assessment through hybrid in vitro/in silico approach: The case of zearalenone. *ALTEX* 2015;32(4):275–286.

Fleming E, Ma P. Drug life-cycle technologies. *Nat Rev Drug Discov* 2002;1(10):751–752.

Fuller SS, Revere D, Bugni PF, Martin GM. A knowledgebase system to enhance scientific discovery: Telemakus. *Biomed Digit Libr* 2004;1(1):2.

Giudetti AM, Leo M, Siculella L, Gnoni GV. Hypothyroidism down-regulates mitochondrial citrate carrier activity and expression in rat liver. *Biochim Biophys Acta* 2006;1761:484–491.

Goldstein MR, Mascitelli L. Do statins cause diabetes? *Curr Diab Rep* 2013;13:381–390.

Gronich N, Deftereos SN, Lavi I, Persidis AS, Abernethy DR, Rennert G. Hypothyroidism is a risk factor for new-onset diabetes: A cohort study. *Diabetes Care* 2015;38(9):1657–1664.

Harpaz R, Chase HS, Friedman C. Mining multi-item drug adverse effect associations in spontaneous reporting systems. *BMC Bioinform* 2010;11(Suppl. 9):S7.

Held KD, Epp ER, Clark EP, Biaglow JE. Effect of dimethyl fumarate on the radiation sensitivity of mammalian cells in vitro. *Radiat Res* 1988;115(3):495–502.

Hervé C, Bergot E, Veziris N, Blanc FX. Tuberculosis in 2015: From diagnosis to the detection of multiresistant cases. *Rev Mal Respir* 2015;32(8):784–790.

Huang CH, Chang PM, Lin YJ, Wang CH, Huang CY, Ng KL. Drug repositioning discovery for early- and late-stage non-small-cell lung cancer. *Biomed Res Int* 2014;2014:193817.

Iorio F, Rittman T, Ge H, Menden M, Saez-Rodriguez J. Transcriptional data: A new gateway to drug repositioning? *Drug Discov Today* 2013;18(7–8):350–357.

Jahchan NS et al. A drug repositioning approach identifies tricyclic antidepressants as inhibitors of small cell lung cancer and other neuroendocrine tumors. *Cancer Discov* 2013;3(12):1364–1377.

Jensen LJ, Saric J, Bork P. Literature mining for the biologist: From information retrieval to biological discovery. *Nat Rev Genet* 2006;7(2):119–129.

Jiang Z, Li L, Huang D. An unsupervised graph based continuous word representation method for biomedical text mining. *IEEE/ACM Trans Comput Biol Bioinform* 2016;13(4):634–642.

Jones G, Willett P, Glen R, Leach A, Taylor R. Development and validation of a genetic algorithm for flexible docking. *J Mol Biol* 1997;267:727–748.

Kamphuis J, Taxis K, Schuiling-Veninga CC, Bruggeman R, Lancel M. Off-label prescriptions of low-dose quetiapine and mirtazapine for insomnia in the Netherlands. *J Clin Psychopharmacol* 2015;35(4):468–470.

Kastrin A, Rindflesch TC, Hristovski D. Link prediction in a MeSH co-occurrence network: Preliminary results. *Stud Health Technol Inform* 2014;205:579–583.

Kerns EH, Di L. *Drug-Like Properties: Concepts, Structure Design and Methods: From ADME to Toxicity Optimization.* Academic Press, Burlington, VT, 2008.

KETEK summary of product characteristics. Available at: http://www.ema.europa.eu/docs/en_GB/document_library/EPAR_-_Product_Information/human/000354/WC500041895.pdf [Accessed April 7, 2017].

Khurana V, Tardiff DF, Chung CY, Lindquist S. Toward stem cell-based phenotypic screens for neurodegenerative diseases. *Nat Rev Neurol* 2015;11(6):339–350.

Kvetny J, Wilms L, Pedersen PL, Larsen J. Subclinical hypothyroidism affects mitochondrial function. *Horm Metab Res* 2010;42:324–327.

Leeson P. Drug discovery: Chemical beauty contest. *Nature,* 2012;481(7382):455–456.

Lekka E, Deftereos SN, Persidis A, Persidis A, Andronis C. Literature analysis for systematic drug repurposing: A case study from Biovista. *Drug Discov Today* 2011;8(3–4):103–108.

Lindsay RK, Gordon MD. Literature-based discovery by lexical statistics. *J Am Soc Inform Sci* 1999;50:574–587.

Liu CC et al. DiseaseConnect: A comprehensive web server for mechanismbased disease–disease connections. *Nucleic Acids Res* 2014;42(Web Server issue):W137–W146.

López Pineda A, Ye Y, Visweswaran S, Cooper GF, Wagner MM, Rich Tsui F. Comparison of machine learning classifiers for influenza detection from emergency department free-text reports. *J Biomed Inform* 2015;58:60–69.

Maratou E et al. Studies of insulin resistance in patients with clinical and subclinical hypothyroidism. *Eur J Endocrinol* 2009;160:785–790.

Matthews L et al. Reactome knowledgebase of human biological pathways and processes. *Nucleic Acids Res* 2009a;37(Database issue):D619–D622.

Matthews EJ et al. Identification of structure activity relationships for adverse effects of pharmaceuticals in humans: C. Use of QSAR and an expert system for the estimation of the mechanism of action of drug-induced hepatobiliary and urinary tract toxicities. *Regul Toxicol Pharmacol* 2009b;54:43–65.

Matthews EJ, Frid AA. Prediction of drug-related cardiac adverse effects in humans—A: Creation of a database of effects and identification of factors affecting their occurrence. *Regul Toxicol Pharmacol* 2010;56(3):247–275.

Melior Discovery. Available at: http://www.meliordiscovery.com/ [Accessed April 7, 2017].

Morris G et al. Automated docking using a Lamarckian genetic algorithm and an empirical binding free energy function. *J Comput Chem* 1998;19:1639–1662.

Mrowietz U, Altmeyer P, Bieber T, Röcken M, Schopf RE, Sterry W. Treatment of psoriasis with fumaric acid esters (Fumaderm). *J Dtsch Dermatol Ges* August 2007;5(8):716–717.

Prescrire Editorial Staff. Telithromycin: Review of adverse effects. *Prescrire Int* 2014;23(154):264–266.

PricewaterhouseCoopers. Pharma 2005 Silicon Rally: The Race to e-R&D. Paraxel's Pharmaceutical R&D Statistical Sourcebook, 2002/2003, PricewaterhouseCoopers.

Qu XA, Gudivada RC, Jegga AG, Neumann EK, Aronow BJ. Inferring novel disease indications for known drugs by semantically linking drug action and disease mechanism relationships. *BMC Bioinform* 2009;10(Suppl. 5):S4.

Quan C, Wang M, Ren F. An unsupervised text mining method for relation extraction from biomedical literature. *PLoS One* 2014;9(7):e102039.

Rarey M, Kramer B, Lengauer T, Klebe G. A fast flexible docking method using an incremental construction algorithm. *J Mol Biol* 1996;261:470–489.

Remeron Package Insert. Available at: https://www.merck.com/product/usa/pi_circulars/r/remeron/remerontablets_pi.pdf [Accessed October 19, 2015].

Roos A, Bakker SJ, Links TP, Gans RO, Wolffenbuttel BH. Thyroid function is associated with components of the metabolic syndrome in euthyroid subjects. *J Clin Endocrinol Metab* 2007;92:491–496.

Rose PW et al. The RCSB Protein Data Bank: Views of structural biology for basic and applied research and education. *Nucleic Acids Res* 2015;43(Database issue):D345–D356.

Sattar N et al. Statins and risk of incident diabetes: A collaborative meta-analysis of randomised statin trials. *Lancet* 2010;375(9716):735–742.

Schiapparelli P et al. Non-pharmacological approach to migraine prophylaxis: Part II. *Neurol Sci* 2010;31(Suppl. 1):S137–S139.

Schilling S, Goelz S, Linker R, Luehder F, Gold R. Fumaric acid esters are effective in chronic experimental autoimmune encephalomyelitis and suppress macrophage infiltration. *Clin Exp Immunol* 2006;145(1):101–107.

Schimrigk S, Brune N, Hellwig K, Lukas C, Bellenberg B, Rieks M, Hoffmann V, Pöhlau D, Przuntek H. Oral fumaric acid esters for the treatment of active multiple sclerosis: An open-label, baseline controlled study. *Eur J Neurol* June 2006;13(6):604–610.

Sheskin J. Thalidomide in the treatment of lepra reaction. *Clin Pharmacol Ther* 1965;6:303–306.

Shivade C, Hebert C, Lopetegui M, de Marneffe MC, Fosler-Lussier E, Lai AM. Textual inference for eligibility criteria resolution in clinical trials. *J Biomed Inform* 2015;58(Suppl):S211–S218.

Smalheiser NR, Torvik VI, Zhou W. Arrowsmith two-node search interface: A tutorial on finding meaningful links between two disparate sets of articles in MEDLINE. *Comput Methods Programs Biomed* May 2009;94(2):190–197.

Smalheizer NR, Swanson DR. Indomethacin and Alzheimer's disease. *Neurology* 1996a;46:583.

Smalheizer NR, Swanson DR. Linking estrogen to Alzheimer's disease: An informatics approach. *Neurology* 1996b;47:809–810.

Sukumar N, Krein M, Breneman CM. Bioinformatics and cheminformatics: Where do the twain meet? *Curr Opin Drug Discov Develop* 2008;11(3):311–319.

Swanson DR. Complementary structures in disjoint science literatures. In: Bookstein A, Chiaramella Y, Salton G, Raghavan VV (eds.), *Proceedings of the 14th Annual International ACM/SIGIR Conference*. ACM Press, New York, 1991, pp. 280–289.

Swanson DR. Fish oil, Raynaud's syndrome, and undiscovered public knowledge. *Perspect Biol Med* 1986;30(1):7–18.

Swanson DR. Migraine and magnesium: Eleven neglected connections. *Perspect Biol Med* 1988;31(4):526–557.

Takarabe M, Kotera M, Nishimura Y, Goto S, Yamanishi Y. Drug target prediction using adverse event report systems: A pharmacogenomic approach. *Bioinformatics* 2012;28(18):i611–i618.

Teixeira SS, Tamrakar AK, Goulart-Silva F, Serrano-Nascimento C, Klip A, Nunes MT. Triiodothyronine acutely stimulates glucose transport into L6 muscle cells without increasing surface GLUT4, GLUT1, or GLUT3. *Thyroid* 2012;22:747–754.

Terry M. AbbVie (ABBV) uses patents to Ambush Gilead (GILD). *BioSpace*, 2014. Available at: http://www.biospace.com/News/abbvie-uses-patents-to-ambush-gilead-sciences-inc/354457 [Accessed April 7, 2017].

The HP, González-Álvarez I, Bermejo M, Mangas Sanjuan V, Centelles I, Garrigues TM, Cabrera-Pérez MA. In silico prediction of Caco-2 cell permeability by a classification QSAR approach. *Mol Inform* 2011;30(4):376–385.

UniProt Consortium. The Universal Protein Resource (UniProt) in 2010. *Nucleic Acids Res* 2010;38(Database issue):D142–D148.

Ursem CJ, Matthews EJ, Kruhlak NL, Contrera JF, Benz RD. Identification of structure–activity relationships for adverse effects of pharmaceuticals in humans A: Use of FDA post market reports to create a database of hepatobiliary and urinary tract toxicities. *Regul Toxicol Pharmacol* 2009b;54:1–22.

van der Eijk CC, van Mulligen EM, Kors JA, Mons B, van den Berg J. Constructing an associative concept space for literature-based discovery. *J Am Soc Inform Sci Technol* 2004;55(5):436–444.

Vilar S, Ryan PB, Madigan D, Stang PE, Schuemie MJ, Friedman C, Tatonetti NP, Hripcsak G. Similarity-based modeling applied to signal detection in pharmacovigilance. *CPT Pharmacomet Syst Pharmacol* 2014;3:e137.

Vincent F et al. Developing predictive assays: The phenotypic screening "rule of 3". *Sci Transl Med* June 24, 2015;7(293):293ps15.

Vosgerau H. Migraine therapy with magnesium glutamate. *Ther Ggw* April 1973;112(4):640 passim.

Wang SM, Lee HK, Kweon YS, Lee CT, Lee KU. Overactive bladder successfully treated with duloxetine in a female adolescent. *Clin Psychopharmacol Neurosci* August 31, 2015;13(2):212–214.

Wang Y et al. In silico ADME/T modelling for rational drug design. *Quart Rev Biophys* 2015;48(4):488–515.

Weeber M, Vos R, Klein H, De Jong-Van Den Berg LT, Aronson AR, Molema G. Generating hypotheses by discovering implicit associations in the literature: A case report of a search for new potential therapeutic uses for thalidomide. *J Am Med Inform Assoc* 2003;10(3):252–259.

Wei CH, Kao HY, Lu Z. GNormPlus: An integrative approach for tagging genes, gene families, and protein domains. *Biomed Res Int* 2015;2015:918710.

Weinstein SP, O'Boyle E, Fisher M, Haber RS. Regulation of GLUT2 glucose transporter expression in liver by thyroid hormone: Evidence for hormonal regulation of the hepatic glucose transport system. *Endocrinology* 1994;135:649–654.

Wren JD, Bekeredjian R, Stewart JA, Shohet RV, Garner HR. Knowledge discovery by automated identification and ranking of implicit relationships. *Bioinformatics* 2004;20(3):389–398.

Xu R, Wang Q. Large-scale extraction of accurate drug-disease treatment pairs from biomedical literature for drug repurposing. *BMC Bioinform* 2013;14:181.

Yang L et al. Exploring off-targets and off-systems for adverse drug reactions via chemical-protein interactome—Clozapine-induced agranulocytosis as a case study. *PLoS Comput Biol* 2011;7(3):e1002016.

Yetisgen-Yildiz M, Pratt W. Using statistical and knowledge-based approaches for literature-based discovery. *J Biomed Inform* 2006;39(6):600–611.

Zerbini LF, Bhasin MK, de Vasconcellos JF, Paccez JD, Gu X, Kung AL, Libermann TA. Computational repositioning and preclinical validation of pentamidine for renal cell cancer. *Mol Cancer Ther* 2014;13(7):1929–1941.

Zhu J, Wang J, Yu H, Li Y, Hou T. Recent developments of in silico predictions of oral bioavailability. *Comb Chem High Throughput Screen* 2011;14(5):362–374.

Zweigenbaum P, Demner-Fushman D, Yu H, Cohen KB. Frontiers of biomedical text mining: Current progress. *Brief Bioinform* 2007;8(5):358–375.

5 Technical Tools for Computational Drug Repositioning

Francesco Napolitano

CONTENTS

5.1 INTRODUCTION

The large amount of data currently available on small molecules and their biological activity at various levels is a fundamental resource both in the search for drug candidates and to better understand the mechanisms behind their effects. Neuropharmacology, often dealing with some of the most complex mechanisms in human pathology, could particularly benefit from the availability of system-wide molecular data. Moreover, the neurochemical space (the space of small molecules that could have neurological activity) has been estimated to possibly include as many as about 6×10^{15} different molecules (Weaver and Weaver, 2011). For such reasons, techniques from statistics and Machine Learning (ML), especially suited to deal with complexity and large samples, have been widely adopted in the modern drug repositioning research (Murphy, 2011; Lavecchia, 2015).

Based on Structure–Activity Relationships (SARs), the search for increasingly efficient and safe drugs has long been taking advantage of the intrinsic physico-chemical properties of bioactive compounds (Topliss, 2012). Rational Drug Design (RDD) principles guide modern drug discovery from the identification of a target ligand to the design and optimization of small molecules that are able to bind it (Silverman and Holladay, 2014). Chemoinformatics plays an important role along this pipeline today: the ability of screening large libraries of compounds solely based on their computational models permits to readily discard hundreds of thousands of unlikely hits on the basis of computer simulations (Kitchen et al., 2004; Cheng et al., 2012). Such virtual screening processes rely on the same SAR principles that form the basis of RDD and predate the advent of computers (Selassie et al., 2003) but formalize them into a computational framework where drug–drug physicochemical similarities and ligand-protein affinities can be predicted in a completely digital fashion.

On the other hand, new approaches in RDD are recently emerging, driven by the innovative perspective on the assessment of drug effects that was provided by systems biology (Butcher et al., 2004). Systems biology aims at studying organisms as a whole, rather than a sum of independent constituents, thus requiring the simultaneous modeling of all the interacting biological entities involved in the response to a given cellular perturbation (Ideker et al., 2001). The sheer complexity of even the simplest living organisms has held back any holistic approach to computational biological research until recently, but current technologies in the area of biological data production and analysis have started to make this possible today. With the advent of microarrays in the early 2000s and their subsequent evolution toward next-generation sequencing technology, it became suddenly possible to measure the entire transcriptome of a cell at a given time point in a single assay (Hoheisel, 2006). Although technical difficulties were present in both the measurement and the analysis of genome-wide expression data (Draghici et al., 2006), the new technology immediately brought an important cultural switch. Besides studying the most likely molecular targets of a drug, it is today commonplace to observe the whole network of transcriptional effects caused to a treated cell. Transcriptional profiles of many cell types in a plethora of different conditions are today publicly available from dedicated databases (Edgar et al., 2002; Brazma et al., 2003; Lamb et al., 2006). Together with proper data analysis techniques, they possibly constitute a new gateway to drug repositioning (Iorio et al., 2013).

This chapter is devoted to the presentation of some of the main technical tools currently used in computational drug repositioning (CDR). As vast as the area is, an exhaustive coverage of the topic falls largely out of the scope of this book. However, the chosen techniques should provide enough insight into the field for the reader to get started and figure out its main potentials and limitations. Toward this aim, the presentation focuses on the most data-centric areas of CDR: those involving ML and statistics. The first part of this chapter gently introduces the fundamentals of ML, followed by concepts related to the use of CDR-relevant publicly available resources. Such topics are preparatory to the second part of this chapter, where a selection of the state-of-the-art CDR methodologies with applications to neurotherapeutics is presented.

5.2   TECHNICAL BACKGROUND

CDR takes advantage of tools from computer science and statistics literature. This section is meant to provide a brief introduction to some of the fundamental concepts from such areas, as a preparatory overview before the presentation of CDR applications to neurotherapeutics.

5.2.1   Machine Learning

ML is a prominent field of artificial intelligence whose final aim is the use of computers to build models of reality by deduction from examples (Bishop, 2006). A properly programmed machine is able to distinguish between different types of objects basing on repeated observations in a process that is conceptually similar to the way living creatures learn. A number of different approaches to implement ML processes exist, with varying degrees of relatedness to the biological intelligence that we are all familiar with: from algorithms trying to mimic neural functions and interconnections to pure mathematical models having no mechanistic counterpart in nature.

From an application point of view, two main areas of ML have been formalized in literature, both of which find a wide range of applications in the context of CDR: unsupervised and supervised learning (Bishop, 2006). In this section, the basic concepts of both worlds will be briefly reviewed. Literature of ML methodologies is vast and varied, with no absolute winner across the spectrum of possible approaches, thus a focused selection of the major players will be mentioned with reminders to relevant publications. The application of each general concept introduced in this overview is then exemplified in subsequent sections of this chapter through the presentation of CDR applications.

5.2.2   Supervised Learning

The concept of supervised learning can be exemplified by a human-learning process including a lazy supervisor. The supervisor's contribution is limited to the provision of question–answer pairs to the student and the verification of his level of accuracy in answering new questions. More formally, given a set of questions numbered 1 to n, the set of questions to learn from is defined as

$$X = \left\{ x_i, x_{i+1}, \ldots, x_n \right\}$$

X is said to be a collection of *training examples*. If the lesson is about zoology, each x_i could be the picture of an animal to be recognized. However, in a formal context, the ith picture must be represented through a mathematical entity, that is, the vector x_i. Each x_i is like a set of measurements of those features that are deemed relevant in characterizing the ith object. This is why the x_i-s are also called *feature vectors*. In a real-world context, for example, x_i could be a collection of physicochemical attributes for a given drug. All x_i-s must refer to the same set of features, thus in principle the physicochemical attributes must be available for all the drugs in the training set, although techniques exist aimed at filling in the missing values (Tshilidzi, 2009).

In addition, a set of answers y_i is provided such that a pair (x_i, y_i) means that the animal x_i belongs to the species y_i, or that the small molecule x_i belongs to the category of those known to be toxic to humans. A collection of such pairs

$$D = \left\{ (x_1, y_1), (x_2, y_2), \ldots, (x_n, y_n) \right\}$$

constitutes a supervised data set. In general, the y_i-s can be real valued, but for the purposes of this chapter they will be considered as categorical variables identifying a class. In ML literature, these kinds of variables are termed *class labels* and the corresponding task is known as *classification*.

The aim of supervised learning algorithms is to build a model of the underlying unknown relation that ties each feature vector x_i to the corresponding class label y_i, that is how the student recognizes each animal. In mathematical terms, this corresponds to finding a function f such that $f(x_i) = y_i \forall (x_i, y_i) \in D$. Once such a function is found, it is reasonable to expect that it will be able to provide generally correct predictions also for $x_i \notin D$, that is to guess the species of animals portrayed in previously unseen pictures.

A popular example of a supervised learning data set in ML literature is the Fisher's Iris data set (Fisher, 1936). Despite being a very simple collection, it poses some interesting learning challenges that will be useful to clarify many concepts throughout this chapter. For the Iris data set, X is defined as a collection of $n = 150$ flowers, each one characterized through four measures: sepal width, sepal length, petal width, and petal length. This means that each feature vector x_i is an ordered list of four numbers. Each of the 150 flowers belongs to one of three possible species: *Iris setosa*, *Iris virginica*, and *Iris versicolor*. Or, in formal and more synthetic terms, $y_i \in \{1, 2, 3\}$. The learning problem is thus to find a function f that is able to guess the species y_i of an Iris flower x_i.

Figure 5.1a shows a two-dimensional reduction of the Iris data set obtained through Principal Component Analysis (PCA) (Bishop, 2006), with different colors corresponding to different species. In that representation, each point represents a flower and close points represent sets of flowers modeled by similar feature vectors. Naturally, the similarity of the feature vectors implies the similarity of the corresponding flowers if the features were chosen carefully enough. A geometrical way of looking at the classification problem is thus to define regions of the space that are associated with a given class. Although our brain is very powerful in performing similar tasks, feature spaces are usually more (or much more) than three-dimensional, while low-dimensional projections are visual approximations with limited accuracy (Young, 2013). Thus, a formal approach is necessary to face the problem in the original feature space. However, a good starting point to better understand the principles of a classification task is to imagine some reasonable hand-made geometric solutions. Figure 5.1 shows three such examples.

The first example as shown in Figure 5.1b is an attempt to isolate the 3 Iris classes using just straight lines, that is, a *linear model*. Linear models are very simple and, depending on the context, can be very effective (Support Vector Machines [SVMs], described in the next sections, are a popular example of the linear classifier). From Figure 5.1b, it is clear that the points belonging to the class *setosa* can easily be separated from all the others by a straight line. However, this is not possible for all the points belonging to the other two classes, which are thus said to be *not linearly separable*.

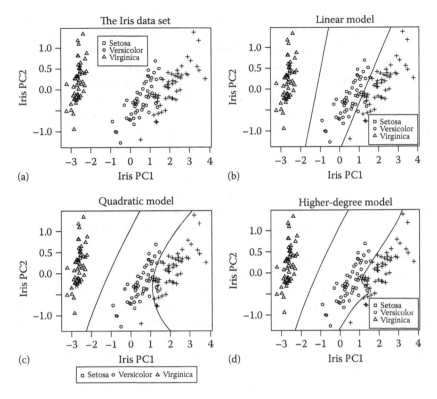

FIGURE 5.1   (a) A visualization of the Iris data set. Different shapes represent different species. (b–d) Three ways of separating the three species by using models (linear, quadratic, and higher degree) of increasing complexity.

In this example, the linear model assigns the label *virginica* to four flowers actually belonging to the class *versicolor*, thus making four *classification errors*. As shown in Figure 5.1c, a better performance for this situation is obtained by a slightly more complex solution: a *quadratic model*. Thanks to its higher flexibility, this model produces only three classification errors, thus beating the linear model. Indeed, one could be mistakenly induced into thinking that the more flexible the model, the higher the accuracy, a hypothesis that seems to be corroborated by Figure 5.1d, showing a high-degree model yielding perfect separation. However, as explained later in this section, this better performance will turn out to be ephemeral.

The problem of automatically creating a decision boundary that correctly separates classes of objects lying in an *N*-dimensional space covers a large slice of the entire ML literature (Bishop, 2006). The first steps into the field of ML were taken back in the early 1940s (McCulloch and Pitts, 1943), when the first computational model of a neural network was proposed, and before computers actually able to train it even existed. Inspired by biological mechanisms underlying the human brain functions, *Artificial Neural Networks* (ANN or simply NN) became widely popular in data analysis research thanks to modern personal computers, and their recent developments, like *Deep Neural Networks*, still represent state-of-the-art tools in ML (Schmidhuber, 2015). An example of a widely used ML technique that is rather far from any biological inspiration is instead the Random Forests (RF) algorithm (Ho, 1998), an efficient development of the Decision Tree (DT) model (Rokach and Maimon, 2008). The DT builds a hierarchical tree of the training data by subsequently splitting them in half according to the value of one variable at a time. Leaves of the tree are associated with the classes y of the training samples. The tree can be used to classify new (testing) samples by following a path down the tree basing on subsequent values of each variable in turn until a class-labeled leaf is reached. Simply put, an RF is a collection of DTs built on random subsets of the training set. The final classification is based on a consensus among different DTs. For this reason, RFs belong to the broader category of *ensemble methods* (Dietterich, 2000), which build *strong classifiers* basing on a large number of *weak classifiers*. The last example of a supervised classification method will be used later in this chapter to demonstrate a neuropharmacological application and is briefly reviewed in the next section.

5.2.2.1 Support Vector Machines

As synthetic as it may be, a review of current major methods in ML cannot miss SVMs (Steinwart and Christmann, 2008). While NNs are biologically inspired and RFs are based on computational logic, SVMs rely on pure mathematical tools. In particular, SVMs build on the concept of *maximum margin* (Boser et al., 1992), which refers to a particular linear classification technique. Given two linearly separable classes of points, the *maximum margin* approach looks for a hyperplane separating the two classes of objects in such a way that the closest point to the margin is as far as possible. Such points are called support vectors and the maximum margin hyperplane will intuitively provide better generalization than the hyperplanes providing smaller margins. Formally, any hyperplane is defined by the following equation:

$$w \cdot x - b = 0$$

With opportune scaling, the hyperplanes crossing the two closest points on either sides of the separating plane can be defined as

$$w \cdot x - b = 1$$

$$w \cdot x - b = -1$$

Their distance is $b/\|w\|$, which also defines the *margin*. Points above and below the margin must belong to the two different classes, or obey the following:

$$w \cdot x_i - b \geq 1$$

$$w \cdot x_j - b \leq -1$$

where all x_i-s belong to one class and all x_j to the other class. Given these constraints, the maximum margin hyperplane is found by minimizing $\|w\|^2$ by means of quadratic optimization techniques.

Because of the linear separability constraint, in the case of the Iris data set previously presented, SVMs in this form would only be able to separate the virginica species from each of the other two. However, SVMs ingenuously support nonlinear classification by means of the so-called *kernel trick* (Steinwart and Christmann, 2008), as explained in the following.

Kernel-based methods use pairwise similarity values between points as opposed to their feature vectors. In other words, they rely on the values $k(x_i, x_j)$ as opposed to the vectors x_i and x_j themselves, with k being a distance function (having a particular mathematical property known as Mercer's condition or a slightly less stringent condition known as positive definiteness) called *kernel*. The term kernel is often used with fewer mathematical implications, which is also justified by the fact that nonpositive definite kernels are known to work well in practice (Haasdonk, 2005). Therefore, replacing the feature vectors with the table of all their pairwise similarities is often enough to apply kernel-based methods, which provide both technical and methodological advantages, as summarized in the following.

The main technical advantage of using kernels is what makes nonlinear SVMs possible. The crucial intuition is the following: point sets that are not linearly separable in the feature space can become linearly separable in a higher-dimensional space. SVMs exploit this principle through a kernel-based formulation of the maximum margin problem, where a nonlinear kernel is used instead of raw data point values. A linear separation in this transformed space constitutes a nonlinear separation in the feature space. A constraint relaxation (the *soft margin*) allowing for imperfect class separation, together with the extension to more than two classes, completes the main features of common SVMs' implementations (Steinwart and Christmann, 2008).

But from a methodological point of view, kernel methods, including SVMs, are particularly useful when dealing with complex objects like biochemical entities. Chemical properties like three-dimensional structures of drugs, or cellular features like entire transcriptomes, cannot easily be represented as vectors of numbers in an efficient way. Moreover, the integration of such diverse entities is even harder. However, the similarities among chemical structures or among transcriptomes can be more easily summarized and expressed through kernels. Kernels in turn, being just tables with the same number of rows and columns, can be more easily integrated. These principles will be exploited in the last part of this chapter in order to compare small molecules using both their structures and their biological effects. Note that, although nonlinear SVMs are natively kernel-based and constitute the most popular method within the class of kernel-based methods, kernel versions of many non-kernel methods exist (see, e.g., Scornet, 2016 for kernel Random Forests).

5.2.2.2 The Overfitting Problem

A fundamental concept in ML is "overfitting" (Bishop, 2006). Along the lines of the human-learning metaphor, overfitting a model is the equivalent of memorizing correct solutions to a number of problems without proper comprehension of the topics. The obvious consequence is the inability to generalize the learnt solutions to new problems. In terms of ML models, overfitting occurs when an excessively complex model is able to fit a training sample near perfectly but performs poorly on other samples extracted from the same distribution. The notion of an "excessively complex" model is linked not only to the distribution of training samples, but also to their number: the more complex the model, the higher the number of training samples necessary to avoid overfitting.

The occurrence of overfitting can be detected by fitting the model on a subsample of the training samples and subsequently checking its performance on the remaining points ("validation samples"). Intuitively, the simple linear and quadratic models of Figure 5.1b and c will likely remain similar after removing a few random points. Conversely, the more complex model of Figure 5.1d will be much more influenced by the removal of any particular subsample, which is a clear sign of overfitting. The model parameters are thus usually optimized by means of subsampling techniques, one of the most popular being the *k-fold cross-validation*. Given any integer $k \in 1, \dots, n$, the k-fold algorithm starts by randomly partitioning the data set into k bins. The model is then trained on the data from $k-1$ bins and validated using the remaining one. The process is iterated for all the possible analogous splits of the data set (there exactly k such splits), and the average number of misclassifications is finally used to define the overall performance of the classifier. Another well-known technique, the *leave-one-out* cross-validation, is actually a special case of the k-fold, where $k=n$, with n being the number of training points.

5.2.3 UNSUPERVISED LEARNING

Sometimes labels expressing a ground-truth association between training data and predefined classes simply don't exist. Market research, for example, tries to partition customers into a small number of typologies in order to better tune their products toward a defined set of targets. However, there exist no predefined set of classes to fit people into. The definition of such classes is in fact an unsupervised learning problem. Unsupervised learning problems include a number of approaches to mine knowledge from data by relying only on their features. The problem of building a taxonomy of drugs basing on their physicochemical properties, for example, can be modeled as an unsupervised learning task.

Two very popular unsupervised approaches are PCA (Bishop, 2006) and Clustering Analysis (Theodoridis and Koutroumbas, 2008). A PCA-related technique, for example, has been used to extract two-dimensional features representing the Iris data in Figure 5.1 starting from four-dimensional vectors (*dimensionality reduction*, see Bishop, 2006). The clustering problem and one corresponding solution, instead, are presented with more details in this section.

Given a collection of objects $X = x_1, x_2, \dots, x_n$ and a set of class labels $y_1, y_2, \dots, y_k$, clustering is the process of assigning a class label y_j to each point x_i in such a way that objects within the same class are "similar" and objects in different classes are "different." As opposed to the supervised framework seen in the previous section, clustering cannot rely on any set of (x_i, y_j) pairings known beforehand.

A clustering solution is defined as a partition C of X, $C = \{C_1, C_2, \dots, C_k\}$, with each C_i being a set of points disjoint from any other $C_j \in C$. The concept of dissimilarity, as in the case of supervised learning, is modeled through a function d defined for pairs of vectors (x_i, x_j). Clustering algorithms are often divided into two broad categories: hierarchical and partitional. Hierarchical clustering (Theodoridis and

Koutroumbas, 2008) techniques build a hierarchical tree of classes and subclasses, from which a partition can be derived thereafter. Partitional algorithms, on the other hand, aim at producing the partition directly.

The performance of a supervised classifier is intuitively related to its accuracy in assigning correct labels to test points. Assessing the performance of an unsupervised method, instead, is less straightforward. A popular measure of *fitness* for a clustering solution is the distortion function (Theodoridis and Koutroumbas, 2008):

$$\sum_{j=1,\dots,k} \frac{1}{|C_j|} \sum_{x_i \in c_j} |x_i - c_j|$$

where c_j is the centroid of the objects belonging to class C_j:

$$c_j = \frac{1}{|C_j|} \sum_{x_i \in C_j} x_i$$

Informally, distortion quantifies how much it is wrong to use the center of a set of points to approximate any point belonging to the set. *K-means*, a classical clustering algorithm, tries to minimize this error.

Note that the definitions mentioned earlier assume that at least the universe of possible classes is known in advance, that is, the number k. Many techniques exist to estimate the value of k from data; however, a number of algorithms have been proposed that are able to find both k and the set partitioning at the same time. One such technique, used later in this chapter, is the Affinity Propagation (AP) algorithm (Frey and Dueck, 2007). AP is said to be a *message passing* algorithm, as it is based on propagating information between data points. The class assignment is performed by choosing "exemplar" points representing classes and associating neighboring points to the best exemplar. Two types of "communication" occur between each pair of points (x, y), with y being a candidate exemplar for x. First, the point x assesses how much the point y is a good exemplar for it as compared to other potential exemplars ("responsibility"). Then the point y evaluates how much the point x can be appropriately assigned to it as compared to other points ("availability"). All points are evaluated as both possible exemplars and regular points. Since responsibility and availability are functions of each other, an iterative procedure is used to update them until convergence. Finally, availability and responsibility values are used to define the role of each point, thus determining the clusters' number and content.

5.3 APPLICATIONS TO DRUG REPOSITIONING

A synthetic overview of a few major ML concepts and techniques was provided in the previous section. This section is devoted to their specialization and application to the field of CDR. Some additional tools are introduced first, particularly covering publicly available data sets that are relevant to CDR applications. Along the lines of the previous section, both supervised and unsupervised applications for drug repositioning are presented thereafter. Finally, a different approach is introduced, the aim of which is to help in understanding the biological mechanisms involved in drug efficiency, thus helping to build rationales for subsequent CDR.

5.3.1 Selected Publicly Available Resources

CDR often takes advantage of large databases of information about small molecules and biological mechanisms involving their targets. While a number of relevant reviews exist (Moreau and Tranchevent, 2012; Villoutreix et al., 2013; Henry et al., 2014; Ding et al., 2014), the quantity of publicly available resources in this area can hardly be covered in a short summary. Here, a small selection of those relevant to the focus of this chapter is briefly mentioned.

Large databases of small molecules exist, such as PubChem (Wang et al., 2009), ChemBank (Seiler et al., 2008), ChemDB (Chen et al., 2007), and DrugBank (Law et al., 2014), which collect both structural and bioactivity-related data for a number of small molecules that are often in the order of millions. Chemoinformatic tools are often provided alongside with data, allowing the user to link drugs among them basing on their structure, or against cell states basing on experimental evidence.

A useful categorization of the most approved drugs is provided by the World Health Organization through the Anatomical Therapeutic Chemical (ATC) Classification (WHO Collaborating Centre for Drug Statistics Methodology, 2016), a curated ontology that hierarchically classifies drugs according to their clinical uses.

On the interface between drugs and cells lie drug-target databases, such as the Search Tool for Interactions of Chemicals (STITCH) (Kuhn et al., 2008), collecting predicted or experimental information about the interaction between chemicals and proteins.

Biological data are a fundamental complement to chemical data in CDR. The Online Mendelian Inheritance in Man (OMIM) (Amberger et al., 2014) database can be used to connect drug targets and disease genes. The database stored at the Protein Interaction Network Analysis (PINA) platform (Cowley et al., 2012) collects protein–protein information that helps in reconstructing the effects of small molecules downstream of their direct targets. Gene set–oriented databases such as Molecular Signatures Database (MSigDB) (Liberzon et al., 2011) and Gene Ontology (GO Consortium, 2004) can be used to extend the analysis of single targets to pathways, protein complexes, and related genes in general.

It is worth mentioning at this point that the MSigDB has been developed together with a widely used technique to analyze gene sets, the Gene Set Enrichment Analysis (GSEA) (Subramanian et al., 2005). Basing on a weighted Kolmogorov–Smirnov-like statistic (Hollander and Wolfe, 1999) acting on ranked lists of genes, the GSEA assesses how much the genes in a set (such as one of those included in the MSigDB) are significantly present among the top (or bottom) expressed genes in a genome-wide profile.

While on the subject, expression profiles as well can be found in *functional genomics* databases, with ArrayExpress (Brazma et al., 2003) and Gene Expression Omnibus (GEO) (Edgar et al., 2002) probably being the most popular collections.

However, the most significant effort to date in producing drug-focused microarray data in a systematic way has been the Connectivity Map (Cmap) (Lamb et al., 2006). The Cmap 2.0 includes 7056 gene expression profiles derived by the treatment of 5 cell types with 1309 small molecules. The profiles include controls, replicates, and treatments at different doses. Having been produced systematically within the same research project, these data are computationally valuable especially because of their controlled technical bias. A new release of the Cmap (within the Library of Integrated Network-Based Cellular Signatures Project, or LINCS) is currently being developed, using dedicated technologies to produce a number of profiles, that is two orders of magnitude bigger than the previous version. Although a first version of the data has been released on the GEO website,* a related paper has not been published at the moment of writing this chapter.

5.3.2 Computational Models of Drugs

The first step of any CDR analysis is to choose a strategy for the computational modeling of small molecules. The aim of this phase is to systematically collect information about a possibly large number of small molecules to be used for automated analysis. The information collected about each small molecule can be used directly as its computational model or indirectly to compare drugs without explicitly modeling them. The first approach is useful to apply computational methods that make use of *feature vectors*, while the latter is suitable for *kernel* approaches (see the previous section). We will sometimes use the term "drug model" in both cases, with the *kernel function* improperly representing the drug model in the cases when it is in fact not directly defined.

A classical and rather intuitive approach to compare small molecules is based on the comparison of their chemical structures. The fundamental notion behind the efficacy of this approach in pharmacology is the so-called SAR, implying that structurally similar small molecules are likely to share some biological effects. This concept has also been applied to the specific context of neuropharmacology (e.g., see Behl et al., 1997).

Structural information about small molecules is mainly derived by the spatial arrangement of atoms in a molecule together with their chemical bonds. Different measures to compare molecular structures exist (Sliwoski et al., 2014), taking into account their 2D topology (pairwise bonding of the atoms), their 3D geometry (the relative position of atoms in space), or even their 4D dynamical structure (the relative position of their atoms for different conformations of the same molecule).

An example of simple and widely used structural similarity is based on *binary fingerprint* representations of the 2D molecular geometry (Sliwoski et al., 2014). Chemical binary fingerprints are 0–1 vectors, whose length is usually in the order of thousands, with each bit usually reporting the presence or absence of a structural pattern in the molecule. The Tanimoto similarity between two molecules is often used to compute a simple but cost-effective binary-fingerprint-based structural similarity score. It is given by the number of 1-bits falling in the same positions in two fingerprints, divided by the number of 1-bits in any position. In other words, it is the number of shared features normalized by the number of shared and unshared features.

---

* http://www.ncbi.nlm.nih.gov/geo/query/acc.cgi?acc=GSE70138.

With the aim of focusing on molecular bioactivity, recent research has defined drug models by the direct collection of pharmacological effects of small molecules after treatment of living organisms. For the purpose of this chapter, such approaches will be divided into two main categories: *knowledge-based* and *data-based*. By knowledge-based models, we mean those making use of curated information about pharmacological effects of a drug at the clinical level. This may include therapeutical applications or known side effects of a small molecule. Data-based models, on the other hand, rely solely on raw information collected through experimental assays, subject to interpretation and potentially leading to new knowledge. Public databases for both kinds of approaches exist and have been cited in the previous section: an example of the former is the ATC classification by the World Health Organization, a curated ontology that hierarchically classifies drugs according to their clinical uses. The collections of genome-wide expression data cited in the previous section (ArrayExpress, GEO, and Cmap), on the other hand, can be used for data-based models.

In summary, the similarity between two drugs can be assessed basing on intrinsic chemical features or derived biological effects. Biological effects can include well-assessed knowledge or raw experimental data.

An obvious but fundamental difference exists between knowledge-driven and data-driven analysis: while the former requires the existence of well-understood knowledge about the biological effects of the molecules to be modeled, the latter can be applied to any molecule producing a measurable biological effect. The practical consequence is that knowledge-driven techniques are well suited to analyze known molecules, thus proving their efficacy mainly in the context of drug repositioning. On the other hand, data-driven techniques can be applied on molecules with completely unknown clinical effects, thus potentially aiding through a drug discovery process.

A final note on the subject of drug modeling must be considered concerning the problem of data integration. Drug models can be built using data of very different nature, thus multiplying the opportunities to look at a given drug repositioning application from the most effective perspective. This implies the development of tools that are able to deal with multiple data sources at the same time. The first section of this chapter posed a particular focus on kernel methods as they can easily overcome the problem of integrating heterogeneous data (Napolitano et al., 2013). The actual drug model used does not really matter to kernel methods, as long as a similarity score between two drugs under the same model can be provided. Integrating kernels obtained with different models is thus a matter of integrating similarity scores, which, at a bare minimum, can be as simple as computing their average.

5.3.3 COMPUTATIONAL DRUG REPOSITIONING USING SUPERVISED MACHINE LEARNING

Theoretically, the perfect supervised machine for CDR would be able to predict what clinical application a molecule may have, basing on the known applications of other molecules. An attempt to build such a machine was proposed in Gottlieb et al. (2014), where drug models were used to predict disease models. Drug models were built basing on structural, clinical, and molecular data, while disease models were mainly based on the previously mentioned OMIM database (Amberger et al., 2014). Chen et al. (2012) tried to directly predict ATC codes (see the previous section) basing on both structural similarity between molecules and chemical interaction data from the STITCH (Kuhn et al., 2008) database.

The relative importance of different data sources is not always clear. However, the method described by Napolitano et al. (2013) provides a useful example of the state-of-the-art ML techniques and integration of heterogeneous drug models for CDR, in which the importance of both individual and integrated sources has been investigated. For this reason, it will be described in more detail as a good example of the application of ML principles described in previous sections.

The drug model proposed by Napolitano et al. is a composition of three different models: one based on chemical structure similarity, another one on genome-wide transcriptional similarity, and a third on common molecular targets. The three kernels are built on a set of 410 small molecules for which all the three types of data were available.

The chemical-structure-based drug model was based on molecular binary fingerprints, as mentioned in the previous section. The fingerprints were obtained by processing structural information obtained through the public database DrugBank (Law et al., 2014) in the form of SMILES string (Nič et al., 2009). The Tanimoto similarity (see the previous section) and other analogous scores were used and averaged over to obtain the final pairwise similarities, that is, the *structure-based kernel*.

The transcriptional-data-based drug model was calculated using the gene expression profiles included in the Cmap (Lamb et al., 2006), for which a weighted Manhattan score was computed as follows. The gene expression profiles are first ranked from the most upregulated to the most downregulated. Given two profiles, the Manhattan distance is simply the sum of the (absolute) differences between the ranks of each gene in the two profiles. *P*-values of the expression fold change (actually their complement to 1) were used to weight the ranks. The table of all the pairwise distances computed in this way constitutes the *transcription-based kernel*.

Finally, a target-based drug model was built by averaging over two scores. The first is the Tanimoto similarity computed on the target sets of each drug (i.e., the number of targets shared by a pair of molecules divided by the total number of different targets of the pair). The second is the average number of protein–protein interactions that are necessary to reach each target of one drug from each target of the other drug in the pair, by traversing the PINA protein–protein interaction network (Cowley et al., 2012).

The three kernels are mathematically summarized by three 410×410 symmetric matrices. The final *integrated* kernel is obtained by the element-wise average of the three matrices. This kernel represents the similarity of each drug pair, basing on three different types of information at the same time. Note that, while in Napolitano et al. (2013) a simple averaging was chosen to integrate the three kernels, more complex techniques are available (Zien and Ong, 2007).

In order to train a supervised model, an *a priori* classification must be provided to build a training set. In Napolitano et al. (2013), ATC codes were used as such classes. An SVM classifier was trained to predict ATC codes and a *k*-fold cross-validation (see Section 5.1) was used to assess its performance, which was finally reported as

78% of correct predictions on average. Importantly, the study showed that the kernel integrating information from chemical structures, molecular targets, and transcriptional response was more efficient at predicting drug therapeutical applications than each kernel used alone.

The trick used to obtain hints for drug repositioning out of the classifier was to focus on the remaining 22% molecules for which the predicted therapeutical application was in disagreement with the known associated ATC codes. In particular, the cases in which the classifier insisted on the same alternative ATC code after reiterated perturbation of the data set were considered as potential repositioning hints. The tool was able to identify known effects that were not annotated in the ATC classification. For example, the ophthalmologicals levobunolol and sulfacetamide were respectively reclassified as beta-blocking agents and antibacterials. Interestingly, chlorphenamine and thiethylperazine, two antihistamines known to have antipsychotic effects, were respectively reassigned to the psychoanaleptics and psycholeptics classes. The inverse reclassification happened to the antihistamine hydroxyzine, classified as a psycholeptic by the ATC system and reassigned to the antihistamines category by the classifier.

Supervised approaches such as those described in Chen et al. (2012) and Napolitano et al. (2013) are very ambitious in trying to directly and blindly predict repositioning opportunities. They constitute important proofs of principle, the results of which must be considered as hints to guide further analyses. The next two sections will show examples of tools that explicitly implement such an exploratory approach.

5.3.4   Computational Drug Repositioning Using Unsupervised Machine Learning

Unsupervised computational tools for drug repositioning have two important advantages: first, they do not need prior knowledge to be trained with; second, they are not biased toward such knowledge, which could be partial. Of course, they cannot be used for automatic class prediction.

Many unsupervised methods for CDR implicitly or explicitly build virtual spaces of chemical and/or biological entities. For example, many chemical databases, like the previously mentioned PubChem, ChemBank, ChemDB, and DrugBank, provide tools to explore molecules that are similar to an input molecule, thus implicitly searching a chemical space. The STITCH database explicitly builds networks of chemical–protein interactions. A largely cited method in this context was described in Iorio et al. (2010) and implemented as the Mode of Action by NeTwoRk Analysis (MANTRA) tool, the latest version of which is described in Carrella et al. (2014). MANTRA is a network analysis approach to drug discovery and repositioning that exploits a genome-wide transcriptional-data-based kernel. For its explicit modeling of a transcriptional-data-based drug space as a tool for CDR and the application of unsupervised techniques for the analysis of such a space, it makes a perfect candidate to be further detailed in this section.

The drug model in MANTRA is built basing on two steps. The first step collects and merges together genome-wide transcriptional data induced by drug treatments.

Raw data are obtained from the Cmap database (Lamb et al., 2006), which includes a number of microarray experiments for each small molecule in a set of 1309. All profiles are ranked such that the most upregulated genes are ranked at the top of the profile, while the most downregulated genes are ranked at the bottom. All the ranked profiles originating from treatments with the same small molecule are merged together by means of an average-like method to build a single Prototype Ranked List (PRL). The set of 1309 PRLs is the final output of the first step toward the construction of the drug model. The second step is meant to reduce noise in the PRLs while building the MANTRA *kernel* at the same time. The key intuition behind the kernel is simple: compute a pairwise similarity that gives particular importance to the most dysregulated genes in the PRLs and less importance to the less dysregulated or not dysregulated. The weighted similarity measure is obtained by exploiting the GSEA, described in the previous section. The GSEA version used in MANTRA is rank-based only and applied to the PRLs as follows. Given two gene expression profiles A and B, MANTRA computes how much the 250 most upregulated genes in A are also upregulated in B using the GSEA, then computes the same for the 250 most downregulated genes. The process is applied again switching A and B, and the results are averaged over. Iterating over all the PRL pairs, the final kernel is obtained.

The MANTRA kernel is used for two different purposes: to obtain drug clusters (or "communities"), and to produce a visual representation of the drug space. The drug communities are obtained by means of the AP clustering algorithm (see the previous section) and are used to characterize regions of the drug space according to local enrichment for known modes of action. Concerning the visual and most important part of the tool, network visualization was chosen. Drugs are visualized as nodes in a network, with edges linking together drugs that induce a similar transcriptional profile. This is obtained by drawing an edge between a pair of drugs A and B if their distance is within the 5% of the smallest distances in the kernel. Note that this step constitutes an alternative clustering process, which is considered easier to visualize than the AP clustering, the communities of which are instead used to assign colors to the network nodes.

The MANTRA network can be used for drug repositioning by exploiting drug neighborhoods. Given a drug D that is known to induce a particular effect at any given biological level (molecular, cellular, and clinical), it can be hypothesized that the neighbors of D in the network could be able to induce the same effect. This simple approach can be used to select a set of candidate drugs for further investigation. Iorio et al. showed how the network was able to suggest a correct repositioning of the Rho-kinase inhibitor and vasodilator Fasudil as an autophagy inhibitor.

It is worth noting that this use of the network is conceptually supervised (or, more precisely, semi-supervised). Moreover, the drug communities were characterized a posteriori basing on the estimated enrichment of known mechanisms of action. However, the exploratory nature of the tool and the absence of predefined prior knowledge in its definition make it technically unsupervised. In fact, this is a good example to demonstrate the existence of a gray area between supervised and unsupervised methods.

A MANTRA-based example of exploratory analysis for CDR in a neuropharmacological context can be obtained as follows. One of the largest drug communities found in MANTRA is characterized by nervous system–related therapeutical applications (community number 100 in Iorio et al., 2010). By using ATC codes, it can be seen that more than half of the small molecules included in this community (26 out of 51 having at least one ATC code assigned) are known to have therapeutical applications on the nervous system. This is a significant enrichment compared with the overall proportion of nervous system–related drugs in the Cmap (125 out of 730 small molecules having at least one ATC code assigned, i.e., 17%). For this reason, looking at the 25 drugs of the community that have not been associated with a nervous system–related ATC code could be of interest. In order to prioritize such drugs, their degree of connectivity within the subnetwork (i.e., the number of connected neighbors) at a stringent threshold can be employed. For example, the MANTRA online tool shows that the subnetwork generated by searching for the drug amoxapine (setting depth = 2) includes 51 nodes at a distance threshold of 0.70 and an empty network at any more stringent threshold (see Figure 5.2).

The most connected nodes in this view of the network are trifluoperazine, perphenazine, and loperamide. Trifluoperazine and perphenazine are known antipsychotics, so prior knowledge confirms their significance in the community. Loperamide, on the other hand, is classified as an antipropulsive by the ATC classification, belonging to the larger class of drugs acting on the alimentary tracts or metabolism. However, loperamide is an opioid receptor agonist and is able to exert central opiate-like effects

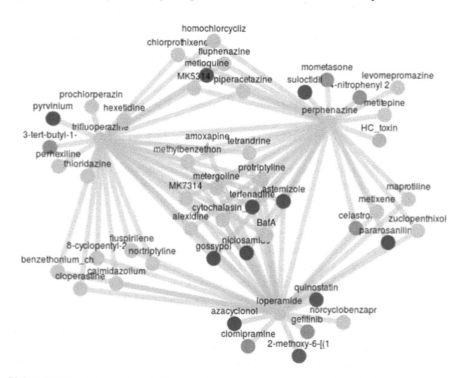

FIGURE 5.2   A drug network. Each node represents a small molecule and edges connect small molecules inducing a similar transcriptional profile. *Note*: The full name for 2-methoxy-6-[(1 is 2-methoxy-6-[(1Z)-[2-(pyridin-2-yl)hydrazin-1-ylidene]methyl]phenol.

if given intravenously (Niemegeers et al., 1979), a condition that is compatible with the *in vitro* high dosages used in Cmap cell lines. Thus, the neuroactivity of the compound could have been hypothesized just basing on drug network analysis.

As an additional example of the use of MANTRA in a neuropharmacological context, Siavelis et al. (2016) found 27 potential candidates for repositioning as anti-Alzheimer agents by combining the MANTRA analysis of publicly available transcriptomic data with other knowledge-based tools.

Of course, the most unexpected connections are often the most intriguing and the ones deserving further investigation, which falls largely out of the scope of this chapter. Network analysis and other kinds of computational tools together with the researcher's intuition and experience may guide through the selection of a set of drug candidates to be brought from *in silico* to *in vitro* experimentation.

5.3.5   COMPUTATIONAL KNOWLEDGE DISCOVERY FOR DRUG REPOSITIONING

Current automated tools for CDR are usually not able to produce a reliable prediction in a completely blind fashion. Prior knowledge about disease and drug mechanisms is usually needed in order to improve the efficacy of computational tools through a more focused and targeted use. This is why knowledge discovery tools can help toward the identification of promising candidate compounds for repositioning. A specific step in this direction has been taken for example in Napolitano et al. (2015), where the Drug Set Enrichment Analysis (DSEA) method was described. As the name suggests, the DSEA is a drug-focused version of the GSEA, methodologically representing its dual. Indeed, while the GSEA is aimed at assessing whether a set of genes tend to be dysregulated by a drug, the DSEA estimates whether a set of drugs tend to dysregulate a given gene. In particular, the method is applied to Cmap gene expression profiles after conversion to "pathway expression profiles." This conversion is performed by the usual GSEA applied to each gene set obtained from a publicly available collection against each PRL built from the Cmap (see the previous section). Once all the pathway expression profiles are computed, all the Cmap drugs are sorted for each pathway, from those most upregulating the pathway to those most downregulating it. At this point, given a set of drugs, the same KS statistic used by the GSEA can be applied to assess how much the drugs in the set tend to appear at the top or bottom of the ranked list of drugs for a pathway. This way DSEA is actually able to estimate if a pathway tends to be dysregulated by most drugs in a set.

The tool is used to help in elucidating the relevant mechanisms playing a role in the therapeutical efficacy of a set of drugs, which are not necessarily linked to the main effects of each individual drug. This is a major problem, for example, when a set of heterogeneous small molecules is found to be effective against a desired phenotype by means of automated screening. In such cases, the heterogeneity of the set leaves with little clue about any particular mechanism of action explaining the efficacy of the drugs, which would in turn be fundamental knowledge in guiding optimization toward final candidates. The principle behind the DSEA is that if a set of small molecules are known to exert a therapeutic effect, then their efficacy should be explained by their common side effects, as opposed to their main individual effects. A crucial point proven in the study is that the simultaneous analysis of the effects of multiple related drugs is able to highlight transcriptionally relevant pathways that would be impossible to detect from individual drugs.

A test drug set for a DSEA application example can be obtained by further restricting the previously described MANTRA network of amoxapine neighbors. Setting the depth parameter to 1, four drugs remain: amoxapine itself, loperamide, perphenazine, and trifluoperazine. A DSEA analysis of this drug set suggests a possible common mode of action through down-regulation of the "Neurotransmitter catabolism process," ranked fifth in the Gene Ontology Biological Process category, and "GABA receptor binding," ranked third in the Gene Ontology Molecular Function category. More specific pathways found by this kind of analysis could reveal relevant modes of action to look for into other drugs for repositioning.

5.4 CONCLUSIONS

CDR is today an assessed research field taking advantage of contributions from a wide range of computer science, mathematics, and statistics areas, with proven efficacy in different therapeutical contexts including neuropharmacology. This chapter has been devoted to an introductory presentation of ML tools for CDR, basing on the perspective of an ever growing availability of biochemically relevant data and efficient algorithms for their analysis. While technology continues to improve our ability to study disease mechanisms from a systemic perspective, computational tools are likely to play a central role in helping researchers to mine knowledge out of present and future large-scale databases. It is therefore reasonable to expect that the basic knowledge of ML concepts and tools such as those mentioned in this chapter will soon be of help to all researchers in the field of computer-aided drug repositioning. With neurological disorders being dependent on biological mechanisms among the most complex, neuropharmacology could find an important lever in CDR techniques.

REFERENCES

Amberger JS, Bocchini CA, Schiettecatte F, Scott AF, Hamosh A. OMIM.org: Online Mendelian Inheritance in Man (OMIM®), an online catalog of human genes and genetic disorders. *Nucleic Acids Res* (2014) 43:789–798.

Behl C, Skutella T, Lezoualc'H F, Post A, Widmann M, Newton CJ, Holsboer F. Neuroprotection against oxidative stress by estrogens: Structure–activity relationship. *Mol Pharmacol* (1997) 51:535–541.

Bishop C. *Pattern Recognition and Machine Learning*. Springer, New York (2006).

Boser BE, Guyon IM, Vapnik VN. A training algorithm for optimal margin classifiers, in: *Proceedings of the Fifth Annual Workshop on Computational Learning Theory, COLT'92*. ACM, New York (1992), pp. 144–152.

Brazma A, Parkinson H, Sarkans U, Shojatalab M, Vilo J, Abeygunawardena N, Holloway E et al. ArrayExpress—A public repository for microarray gene expression data at the EBI. *Nucleic Acids Res* (2003) 31:68–71.

Butcher EC, Berg EL, Kunkel EJ. Systems biology in drug discovery. *Nat Biotechnol* (2004) 22:1253–1259.

Carrella D, Napolitano F, Rispoli R, Miglietta M, Carissimo A, Cutillo L, Sirci F, Gregoretti F, di Bernardo D. Mantra 2.0: an online collaborative resource for drug mode of action and repurposing by network analysis. *Bioinformatics* (2014) 30:1787–1788.

Chen JH, Linstead E, Swamidass SJ, Wang D, Baldi P. ChemDB update—Full-text search and virtual chemical space. *Bioinformatics* (2007) 23:2348–2351.

Chen L, Zeng W-M, Cai Y-D, Feng K-Y, Chou K-C. Predicting anatomical therapeutic chemical (ATC) classification of drugs by integrating chemical–chemical interactions and similarities. *PLoS ONE* (2012) 7:e35254.

Cheng T, Li Q, Zhou Z, Wang Y, Bryant SH. Structure-based virtual screening for drug discovery: A problem-centric review. *AAPS J* (2012) 14:133–141.

Cowley MJ, Pinese M, Kassahn KS, Waddell N, Pearson JV, Grimmond SM, Biankin AV, Hautaniemi S, Wu J. PINA v2.0: Mining interactome modules. *Nucleic Acids Res* (2012) 40:862–865.

Dietterich TG. Ensemble methods in machine learning. In: *Multiple Classifier Systems*. Lecture Notes in Computer Science. Springer, Berlin, Germany (2000), Vol. 1857, pp. 1–15.

Ding H, Takigawa I, Mamitsuka H, Zhu S. Similarity-based machine learning methods for predicting drug–target interactions: A brief review. *Brief Bioinform* (2014) 15:734–747.

Draghici S, Khatri P, Eklund AC, Szallasi Z. Reliability and reproducibility issues in DNA microarray measurements. *Trend Genet* (2006) 22:101–109.

Edgar R, Domrachev M, Lash AE. Gene Expression Omnibus: NCBI gene expression and hybridization array data repository. *Nucleic Acids Res* (2002) 30:207–210.

Fisher RA. The use of multiple measurements in taxonomic problems. *Annals Eugen* (1936) 7:179–188.

Frey BJ, Dueck D. Clustering by passing messages between data points. *Science* (2007) 315:972–976.

GO Consortium. The gene ontology (GO) database and informatics resource. *Nucleic Acids Res* (2004) 32:258–261.

Gottlieb A, Stein GY, Ruppin E, Sharan R. PREDICT: A method for inferring novel drug indications with application to personalized medicine. *Mol Syst Biol* (2014) 7:496.

Haasdonk B. Feature space interpretation of SVMs with indefinite kernels. *IEEE Trans Pattern Anal Mach Intell* (2005) 27:482–492.

Henry VJ, Bandrowski AE, Pepin A-S, Gonzalez BJ, Desfeux A. OMICtools: An informative directory for multi-omic data analysis. *Database* (2014) 2014: article ID bau069; doi:10.1093/database/bau069.

Ho TK. The random subspace method for constructing decision forests. *IEEE Trans Pattern Anal Mach Intell* (1998) 20:832–844.

Hoheisel JD. Microarray technology: Beyond transcript profiling and genotype analysis. *Nat Rev Genet* (2006) 7:200–210.

Hollander M, Wolfe DA. *Nonparametric Statistical Methods*, 2nd edn. Wiley-Interscience, New York (1999).

Ideker T, Galitski T, Hood L. A new approach to decoding life: Systems biology. *Annu Rev Genom Human Genet* (2001) 2:343–372.

Iorio F, Bosotti R, Scacheri E, Belcastro V, Mithbaokar P, Ferriero R, Murino L et al. Discovery of drug mode of action and drug repositioning from transcriptional responses. *Proc Natl Acad Sci* (2010) 107:14621–14626.

Iorio F, Rittman T, Ge H, Menden M, Saez-Rodriguez J. Transcriptional data: A new gateway to drug repositioning? *Drug Discov Today* (2013) 18:350–357.

Kitchen DB, Decornez H, Furr JR, Bajorath J. Docking and scoring in virtual screening for drug discovery: Methods and applications. *Nat Rev Drug Discov* (2004) 3:935–949.

Kuhn M, von Mering C, Campillos M, Jensen LJ, Bork P. STITCH: Interaction networks of chemicals and proteins. *Nucleic Acids Res* (2008) 36:684–688.

Lamb J, Crawford ED, Peck D, Modell JW, Blat IC, Wrobel MJ, Lerner J et al. The connectivity map: Using gene-expression signatures to connect small molecules, genes, and disease. *Science* (2006) 313:1929–1935.

Lavecchia A. Machine-learning approaches in drug discovery: Methods and applications. *Drug Discov Today* (2015) 20:318–331.

Law V, Knox C, Djoumbou Y, Jewison T, Guo AC, Liu Y, Maciejewski A et al. DrugBank 4.0: Shedding new light on drug metabolism. *Nucleic Acids Res* (2014) 42:1091–1097.

Liberzon A, Subramanian A, Pinchback R, Thorvaldsdóttir H, Tamayo P, Mesirov JP. Molecular signatures database (MSigDB) 3.0. *Bioinformatics* (2011) 27:1739–1740.

McCulloch WS, Pitts W. A logical calculus of the ideas immanent in nervous activity. *Bull Math Biophys* (1943) 5:115–133.

Moreau Y, Tranchevent L-C. Computational tools for prioritizing candidate genes: Boosting disease gene discovery. *Nat Rev Genet* (2012) 13:523–536.

Murphy RF. An active role for machine learning in drug development. *Nat Chem Biol* (2011) 7:327–330.

Napolitano F, Zhao Y, Moreira VM, Tagliaferri R, Kere J, D'Amato M, Greco D. Drug repositioning: A machine-learning approach through data integration. *J Cheminform* (2013) 5:30.

Napolitano F, Sirci F, Carrella D, di Bernardo D. Drug-set enrichment analysis: A novel tool to investigate drug mode of action. *Bioinformatics* (2015) 32(2):235–241. doi:10.1093/bioinformatics/btv536.

Nič M, Jirát J, Košata B, Jenkins A, McNaught A. *IUPAC Compendium of Chemical Terminology: Gold Book*, 2.1.0 edn. IUPAC, Research Triangle Park, NC (2009).

Niemegeers CJ, McGuire JL, Heykants JJ, Janssen PA. Dissociation between opiate-like and antidiarrheal activities of antidiarrheal drugs. *J Pharmacol Exp Ther* (1979) 210:327–333.

Rokach L, Maimon O. *Data Mining with Decision Trees: Theory and Applications*. World Scientific Publishing Co., Inc., River Edge, NJ (2008).

Schmidhuber J. Deep learning in neural networks: An overview. *Neural Networks* (2015) 61:85–117.

Scornet, E. Random forests and kernel methods. *IEEE Transactions on Information Theory* (2016) 62:1485–1500.

Seiler KP, George GA, Happ MP, Bodycombe NE, Carrinski HA, Norton S, Brudz S et al. ChemBank: A small-molecule screening and cheminformatics resource database. *Nucleic Acids Res* (2008) 36:351–359.

Selassie C, Verma RP, Abraham DJ. History of quantitative structure–activity relationships. In: *Burger's Medicinal Chemistry and Drug Discovery* (DJ Abraham, ed.). John Wiley & Sons, Inc. (2003) pp. 1–48.

Siavelis JC, Bourdakou MM, Athanasiadis EI, Spyrou GM, Nikita KS. Bioinformatics methods in drug repurposing for Alzheimer's disease. *Brief Bioinform* (2016) 17(2):322–335.

Silverman RB, Holladay MW. *The Organic Chemistry of Drug Design and Drug Action*, 3rd edn. Academic Press, Amsterdam, the Netherlands (2014).

Sliwoski G, Kothiwale S, Meiler J, Lowe EW. Computational methods in drug discovery. *Pharmacol Rev* (2014) 66:334–395.

Steinwart I, Christmann A. *Support Vector Machines*. Springer Science & Business Media (2008).

Subramanian A, Tamayo P, Mootha VK, Mukherjee S, Ebert BL, Gillette MA, Paulovich A et al. Gene set enrichment analysis: A knowledge-based approach for interpreting genome-wide expression profiles. *Proc Natl Acad Sci USA* (2005) 102:15545–15550.

Theodoridis S, Koutroumbas K. *Pattern Recognition*, 4th edn. Academic Press (2008).

Topliss J. *Quantitative Structure-Activity Relationships of Drugs*. Elsevier (2012).

Tshilidzi M. *Computational Intelligence for Missing Data Imputation, Estimation, and Management: Knowledge Optimization Techniques: Knowledge Optimization Techniques*. IGI Global (2009).

Villoutreix BO, Lagorce D, Labbé CM, Sperandio O, Miteva MA. One hundred thousand mouse clicks down the road: Selected online resources supporting drug discovery collected over a decade. *Drug Discov Today* (2013) 18:1081–1089.

Wang Y, Xiao J, Suzek TO, Zhang J, Wang J, Bryant SH. PubChem: A public information system for analyzing bioactivities of small molecules. *Nucleic Acids Res* (2009) 37:W623–W633.

Weaver DF, Weaver CA. Exploring neurotherapeutic space: How many neurological drugs exist (or could exist)? *J Pharm Pharmacol* (2011) 63:136–139.

WHO Collaborating Centre for Drug Statistics Methodology. Anatomical Therapeutic Chemical (ATC) classification system. https://www.whocc.no/atc. Accessed on September 19, 2016.

Young FW. *Multidimensional Scaling: History, Theory, and Applications*. Psychology Press (2013).

Zien A, Ong CS. Multiclass multiple kernel learning, in: *Proceedings of the 24th International Conference on Machine Learning* (2007), pp. 1191–1198.

6 RNAi Screening toward Therapeutic Drug Repurposing

Nichole Orr-Burks, Byoung-Shik Shim,
Olivia Perwitasari, and Ralph A. Tripp

CONTENTS

6.1 INTRODUCTION

As a result of large-scale genome-wide screens, thousands of genes have been discovered, but many of their functions remain unknown. Initially, functional analysis relied on the characterization of mutant phenotypes, a generally slow and tedious practice. At present, reverse genetics approaches are considered the most effective way to characterize gene function. Loss-of-function reverse genetics methods involve targeting genes *via* homologous recombination, utilize antisense oligonucleotides, ribozyme technologies, or a combination of these and related technologies, but their implementation is a cumbersome and costly procedure. In contrast, RNA interference (RNAi), specifically using short interfering RNAs (siRNAs), have revolutionized the field providing multifaceted, relatively quick, and cost-effective methods to characterize traditional gene function, as well as their function under certain conditions, such as viral infection. siRNAs, when used to exploit the cells machinery, have the ability to systematically uncover the function and interactions of most vertebrate genes (Elbashir 2002).

RNAi is a conserved cellular process that regulates gene expression at the post-transcriptional level providing transient and specific knockdown of cellular messenger RNA (mRNA) and proteins, features affecting cellular pathways. The RNA–mediated knockdown is sequence-directed, and siRNA:mRNA pairing results in targeted nucleolytic cleavage of the mRNA or translational impediment. As illustrated in Figure 6.1, microRNAs (miRNAs) are endogenous, small noncoding

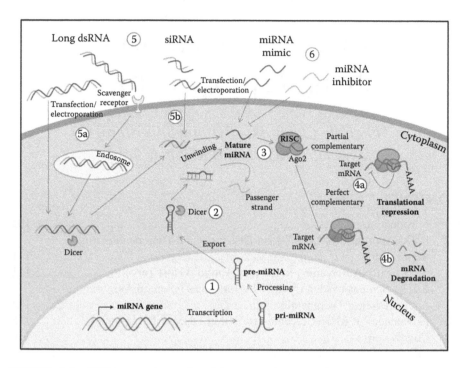

FIGURE 6.1 RNAi strategies to knockdown protein expression by harnessing cellular miRNA pathway. miRNAs are endogenous small noncoding RNAs that regulate gene expression at the post-transcriptional level. (1) RNA polymerase II mediates transcription of miRNA genes to produce pri-miRNAs. Pri-miRNAs are processed by Drosha and Pasha to generate pre-miRNAs, which are subsequently transported to the cytoplasm by exportin-5. (2) Once in the cytoplasm, pre-miRNAs are further processed by dicer into ~22 nt double-stranded miRNA duplexes. miRNA duplexes unwind into single-stranded mature miRNAs, which are then loaded on RISC-containing Ago2 (3). Ago2 guides loaded RISC to complementary miRNA recognition element (MRE) sequences in 5′ UTR, coding region, or 3′ UTR of target mRNAs. (4a) miRNAs repress protein translation through partial complementarity with 6–8 nt seed region on target mRNAs. (4b) miRNAs can also bind target mRNAs through perfect sequence complementary, resulting in degradation of target mRNAs. To harness this pathway and achieve protein expression knockdown, long or short synthetic siRNAs can be utilized. (5a) Long synthetic siRNAs can be transfected or electroporated into cells, although they can also be taken up by receptor-mediated endocytosis. Inside the cells, these long dsRNAs are cleaved by dicer to produce short siRNA, which are then unwound and loaded into the RISC. (5b) Alternatively, short synthetic siRNA can also be transfected or electroporated into the cells. (6) Another method for protein expression perturbation by RNAi is by using synthetically made miRNA mimics or inhibitors. As with endogenous miRNAs, mimics are loaded into the RISC to repress target mRNAs. On the other hand, miRNA inhibitors block endogenous miRNAs either by binding to miRNAs through sequence complementary or by masking the MRE on target mRNAs.

RNAs that are involved in governing gene expression at the posttranscriptional level by binding to mRNAs (Bartel 2004, Filipowicz et al. 2008). Typically, miRNA transcripts are transcribed in the nucleus by RNA polymerase II to produce primary miRNAs (pri-miRNAs) of ~2 kilobases (kb) in length (Lee et al. 2004), generating precursor miRNAs (pre-miRNAs), ~70 nucleotides (nt) in length. Pre-miRNAs are transported to the cytoplasm by exportin-5 (Okada et al. 2009), where they are further processed by dicer into ~22 nt double-stranded miRNAs. Double-stranded miRNAs unwind into a single strand, which is then loaded on Argonaute 2 (Ago2) containing RNA-induced silencing complex (RISC) (Chendrimada et al. 2005). The complex guides Ago2 to complementary miRNA recognition element sequences in 5′ UTR, coding region, or 3′ UTR of target mRNAs (Lytle et al. 2007, Rigoutsos 2009, Hafner et al. 2010). miRNAs reduce protein expression mostly through imperfect complementarity with 6–8 nt region called the "seed region." Therefore, binding to a small site enables individual miRNA to potentially target numerous mRNAs (Carthew and Sontheimer 2009).

As opposed to miRNA, siRNAs are exogenous synthetic RNA oligonucleotide duplexes composed of two single-stranded RNA molecules, a guide strand, and a passenger strand, and approximately 20–24 nucleotides (nt) in length having a two nt 3′-overhang. Within the cell, siRNAs are remodeled in the cytoplasm by cellular kinases. As with miRNA, modified duplexes associate with Ago2 and RISC. Ago2 cleaves one strand (the passenger strand) of the RNA duplex, leaving the other strand (guide strand) associated with the RISC complex (Carthew and Sontheimer 2009). Site-directed cleavage of target mRNAs is facilitated by the sequence-dependent recognition of seed region complements within the target mRNA by the RISC Ago2 associate guide strand. siRNAs typically induce gene-specific knockdown *via* full length complementarity with "seed regions" on target mRNA. This sequence is responsible for the specificity of the siRNA as it facilitates the initial recognition and binding of the siRNA to its target. Recognition and binding of the guide strand with the siRNA allows for the direct cleavage of the mRNA *via* the Ago2 catalytic domain resulting in posttranscriptional gene silencing (Elbashir 2002, McManus and Sharp 2002). This route of alteration obviates the need for processes that induce the activation of type I interferons. This natural process can be utilized as an exploratory *in vitro* and *in vivo* tool to characterize host machinery and provide a platform for the discovery of novel drug targets. This section will discuss siRNAs and the intermediate of the RNAi pathway and their role in target discovery and drug repositioning.

There are two popular approaches to harnessing the RNAi pathway. One option is to transfect short siRNAs (21–22 nt) into cells. The other option is to transfect cells with longer siRNAs (25–27 nt). Following transfection, dicer processes these longer "dicer-ready siRNAs." Both methods provide transient knockdown of target mRNA, but for most applications "dicer-ready siRNAs" provide a more potent silencing (Boutros and Ahringer 2008).

A benefit to using siRNAs as an RNAi trigger is that siRNAs are rationally designed to reduce immune stimulation *via* scavenger and Toll-like receptors, promote effective silencing, increase stability, and reduce off-target effects. When designing siRNAs, one should consider the mRNA target sequence, the silencing moiety, whether or not the RNA requires modification to confer better complementarity to its target, the siRNA sequence, the length of RNA, and physical characteristics of the RNA (e.g., the nature of the 5' and 3' ends) (Schwarz 2006). Rational design is important to reduce immune stimulation within the cell. For example, type I interferon and the cytoplasmic retinoic acid–inducible gene I protein can be induced by the 5' triphosphates and blunt ends of RNA, respectively. These issues can be avoided by designing siRNAs to lack the 5' triphosphates, and to contain the 3' overhangs. Furthermore, incorporating 2'-O-methyl–modified purine nucleosides into the passenger strand can reduce IFN triggering while target specificity remains (Watts et al. 2008). Toll-like receptor 3 (TLR3) is an endosomal or cell surface pattern recognition receptor that recognizes dsRNA and 21-mer siRNAs (Watts et al. 2008). Although there are situations innate immune induction may be desirable, for example cancer therapies, generally this stimulation is unwanted. Chemical alteration can reduce TLR activation (Davidson and McCray 2011). The siRNAs share a more perfect sequence complementarity to their target mRNA as compared to miRNAs (Jackson et al. 2006, Carthew and Sontheimer 2009) and thus produce more potent knockdown with less off-target effects. The siRNAs can be further modified to reduce unwanted off-target effects and enhance target specificity. Off-target effects are those knockdown events that occur due to the targeting of mRNAs other than the target mRNA. As previously noted, siRNAs are composed of two strands, that is, the sense strand and the antisense strand. It is not clear how the guide strand is chosen during RISC loading, but the sense strand can be modified using established chemistries to prevent its association with the RISC complex, and thus the antisense strand becomes the guide strand more often than the sense strand (Watts et al. 2008). The antisense strand seed region can be further modified to enhance target specificity (Anderson et al. 2008).

Efficient uptake of siRNAs into cells is required to achieve gene silencing. Chemically modified siRNAs require a delivery system to overcome their net negative charge and size, which impedes them from penetrating the cellular membrane (Schroeder et al. 2010). There are several procedures by which siRNA can be delivered into cells and this includes transfection, electroporation, DNA plasmid, and viral gene transfer (Table 6.1).

The transfection method is an easy and rapid method that is typically used when working with most immortalized adherent cell types. The majority of adherent and immortalized cell lines can be transfected with minimal transfection reagent toxicity. Most commonly, transfection procedures utilize a liposome-based transfection reagent where the cationic head group of the lipid interacts with negatively charged phosphates on the nucleic acid, forming an RNA–liposome complex (Whitehead et al. 2009). These complexes enter the cells by endocytosis or fusion with the cell plasma membrane mediated by the lipid moieties of the liposomes. There are two transfection methods, that is, forward transfection and reverse transfection. Reverse transfection is typically used in format plates, which are commercially available (Figure 6.2). Each transfection method requires the siRNA, lipid-based transfection reagent, and cells. The order, timing, and addition of these components vary with each method. The forward transfection method is a two-step process where plated cells are treated with siRNAs complexed with a lipid-based transfection reagent. The reverse transfection

TABLE 6.1

Method for RNAi Delivery for Transient Gene Knockdown

Method	Advantage	Disadvantage
Lipid-based transfection	• Highly efficient, reproducible, and commercially available from multiple vendors. • Delivery into most secondary and transformed cell lines.	• Primary and nontransformed cells are difficult to transfect. • Limited *in vivo* application.
Electroporation	• Delivery difficult; can overcome hard to transfect primary and nontransformed cells. • Nucleofection can be utilized to deliver RNAi to nondividing cells.	• Significant cell death. • Potential escape of intracellular contents. • Requires cell type–specific protocol optimization.
Viral vector (e.g., adenovirus/AAV, retrovirus, lentivirus)	• Delivery to growth-arrested and contact-inhibited cells. • Integrating viral vectors (retrovirus, lentivirus) allows for stable knockdown. • *In vivo* delivery application.	• Requires generation and titration of virus vectors. • Can induce nonspecific and unintentional mutations on cell genome. • Viral vectors can induce significant immune response *in vivo*.

Sources: Adapted from Jiang, Q. et al. *Pathways*, 9, 10, 2009; Perwitasari, O. et al., *Pharmaceuticals (Basel)*, 6(2), 124, 2013.

method is a three-step process where the siRNA is complexed with the transfection reagent, added to a suspension of cells, and then all components are added to the cell culture plate together. The reverse transfection format plate method is a truncated derivative of the reverse transfection method. This method utilizes a cell culture plate that has already been treated with the siRNA. The transfection reagent is added directly to the plated siRNAs followed by addition of cells. The aforementioned methods require a 2–3-day incubation period in appropriate cell culture conditions but can last up to a week. The method chosen depends on the desired transfection efficiency, cytotoxicity, and rate of cell division.

Electroporation is most often utilized when working with primary cell and suspension cell (Tsong 1991). This method obviates the impermeability of the siRNA *via* manipulation of the semipermeability of the cellular membrane with a brief powerful pulse of electric current. Applying this pulse of energy forces the lipid molecules within the membrane to undergo remodeling *via* reorientation and thermal phase transitioning, inducing the formation of transient hydrophilic pores and localized weak points within the membrane (Andreason and Evans 1989). This process is not the preferred method of delivery due to the potential escape of intracellular contents during membrane disruption, but it can resolve some cell types' specific limitations associated with liposomal transfection methods. Factors to consider when using electroporation include the voltage, quantity, and duration

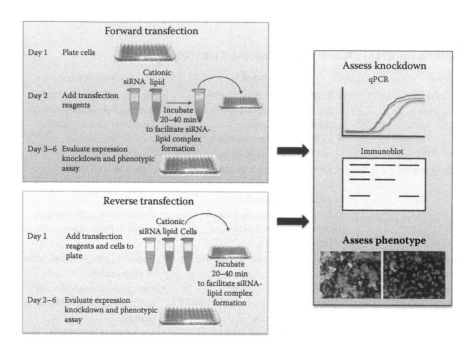

FIGURE 6.2   Transfection protocols for synthetic RNAi (siRNAs, miRNA mimics, or inhibitors) delivery into cells using lipid-based transfection reagent. Traditional forward transfection protocol requires cell plating the day before transfection. The following day, RNA-lipid complexes can be added to the adhered cells. In contrast, reverse transfection protocol permits direct addition of cell suspension into culture plate, which contains preincubated RNA and transfection reagent. Reverse transfection format (RTF) plates are multiwell culture plates readily coated with lyophilized siRNAs and are commercially available. RTF protocol allows for rapid and convenient genome-wide or library-specific RNAi screening. Regardless of the transfection method, knockdown efficiency and desired endpoint phenotypes (infection, apoptosis, etc.) can be evaluated at 24–96 h post-transfection.

of pulses. These parameters should be optimized for different cell types to limit cell mortality. Transfection and electroporation of siRNAs provide transient short-term knockdown. Long-term stable expression of siRNAs can be accomplished *via* DNA expression plasmids expressing short hairpin RNAs (shRNAs) (Taxman et al. 2006). shRNAs are recognized and cleaved by dicer to generate mature siRNAs. Recombinant viral vectors modeling retrovirus, adeno-associated virus (AAV), adenovirus, and lentivirus can be used to deliver and express siRNAs (Sliva and Schnierle 2010). This method can also be used to integrate shRNAs into the cell genome, thus providing stable siRNA expression and long-term knockdown (Whitehead et al. 2009). A variety of methods can be implemented to deliver libraries of siRNAs to cells in culture providing a high-throughput platform for analyzing thousands of gene targets per experiment. There are advantages and disadvantages to each of the methods discussed (Table 6.1). Transfection methods require less effort, are highly efficient, reproducible, and commercially available.

When choosing and optimizing a delivery method, take care to be consistent, use healthy cells, choose appropriate culturing conditions, use siRNAs at their lowest effective concentrations, optimize exposure time, monitor delivery and effectiveness *via* positive, negative, and untreated controls while monitoring knockdown efficiency *via* mRNA and protein quantification.

6.2 RNAi APPROACH FOR DRUG TARGET DISCOVERY AND THERAPEUTIC DRUG REPURPOSING

Traditionally, drug development involves two general approaches, that is, large-scale chemical library screening and structure-based virtual screening. Large-scale screening requires considerable compound libraries, and virtual screening requires the precise target structure. Both processes can be costly and cumbersome and still yield minimal results. These processes have been successful in identifying unique drug targets, but most of the compounds identified never make it to market. Alternative methods are now being used to facilitate drug discovery and repurposing. One such method utilizes siRNAs to screen host genes required by pathogens (e.g., viruses) for replication, which are temporally dispensable to the cell and do not immediately cause cell death when silenced or knocked down. This method circumvents the need for large compound libraries and is a simpler, more efficient way to tease out host factors known to be involved during a particular disease (Perwitasari et al. 2013). For example, influenza viruses, like all viruses, usurp host cell genes to replicate. To inhibit virus replication, one may use siRNA screening and validation approaches to identify host genes and cellular proteins and pathways required by the virus for replication and find drugs that target those genes to be used as countermeasures to prevent disease. Understanding these host factors provides a pool of potential pharmacological targets, while also adding to our knowledge of the molecular mechanisms involved during virus replication (Konig et al. 2010, Meliopoulos et al. 2012).

6.2.1 siRNA SCREENING TO IDENTIFY ANTIVIRAL HOST TARGETS

siRNAs can be used to screen host protein gene families such as kinases, proteases, ion channels, G-protein coupled receptors (GPCRs), etc., to determine if said genes within these families are needed by a virus to replicate (Karlas et al. 2010). Using influenza A virus as an example, one can determine whether a host gene is proviral or antiviral relative to influenza replication. Proviral genes are those genes whose expression is required for efficient virus replication. In contrast, the expression of antiviral genes impedes virus replication. An example of such a genome-wide screen is summarized in Figure 6.3.

Initially, a primary screen of potential host genes is performed using pools of validated siRNAs from a genome-wide siRNA library targeting all or some of the unique human genes in the NCBI RefSeq database, which is commercially available. These pools are composed of four individual siRNAs that target distinct "seed regions" on the target mRNA for each gene to be tested. Pools are necessary for the initial primary screen to ensure the most effective knockdown phenotype.

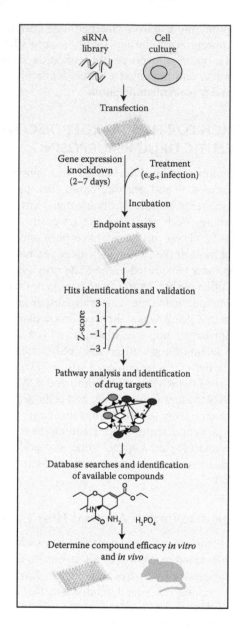

FIGURE 6.3 Overview of siRNA screening for target identification and drug repositioning. A primary screen is completed utilizing an siRNA library targeting genes within a family of genes. Alternatively, a miRNA screen can also be performed using a library of miRNA mimics. Following transfection, cells are treated, incubated, and endpoint assays are completed to determine the knockdown phenotype. These data are then used to identify possible hits for validation. Identified hits are subject to repeated transfection and endpoint assays' experiments to confirm the knockdown phenotype. Validated gene targets are subjected to pathway analysis to determine possible mechanisms of action. Following pathway analysis, compounds are identified through database searches and evaluated for efficacy *in vitro* and *in vivo*.

When designing a screen, careful consideration should be directed toward the selection and implementation of appropriate controls such as a positive, negative, mock, and transfection control. For example, transfection efficiency and host gene silencing can be evaluated with a nontargeting siRNA and siRNA known to cause cell cytotoxicity. A nontargeting siRNA (NTC) does not share sequence complementarity with any known mRNAs and thus should not affect the expression of host genes. This type of siRNA would be considered a negative control. A siRNA that induces cell myotoxicity targets a host gene known to be required for cell survival. Thus, if the transfection procedure is working appropriately, cell death is observed. In the context of the example, influenza virus is known to have cell tropism for respiratory epithelial cells, so an appropriate cell line for screening would be A549 cells, which are type II human alveolar pneumocytes. Pooled siRNA library constituents and controls are reverse transfected into cells as previously discussed. Following transfection, cells are evaluated for cellular toxicity and cytopathic events. The siRNAs that are associated with high cell toxicity and death should be excluded. Accordingly, cells are infected with a low multiplicity of influenza A virus, and 48 h later infectious virus is measured and the results normalized to the NTC transfected cells. One such method of normalizing sets the NTC siRNA to an arbitrary value of 1.0. Genes are then assigned a standard score (z-score) based on the distribution of these values compared to nontargeting control. Z-scores less than zero are considered proviral, conversely, z-scores greater than zero are considered an antiviral hit (Meliopoulos et al. 2012, Bakre et al. 2013).

Data collected during the primary screen can be used to determine which genes are potential "druggable" targets. Negative z-scores are associated with genes required for influenza replication and are thus possible targets for antiviral therapeutics. These proviral genes are then subjected to elimination based on standard deviations from control values from a z-score, redundancy, and the availability of known drug antagonists/inhibitors. Redundancy refers to a gene or genes whose products have similar function, which can partially or completely compensate for the diminished expression and/or knockout of another in the same pathway. Redundancy can be determined by Ingenuity Pathway Analysis software (IPA; Ingenuity systems, http://www.ingenuity.com). IPA software propagates interaction networks and identifies functional groups to which preliminary genes belong. IPA and other database searches are useful tools to determine previously evaluated drugs that target the gene candidates (Gusev 2008, Jimenez-Marin et al. 2009). This process allows the researcher to short-list the best candidates for drug repositioning. Promising candidates undergo additional validation steps to ensure the legitimacy of the knockdown phenotype. Following the primary screen, a secondary validation screen is completed. In this screen, chosen genes are targeted with deconvoluted siRNAs from the pooled siRNAs that showed the desired efficacy. The overall process is similar to the primary screen with the exception that target mRNA transcripts are evaluated by PCR to confirm knockdown of the gene target *via* knockdown of gene specific mRNA transcripts. Knockdown of gene specific mRNA transcripts reduces the targeted gene's expression. Also during the secondary screen, multiple endpoint assays should be implanted to determine phenotype (i.e., viral titer) following transfection to ensure the accuracy and validity of the screen. Genes associated with reduction in viral titer following transfection are confirmed proviral genes and can be now evaluated as possible targets for drug repositioning. After gene targets are identified, possible drug inhibitors and antagonists can be identified through database mining. Particular interests should be assigned to those drugs that have been approved for clinical use or evaluated in clinical safety trials for repurposing. Once drugs are selected, they can be evaluated *in vitro* and subsequently *in vivo* (Perwitasari et al. 2013).

6.2.2   SCREENS TO IDENTIFY miRNA TARGETS AS POTENTIAL ANTIVIRAL DISEASE THERAPEUTIC

In addition to siRNA screens, an miRNA screen can also be utilized to identify the host miRNAs and their downstream factors that regulate the host genes required, for example, in virus infection and replication. Currently, more than 2500 miRNAs have been identified in human cells (Griffiths-Jones et al. 2008). miRNAs are either expressed in specific tissue types, or in certain responses, such as virus infection (Xu et al. 2007). As a result, miRNAs affect various biological processes, including proliferation, apoptosis, differentiation, development, and immune response (Jovanovic and Hengartner 2006, Kedde and Agami 2008, O'Connell et al. 2010). In addition, miRNAs have been known to be a critical factor involved in host–virus interaction during virus infection (Umbach and Cullen 2009, Skalsky and Cullen 2010). Once a virus enters host cells, it may regulate the miRNA expression to support its replication (Jopling et al. 2005, Farberov et al. 2015). Conversely, miRNAs expressed by host cells can inhibit virus replication by targeting the viral genome directly (Zheng et al. 2013, Trobaugh et al. 2014), or by reducing the expression of host factors that are required for virus replication (Fu et al. 2015, Huang et al. 2015).

Several efforts have focused on developing miRNA-based therapeutic interventions against pathogens and diseases. For example, miravirsen, an anti-miR-122 oligonucleotide, has been evaluated in patients chronically infected with hepatitis C virus (HCV) and demonstrated promising reduction of HCV titers in patients' serum (Janssen et al. 2013). In addition, MRX34, a liposome-based miR-34 mimic, has undergone a clinical trial to evaluate its safety as an anticancer therapeutic candidate (Bouchie 2013).

High-throughput screening with commercially available libraries of miRNAs can be utilized to identify miRNAs involved in biological processes as well as host–pathogen interaction. MicroRNA mimics and inhibitors can be used for the analysis of gain or loss of function for miRNAs. However, while siRNA screening can provide direct gene targets related to a specific biological process or disease, miRNA screening requires relatively more robust data processing. Additional analyses are required to identify target genes and/or biological mechanism due to the broad activity of miRNAs. Here, bioinformatics tools can be employed to identify target genes regulated by miRNAs, as summarized in Table 6.2. These tools provide target prediction, typically on the basis of seed match, conservation, site accessibility, and free energy (Peterson et al. 2014). The predicted genes can be further confirmed by overlapping with gene hits identified from RNAi screens for high-confidence hits (Meliopoulos et al. 2012, Bakre et al. 2013). Therefore, correlations between miRNA and RNAi screenings provide an approach to identify high-confidence hits and therapeutic targets.

TABLE 6.2
Bioinformatics Tools for microRNA Target Prediction

Tools	Website	Reference
TargetScan	http://www.targetscan.org/	Lewis et al. (2005)
miRWalk	http://zmf.umm.uni-heidelberg.de/apps/zmf/ mirwalk2/miRretsys-self.html	Dweep et al. (2011)
miRanda	http://www.microrna.org/	John et al. (2004)
starBase	http://starbase.sysu.edu.cn/	Li et al. (2014)
DIANA-microT-CDS	http://www.microrna.gr/microT-CDS	Paraskevopoulou et al. (2013)
miRDB	http://mirdb.org	Wong and Wang (2015)
RNA22	https://cm.jefferson.edu/rna22/	Miranda et al. (2006)
PITA	http://genie.weizmann.ac.il/pubs/mir07/	Kertesz et al. (2007)

6.3 ADVANTAGES OF RNAi SCREENING FOR NOVEL DISEASE THERAPEUTICS

Utilizing RNAi screens such as siRNA and miRNA screens for drug repurposing other than classical approaches can be hugely beneficial. RNAi is typically more cost-effective than large-scale compound library screens, and RNAi procedures require small amounts of reagents lowering the overall cost of each experiment. Unlike compounds that may require a toxicity solvent such as dimethyl sulfoxide (DMSO), RNAi does not. RNAi reagents are easily introduced into cell culture with relatively high efficiency and minimally cellular toxicity due to reagent formulation. Unlike compound libraries, siRNAs and miRNA mimic/inhibitor can be easily generated and manipulated to match one's needs and goals. RNAi screens provide a plethora of information about the target, and siRNAs are specific, thus they provide immediate identification of genes important during a specific process without having a priori knowledge of the protein structure. Although the sequences of all human genes are available, in most cases the structure and function of the proteins expressed are yet to be identified. Virtual screening requires information about the structure of the target where this information may not be available. One can follow the drug or follow the gene. Drug compound screening requires backtracking to determine the protein target as well as the mechanism by which this target is implicated in disease. As previously discussed, siRNAs are rationally designed to specifically target the gene of interest. Virtual screening provides matches based on a set of parameters and algorithms, which must be evaluated to determine if the predictions can be confirmed. Due to their specificity, siRNA screens provide immediate information about the target and its implication during disease (Fougerolles 2007). In the previous example, it was determined whether a particular gene is beneficial or detrimental to influenza replication. Following a similar procedure, compound screening would provide information about whether or not the compound impedes viral replication but would provide little to no information about the mechanism by which the compound affects virus replication. For example, a particular compound may be an agonist, activator, antagonist, inhibitor, etc. Furthermore, while there may not be an available compound inhibitor, siRNAs can be designed to target the majority of genes and can potentially be used as therapeutics. siRNAs provide insight into the mechanics of a particular genetic factor during disease by providing a reduced level of wild-type product. By examining the resulting phenotype, one can gain insight into its role within a particular cell based on the disease process. Unfortunately, siRNA screens are not perfect. Not all genes can be successfully targeted with siRNA. There are cells and tissues that are resistant to transfection. Other genes express proteins will long half-lives reducing the effectiveness of siRNA knockdown due to residual protein. siRNA can be chemically modified, but there is always still a chance of off-target effects during transfection.

6.4 CONCLUSION

In summary, RNAi screening is a viable approach to uncover disease-specific host factor or genes for novel therapeutic applications. As many genes and gene factors have been previously targeted for unrelated disease indications, such as cancer or inflammation, the RNAi approach can identify genes required by the pathogen or disease process to allow drug repurposing of currently approved drugs. It can also reduce costs and time as it allows drugs to be tested for efficacy using drugs having good safety profiles and known pharmacokinetics. Drug repositioning gives life to forgotten or overlooked drugs to provide therapeutics and prophylactics. There are several methods by which one can determine which of these drugs is suitable for repositioning. One such method utilizes siRNA screening to uncover disease-specific host factor candidates for novel target applications. By reducing the time and cost associated with other methods, siRNA screening can provide an easy, efficient way to work forward and not backward for disease therapeutic.

REFERENCES

Anderson, E. M., A. Birmingham, S. Baskerville, A. Reynolds, E. Maksimova, D. Leake, Y. Fedorov, J. Karpilow, and A. Khvorova. 2008. Experimental validation of the importance of seed complement frequency to siRNA specificity. *RNA* 14 (5):853–861. doi:10.1261/rna.704708.

Andreason, G. L. and G. A. Evans. 1989. Optimization of electroporation for transfection of mammalian cell lines. *Anal Biochem* 180:269–275.

Bakre, A., L. E. Andersen, V. Meliopoulos, K. Coleman, X. Yan, P. Brooks, J. Crabtree, S. M. Tompkins, and R. A. Tripp. 2013. Identification of host kinase genes required for influenza virus replication and the regulatory role of microRNAs. *PLoS One* 8 (6):e66796.

Bartel, D. P. 2004. MicroRNAs: Genomics, biogenesis, mechanism, and function. *Cell* 116 (2):281–297.

Bouchie, A. 2013. First microRNA mimic enters clinic. *Nat Biotechnol* 31 (7):577.

Boutros, M. and J. Ahringer. 2008. The art and design of genetic screens: RNA interference. *Nat Rev Genet* 9 (7):554–566.

Carthew, R. W. and E. J. Sontheimer. 2009. Origins and mechanisms of miRNAs and siRNAs. *Cell* 136 (4):642–655.

Chendrimada, T. P., R. I. Gregory, E. Kumaraswamy, J. Norman, N. Cooch, K. Nishikura, and R. Shiekhattar. 2005. TRBP recruits the Dicer complex to Ago2 for microRNA processing and gene silencing. *Nature* 436 (7051):740–744.

Davidson, B. L. and P. B. McCray, Jr. 2011. Current prospects for RNA interference-based therapies. *Nat Rev Genet* 12 (5):329–340.

Dweep, H., C. Sticht, P. Pandey, and N. Gretz. 2011. miRWalk—Database: Prediction of possible miRNA binding sites by walking the genes of three genomes. *J Biomed Inform* 44 (5):839–847.

Elbashir, S. M. 2002. Analysis of gene function in somatic mammalian cells using small interfering RNAs. *Methods* 26:199–213.

Farberov, L., E. Herzig, S. Modai, O. Isakov, A. Hizi, and N. Shomron. 2015. MicroRNA-mediated regulation of p21 and TASK1 cellular restriction factors enhances HIV-1 infection. *J Cell Sci* 128 (8):1607–1616.

Filipowicz, W., S. N. Bhattacharyya, and N. Sonenberg. 2008. Mechanisms of post-transcriptional regulation by microRNAs: Are the answers in sight? *Nat Rev Genet* 9 (2):102–114.

Fougerolles, A. D. 2007. Interfering with disease: A progress report on siRNA-based therapeutics. *Nat Rev Drug Discov* 6:443–453.

Fu, Y. R., X. J. Liu, X. J. Li, Z. Z. Shen, B. Yang, C. C. Wu, J. F. Li et al. 2015. MicroRNA miR-21 attenuates human cytomegalovirus replication in neural cells by targeting Cdc25a. *J Virol* 89 (2):1070–1082.

Griffiths-Jones, S., H. K. Saini, S. van Dongen, and A. J. Enright. 2008. miRBase: Tools for microRNA genomics. *Nucleic Acids Res* 36 (Database issue):D154–D158.

Gusev, Y. 2008. Computational methods for analysis of cellular functions and pathways collectively targeted by differentially expressed microRNA. *Methods* 44 (1):61–72.

Hafner, M., M. Landthaler, L. Burger, M. Khorshid, J. Hausser, P. Berninger, A. Rothballer et al. 2010. Transcriptome-wide identification of RNA-binding protein and microRNA target sites by PAR-CLIP. *Cell* 141 (1):129–141.

Huang, J. Y., S. F. Chou, J. W. Lee, H. L. Chen, C. M. Chen, M. H. Tao, and C. Shih. 2015. MicroRNA-130a can inhibit hepatitis B virus replication via targeting PGC1alpha and PPARgamma. *RNA* 21 (3):385–400.

Jackson, A. L., J. Burchard, D. Leake, A. Reynolds, J. Schelter, J. Guo, J. M. Johnson et al. 2006. Position-specific chemical modification of siRNAs reduces off-target transcript silencing. *RNA* 12 (7):1197–1205.

Janssen, H. L., H. W. Reesink, E. J. Lawitz, S. Zeuzem, M. Rodriguez-Torres, K. Patel, A. J. van der Meer et al. 2013. Treatment of HCV infection by targeting microRNA. *N Engl J Med* 368 (18):1685–1694.

Jiang, Q., Z. Zhang, A. Bank, G. Quellhorst, J. Huang, and R. Medicus. 2009. siRNA delivery methods into mammalian cells: Gene function study guide in stem cells. *Pathways* 9: 10–11.

Jimenez-Marin, A., M. Collado-Romero, M. Ramirez-Boo, C. Arce, and J. J. Garrido. 2009. Biological pathway analysis by ArrayUnlock and Ingenuity Pathway Analysis. *BMC Proc* 3 (Suppl. 4):S6.

John, B., A. J. Enright, A. Aravin, T. Tuschl, C. Sander, and D. S. Marks. 2004. Human MicroRNA targets. *PLoS Biol* 2 (11):e363.

Jopling, C. L., M. Yi, A. M. Lancaster, S. M. Lemon, and P. Sarnow. 2005. Modulation of hepatitis C virus RNA abundance by a liver-specific MicroRNA. *Science* 309 (5740):1577–1581.

Jovanovic, M. and M. O. Hengartner. 2006. miRNAs and apoptosis: RNAs to die for. *Oncogene* 25 (46):6176–6187.

Karlas, A., N. Machuy, Y. Shin, K. P. Pleissner, A. Artarini, D. Heuer, D. Becker et al. 2010. Genome-wide RNAi screen identifies human host factors crucial for influenza virus replication. *Nature* 463 (7282):818–822.

Kedde, M. and R. Agami. 2008. Interplay between microRNAs and RNA-binding proteins determines developmental processes. *Cell Cycle* 7 (7):899–903.

Kertesz, M., N. Iovino, U. Unnerstall, U. Gaul, and E. Segal. 2007. The role of site accessibility in microRNA target recognition. *Nat Genet* 39 (10):1278–1284.

Konig, R., S. Stertz, Y. Zhou, A. Inoue, H. H. Hoffmann, S. Bhattacharyya, J. G. Alamares et al. 2010. Human host factors required for influenza virus replication. *Nature* 463 (7282):813–817.

Lee, Y., M. Kim, J. Han, K. H.'Yeom, S. Lee, S. H. Baek, and V. N. Kim. 2004. MicroRNA genes are transcribed by RNA polymerase II. *EMBO J* 23 (20):4051–4060.

Lewis, B. P., C. B. Burge, and D. P. Bartel. 2005. Conserved seed pairing, often flanked by adenosines, indicates that thousands of human genes are microRNA targets. *Cell* 120 (1):15–20.

Li, J. H., S. Liu, H. Zhou, L. H. Qu, and J. H. Yang. 2014. starBase v2.0: Decoding miRNA-ceRNA, miRNA-ncRNA and protein-RNA interaction networks from large-scale CLIP-Seq data. *Nucleic Acids Res* 42 (Database issue):D92–D97.

Lytle, J. R., T. A. Yario, and J. A. Steitz. 2007. Target mRNAs are repressed as efficiently by microRNA-binding sites in the 5′ UTR as in the 3′ UTR. *Proc Natl Acad Sci USA* 104 (23):9667–9672.

McManus, M. T. and P. A. Sharp. 2002. Gene silencing in mammals by small interfering RNAs. *Nat Rev Genet* 3 (10):737–747.

Meliopoulos, V. A., L. E. Andersen, P. Brooks, X. Yan, A. Bakre, J. K. Coleman, S. M. Tompkins, and R. A. Tripp. 2012. MicroRNA regulation of human protease genes essential for influenza virus replication. *PLoS One* 7 (5):e37169.

Miranda, K. C., T. Huynh, Y. Tay, Y. S. Ang, W. L. Tam, A. M. Thomson, B. Lim, and I. Rigoutsos. 2006. A pattern-based method for the identification of MicroRNA binding sites and their corresponding heteroduplexes. *Cell* 126 (6):1203–1217.

O'Connell, R. M., D. S. Rao, A. A. Chaudhuri, and D. Baltimore. 2010. Physiological and pathological roles for microRNAs in the immune system. *Nat Rev Immunol* 10 (2):111–122.

Okada, C., E. Yamashita, S. J. Lee, S. Shibata, J. Katahira, A. Nakagawa, Y. Yoneda, and T. Tsukihara. 2009. A high-resolution structure of the pre-microRNA nuclear export machinery. *Science* 326 (5957):1275–1279.

Paraskevopoulou, M. D., G. Georgakilas, N. Kostoulas, I. S. Vlachos, T. Vergoulis, M. Reczko, C. Filippidis, T. Dalamagas, and A. G. Hatzigeorgiou. 2013. DIANA-microT web server v5.0: Service integration into miRNA functional analysis workflows. *Nucleic Acids Res* 41 (Web Server issue):W169–W173.

Perwitasari, O., A. Bakre, S. M. Tompkins, and R. A. Tripp. 2013. siRNA genome screening approaches to therapeutic drug repositioning. *Pharmaceuticals (Basel)* 6 (2):124–160.

Peterson, S. M., J. A. Thompson, M. L. Ufkin, P. Sathyanarayana, L. Liaw, and C. B. Congdon. 2014. Common features of microRNA target prediction tools. *Front Genet* 5:23.

Rigoutsos, I. 2009. New tricks for animal microRNAS: Targeting of amino acid coding regions at conserved and nonconserved sites. *Cancer Res* 69 (8):3245–3248.

Schroeder, A., C. G. Levins, C. Cortez, R. Langer, and D. G. Anderson. 2010. Lipid-based nanotherapeutics for siRNA delivery. *J Intern Med* 267 (1):9–21.

Schwarz, D. S. 2006. Designing siRNA that distinguish between genes that differ by a single nucleotide. *PLoS Genetics* 2 (9):1307–1318.

Skalsky, R. L. and B. R. Cullen. 2010. Viruses, microRNAs, and host interactions. *Annu Rev Microbiol* 64:123–141.

Sliva, K. and B. S. Schnierle. 2010. Selective gene silencing by viral delivery of short hairpin RNA. *Virol J* 7:248.

Taxman, D. J., L. R. Livingstone, J. Zhang, B. J. Conti, H. A. Iocca, K. L. Williams, J. D. Lich, J. P. Ting, and W. Reed. 2006. Criteria for effective design, construction, and gene knockdown by shRNA vectors. *BMC Biotechnol* 6:7.

Trobaugh, D. W., C. L. Gardner, C. Sun, A. D. Haddow, E. Wang, E. Chapnik, A. Mildner, S. C. Weaver, K. D. Ryman, and W. B. Klimstra. 2014. RNA viruses can hijack vertebrate microRNAs to suppress innate immunity. *Nature* 506 (7487):245–248.

Tsong, T. Y. 1991. Electroporation of cell membranes. *Biophysiol J* 60:297–306.

Umbach, J. L. and B. R. Cullen. 2009. The role of RNAi and microRNAs in animal virus replication and antiviral immunity. *Genes Dev* 23 (10):1151–1164.

Watts, J. K., G. F. Deleavey, and M. J. Damha. 2008. Chemically modified siRNA: Tools and applications. *Drug Discov Today* 13 (19–20):842–855.

Whitehead, K. A., R. Langer, and D. G. Anderson. 2009. Knocking down barriers: Advances in siRNA delivery. *Nat Rev Drug Discov* 8 (2):129–138.

Wong, N. and X. Wang. 2015. miRDB: An online resource for microRNA target prediction and functional annotations. *Nucleic Acids Res* 43 (Database issue):D146–D152.

Xu, S., P. D. Witmer, S. Lumayag, B. Kovacs, and D. Valle. 2007. MicroRNA (miRNA) transcriptome of mouse retina and identification of a sensory organ-specific miRNA cluster. *J Biol Chem* 282 (34):25053–25066.

Zheng, Z., X. Ke, M. Wang, S. He, Q. Li, C. Zheng, Z. Zhang, Y. Liu, and H. Wang. 2013. Human microRNA hsa-miR-296-5p suppresses enterovirus 71 replication by targeting the viral genome. *J Virol* 87 (10):5645–5656.

7 Phenotypic Screening

Christine M. Macolino-Kane, John R. Ciallella,
Christopher A. Lipinski, and Andrew G. Reaume

CONTENTS

7.1 INTRODUCTION

Since 1950, the FDA has approved 1222 new drugs. Yet despite new technologies, research spending of over $50 billion per year, mergers and acquisitions, and improvements in the pharmaceutical development process, the rate of newly approved drugs is no better than it was 50 years ago (Munos 2009). It can take up to 15 years and over $1 billion to develop a new drug from original design to final product launch, and it is estimated that large pharmaceutical companies would require 2–3 new molecular entities (NMEs) per year to sustain a profitable business. No company is coming close to this productivity (Munos 2014), suggesting that current approaches have reached their innovative capacities. While a slight uptake in pharmaceutical productivity has been noted very recently, an analysis on this topic concluded with the observation that "old fashioned" empirical approaches, such as phenotypic screening and drug repurposing, still have much to contribute to their innovation revival (Munos 2013). Without drastic improvement in research and development techniques, the pharmaceutical industry cannot sustain such a severe loss in revenue and patent expirations for successful products.

Historically, the pharmaceutical industry has used two alternative approaches to drug discovery: phenotypic screening and target-based or hypothesis-driven screening. The former evaluates the effect of a test article on a biological system (isolated cells, isolated organs, or whole organisms), whereas the latter evaluates the effect of the test article on an isolated or purified targeted protein using *in vitro* assays. Over the last 30 years or so, the pharmaceutical industry has increasingly relied on hypothesis-driven and target-based screening strategies (Munos 2009; Pharmaceutical Research and Manufacturers of America 2016; Paul et al. 2010; Hughes et al. 2011; Swinney and Anthony 2011; Sams-Dodd 2013; Vincent et al. 2015). These strategies were fueled by the successful sequencing of the human genome and the belief that one target equates to one drug. Enormous budgets were directed at high-throughput screening (HTS) efforts to identify molecules that bind to a specific target; however, most of these compounds did not produce a desired therapeutic response in preclinical models or clinical trials. It has been estimated that drugs progressing to clinical trials have only an 11.5% probability of approval (Munos 2009). This abysmal attrition rate may be partly due to the underappreciated promiscuity of many drugs and many targets. One report estimates that all approved drugs with known mechanisms of action act through only 324 distinct molecular drug targets (Overington et al. 2006) suggesting that a single target can have multiple effects in different diseases and that many biological pathways and interactions are still unknown.

The focus on target interaction has led to a decreased focus on translatable preclinical *in vivo* models of disease, further contributing to attrition rates of drug candidates. This might be best exemplified by the established trends in the psychotherapeutic area that has never been as well served by translatable animal models as other therapeutic areas due, in part, to the relative uniqueness of the human brain with its expanded prefrontal cortex, compared to rodent brains for example (Hyman 2012). Over the last 30 years, the increasing focus on target-based approaches (versus phenotypic discovery approaches) has further driven up the attrition rate in central nervous system (CNS) drug discovery due to the particular biological complexity associated with individual CNS targets (i.e., psychiatric diseases are often influenced by multiple CNS receptor types) and thereby the especially high rate at which mechanistic hypotheses about CNS disease fail. One potential solution to narrow this gap is to augment target-based screening with phenotypic screening. Even in a therapeutic area such as CNS disease, where animal models are somewhat limiting as mentioned here, phenotypic screening has been used to identify essentially all of the first-in-class CNS therapeutics and will continue to serve the field in the future. These approaches are discussed in detail here as well as the application of these techniques to the discovery of new therapies for CNS disorders.

7.2 DRUG DISCOVERY APPROACHES AND TECHNIQUES

In 2011, David Swinney and Jason Anthony investigated effective strategies for the discovery of new drugs that were approved by the Food and Drug Administration (FDA) between 1999 and 2008. They emphasize two particular approaches to drug discovery: target-based screening and phenotypic screening (Swinney and Anthony 2011).

7.2.1 TARGET-BASED SCREENING

Invariably, a drug's mechanism of action (MOA) involves modulating a target or gene product, resulting in a therapeutic effect (Swinney and Anthony 2011; Swinney 2013a). Target-based screening, also known as the reverse pharmacology method, measures the effect of compounds on an isolated target protein *via in vitro* assays (Kotz 2012). Target-based screening *per se* is independent of an indication endpoint (i.e., one can screen against a target to discover small molecule probes that modulate the target, perhaps by inhibition or stimulation of the target, independently of its biological effect). Once discovered, the probes can be used as chemical biology tools to determine whether a particular mechanism might have relevance to a particular biological process or to a disease state. The chemistry requirements for a tool or probe in the chemical biology sense can differ quite markedly from the chemistry requirements for a drug lead or a drug candidate (Arrowsmith et al. 2015).

Target-based screening can also be conducted with an indication endpoint in mind (i.e., as part of a drug discovery program). Typically, the drug discovery target-based screening is hypothesis driven. For example, the hypothesis might be based on chemistry. Compounds might be chosen for screening because they, or similar compounds, are known to modulate an existing known target that has some similarity to the proposed new target. Perhaps more commonly, the hypothesis comes from the realm of biology. A biologist speculates that based on biological information, modulating a particular target might have beneficial physiological effects. A target-based *in vitro* screen is set up and compounds, perhaps at random, are tested to discover actives that modulate the target in the desired sense. These actives can then be tested in further screens to see if the compound activity is consistent with the underlying biological hypothesis. In both these target-based scenarios, preexisting knowledge at the molecular (chemistry) level or the biological level is applied to the small molecule screening strategy. In the drug discovery mode, target-based screening relies on a hypothesis that a protein specifically targeted with a chemical ligand will exhibit beneficial therapeutic effects. For example, a protein may represent the target and the ligand represents a small molecule. The same principles apply irrespective of whether the target is a protein, a nucleotide, a lipid, or a polysaccharide, or whether the ligand is very small, like a fragment, or very large like an antibody.

Technical developments facilitated the advent of target-based screening in the late 1980s–1990s. For example, the sequencing of the human genome permitted rapid cloning and synthesis of milligram to gram quantities of purified proteins that small molecules could be tested against to see if there were any high affinities to a candidate target protein. Purified protein could be used in experiments to determine if a ligand could be crystallized or soaked into the protein crystal. Solving the x-ray structure of the protein ligand complex leads to "structure-based" drug discovery—a very powerful tool in drug discovery (Lounnas et al. 2013).

Most pharmaceutical companies utilize target-based screening due to the fact that *in vivo* screening approaches (i.e., phenotypic screening) do not provide the target, or MOA, thereby creating possible challenges for optimizing small molecule design by medicinal chemistry. Medicinal chemists may balk at trying to optimize a "black box" mechanism or mechanisms from a phenotypic screen. Much of this reluctance comes from the correct perception that an affinity of a drug to a target in a mechanistic screen might need to improve a thousandfold from a compound identified in an initial high-throughput screen in order to yield a drug-like candidate. However, the phenotypic lead compound, which inherently has been evaluated across a number of drug-likeness parameters in the course of the *in vivo* screen, might need to only improve 20 to 100-fold. This improvement is possible using *in vivo* assays as a screening tool, based on drug discovery phenotypic lead optimization history in the 1970s and 1980s before the widespread advent of mechanistic screening. For example, activity in an *in vivo* screen implicitly means that the combined target affinity and bioavailability, as well as toxicity profile, are all within some range of adequacy. When comparing the medicinal chemistry requirements for each of these two strategies, it is important to also take into consideration that evidence now indicates that more extensively iterative medicinal chemistry processes to improve potency at a target are associated with an output of compounds with poorer physiochemical properties (Gleeson et al. 2011).

In March 2012 at the *"Addressing the Challenges of Drug Discovery" Keystone Symposium* in Lake Tahoe, Joanne Kotz, the senior editor for Science Business eXchange (SciBX) had the opportunity to interview drug developers from academia and industry. She noted that Genentech, a biotechnology company that focuses on oncology, neuroscience, metabolism, infectious disease, and ophthalmology, was focused on target-based screening rather than phenotypic screening strategies in drug development. During her interview with Michael Varney (then senior vice president, co-head of Research and Small Molecule Discovery), he emphasized the dogmatic view that the key hurdle in phenotypic screening was identifying the target of active molecules, a critical step in drug discovery. Varney noted "...even if researchers can find the target, there is no guarantee that it will be a tractable target for lead optimization or that the target will have the right toxicity profile or even elicit the amount of efficacy needed..." implying that phenotypic screening success rate is too low (Kotz 2012). Although target-based molecular screening may reasonably model pharmacokinetic (PK) properties and some *in vivo* pharmacology features, in most cases it may not reflect the phenotypic changes or effects toward pathology of human diseases (Saporito et al. 2012b). While concerns about phenotypic screening do exist, similar concerns are also expressed about mechanistic screening. For example, concerns about the excessive reductionism of mechanistic screens are typified as in this quote from a review on receptor theory: "...a key issue in the elaboration of receptor theory at the molecular level is the need to avoid an overtly reductionistic framework that has little, if any, hierarchal relationship to either native or intact mammalian systems" (Kenakin and Williams 2014). Alternatively, phenotypic screening approaches, although not providing initial target information, may replicate physiopathological effects of a disease, pharmacological responsiveness, and have increased broad-range success in clinical translation (Saporito et al. 2012b).

7.2.2 PHENOTYPIC SCREENING

The alternative to target-based or hypothesis-driven screening is phenotypic screening, in which a drug candidate is evaluated in a biological system, be it a culture, cell system, isolated organ, whole animal, or human. Historically, the phenotypic approach has played a main role in drug discovery where knowledge of the disease mechanism was limited. It is the empirical approach to drug discovery, sometimes called "classical pharmacology," and it does not require an understanding of the drug's method of action.

In Swinney's 2013 analysis of NMEs and the distribution of new drugs discovered between 1999 and 2008, phenotypic screening was the most successful for first-in-class, small molecule drugs while target-based screening was the most successful for follower or second-generation drugs (Swinney and Anthony 2011; Swinney 2013a,b). Once a drug candidate resulted in a therapeutic effect in selected disease areas, chemists or drug developers are able to use that information to develop a second-generation drug with improved capabilities (see Figure 7.1).

Although many researchers and pharmaceutical institutions are reluctant to develop compounds for therapeutic areas unless the target or the mechanisms of action are known, this is not a universally held view. For example, the Eli Lilly and Company has pursued parallel mechanistic and phenotypic drug discovery (termed PD(2)) for over a decade; an initiative has now morphed into a collaboration model with academics (Lee et al. 2011). Not knowing the MOA or the target may be a useful option in opening up therapeutic pathways for diseases with unmet medical need. Such is certainly the case in the discovery of medicines for rare diseases (Swinney

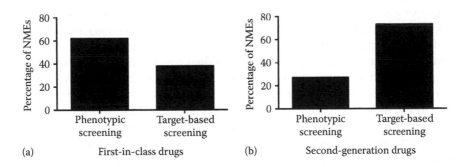

FIGURE 7.1 Percentage of new molecular entities (NMEs) discovered between 1999 and 2008 according to discovery strategy. (a) Phenotypic screening was the most successful approach for first-in-class drugs, whereas (b) target-based screening was the most successful for follower drugs during the period of this analysis. The total number of medicines that were discovered *via* target-based screening was nearly three times higher for follower drugs *versus* first-in-class drugs. (From Swinney, D.C. and Anthony, J., *Nat. Rev. Drug Discov.*, 10(7), 507, 2011; Swinney, D.C., *Clin. Pharmacol. Ther.*, 93(4), 299, 2013.)

and Xia 2014). It is interesting to note that most of the first-in-class molecules discovered between 1999 and 2008 were not found based on target, but rather functional responses associated with diseases (especially metabolic, CNS, pain, and inflammation). Inflammation is a key broad biological area that is related to multiple conditions; therefore, therapeutic areas may be identified utilizing multiplexed phenotypic screening methods (Swinney and Anthony 2011; Saporito et al. 2012b; Swinney 2013b). The starting points for a phenotypic screening program may involve the knowledge of biomarkers, biochemical mechanism and pathways, molecular mechanisms of action, chemical starting points, and last but not least, the benefits of preclinical and/or clinical serendipity (Swinney 2013b). Very often, the findings from an initial phenotypic screen may then lead to new insights into mechanism-pathway or pathway-disease relationships that in turn provide new starting points for target-based strategies. As one exciting recent example, the serendipitous discovery of ketamine, by way of clinical observation, revealed a potentially new class of antidepressants (Nguyen et al. 2015; Tizabi 2015; Xu et al. 2015). In turn, this discovery has helped to underscore the role of the glutamate NMDA receptor in depression and has invigorated a new area of exploration into other NMDA modulators that may be more useful therapeutics than ketamine (Boczek et al. 2015; Caddy et al. 2015; Liu et al. 2015; Nagy et al. 2015; Newport et al. 2015).

7.3   RESURGENCE OF SERENDIPITOUS DRUG DISCOVERY

7.3.1   CLASSICAL PHARMACOLOGY, EMPIRICAL AND SERENDIPITOUS DRUG DISCOVERY

As sustainability remains a current drug discovery crisis, it is useful to dive deeper into classical pharmacology and the history of drug discovery and development. By understanding the history of drug discovery, scientists can use modern techniques and advances to improve upon the classical process.

The origins of the pharmaceutical industry can be traced back to the mid-1800s and started in the dye-making business (Ban 2006). Various artificial fabric dyes were made through modifications of organic chemistry processes and eventually led to the synthesis of pharmaceutical compounds. By the late 1800s, companies such as Bayer and Ciba, that had been manufacturing dyes, turned their attention to developing drugs. Most of the early drugs were for psychiatric indications such as agitation, excitement, seizure disorders, and insomnia and included drugs such as potassium bromide, chloral hydrate, and lithium. Even up through the 1980s, development of almost all new psychotropic drugs was through classical phenotypic pharmacology, usually with an element of serendipity (Ban 2006).

Classical pharmacology is the foundation of drug discovery. In this process, scientists utilize developmental compounds *via in vitro* or *in vivo* assays to explore serendipitous phenotypic changes after treatment. In classical pharmacology, if treatment with the compound resulted in an unexpected phenotypic change during an assay, then it advances to the next step in the developmental process in assessing the biological target. Assessment of the phenotypic changes requires experienced observational skills to identify not only the expected endpoint(s), but also any unexpected results. This was the science of serendipitous discovery.

In 1928 when Alexander Fleming, a well-known scientist actively engaged in influenza research, noticed one of his staphylococcus culture plates was contaminated and had developed a bacteria-free circle of mold, he acted on his microbiologist instincts, isolated the mold, and determined that it was an antibacterial substance. This pure serendipitous observation is what engaged him to develop and optimize penicillin, a world-changing drug (Fleming 1964).

Another great example of serendipitous discovery directly relates to CNS drug discovery and occurred in 1911. Heinrich Hörlein of F. Bayer Company and Co. synthesized phenobarbital, a barbiturate and a hypnotic at F. Bayer Company and Co. Hörlein happened to be a resident physician in psychiatry at the time, treating epileptic patients in Freiburg. Hörlein was having sleepless nights during this period because patients with frequent nocturnal epileptic episodes interrupted his sleep. He administered Luminol® to those patients who had difficulty sleeping due to epileptic restlessness. Not only did Luminol act as a sedative and sleep aid, but it also reduced the number and intensity of seizures, further allowing patients (and their treating physicians) adequate sleep. Originally developed as a hypnotic, this treatment in low doses enhanced the lives of many epileptics in the early 1900s as an efficient anti-seizure medication (Hauptmann 1912; Shorter 1997; López-Muñoz et al. 2012).

The example of Hörlein's use of Luminol illustrates the potential for using an existing drug for a new indication. Another great example of additional potential for an existing drug is aspirin. Aspirin is a simple acetylated derivative of a compound found in willow bark. The bark was used as an antipyretic and reduced pain and inflammation throughout antiquity. In more modern times, the semisynthetic aspirin replaced the bark extract. In 1950, a family doctor, Lawrence Craven, instructed tonsillectomy patients to chew Aspergum, laced with aspirin, to help relieve inflammation and pain. Craven recognized a large number of his patients had been hospitalized due to severe bleeding; these were patients who were using large amounts of Aspergum. Craven realized aspirin's effects on clotting, began prescribing aspirin to his patients at low doses, and noticed that none of these patients developed thrombosis. In the 1980s, aspirin was repurposed for its antithrombotic potential and is currently used worldwide to prevent the incidence of myocardial infarction (Jeffreys 2004; Miner and Hoffhines 2007; Bayer HealthCare LLC 2015).

Clearly, there is more to learn about drugs already on the market, and one way to explore their capabilities is through phenotypic screening. It is important to be aware that many drugs treating the CNS act *via* multiple pathways. These empirical, phenotypic observations drove CNS discovery in areas such as schizophrenia, depression, anxiety, and alertness (Saporito et al. 2012b). In contrast to the earlier time period, the discovery of CNS drugs in the current era has slowed to a crawl with many drug companies abandoning the effort entirely. An insufficient understanding of the basic science in CNS diseases is the cause with companies choosing a time-out until the science becomes clearer (Hyman 2012).

7.3.2 Effective Drug Discovery Strategies

As Swinney and Anthony analyzed the drug discovery strategy between 1999 and 2008, 62% of first-in-class, small molecule drugs were discovered using phenotypic-based approaches *versus* the 38% that were discovered using target-based approaches (see Figure 7.1a; Swinney and Anthony 2011). This should be especially surprising given that the industry was very much entrenched into the dogma of target-based discovery during these years.

What accounts for this surprising disparity? In target-based screening, if there is not enough mechanistic and molecular etiological data to form a hypothesis around a given target, then one may completely overlook the best target. In contrast, phenotypic screening is entirely agnostic as regards to whether a molecular target is known or not. In this way, phenotypic screening can also provide a compliment to the traditional target-based, hypothesis-driven drug discovery paradigm.

Given the historical success rate of phenotypic screening, why does the industry not utilize it more? In the past 5 years, there has been a resurgence in drug repositioning that directly correlates with phenotypic screening discoveries. In a PubMed literature "hits by year" analysis, phenotypic screening discoveries were stagnant from 2000 to 2008. However, from 2009 to 2014, there has been a steady increase in publications each year (see Figure 7.2).

It is important to understand that phenotypic screening and target-based discovery can work together and are not necessarily mutually exclusive. The findings of Swinney and Anthony suggest that an effective strategy might be to use phenotypic screening to identify and to progress a therapeutic approach with the aim of discovering a validated target, and then subsequently to use target-based screening to optimize second-generation drugs (see Figure 7.1b).

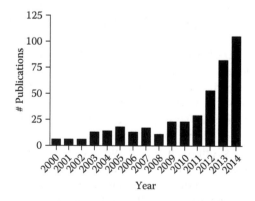

FIGURE 7.2 PubMed literature search for phenotypic screening publications. Phenotypic screening publication activity was nearly stagnant from 2000 to 2008. However, from 2009 to 2014, publications have since nearly doubled or tripled in number. A trend is clearly displayed in utilizing phenotypic screening efforts in drug discovery.

7.4 PHENOTYPIC SCREENING IS AN EFFECTIVE APPROACH TO DRUG DISCOVERY

7.4.1 IN VITRO SCREENING

In vitro screening approaches utilize high-content screening (HCS) processes to identify small molecules (mostly drugs with a low molecular weight), RNAi, or peptides so as to modify a cell's physical or chemical characteristics (Haney 2008; Taylor et al. 2010). These are high-throughput cell-based systems that identify potential therapeutic molecules by measuring changes in cellular responses or biomarkers of disease. HCS systems provide a high-throughput means to screen large numbers of compounds through a disease-relevant biological system, compared to HTS, which filters compounds *via* a target-based binding assay or other cell-free assay. HCS can be a useful process that provides phenotypic information, target-based hit verification, and MOA information (see Table 7.1). Often, HCS is intermediate between HTS and *in vivo* screening in the drug discovery process and can complement and enrich HTS methods through target hypothesis, generation, and confirmation.

7.4.2 IN VIVO SCREENING

One of the benefits of utilizing *in vivo* (animal) models is determining the crucial "go/ no go" developmental decisions that pharmaceutical and biopharmaceutical companies are faced with as a routine part of any drug discovery and development program. Animal models are essential research tools in developing an understanding of the pathology, MOA, efficacy, and potentially safe dosing parameters of a drug candidate (McGonigle 2014). Pathology-based models are designed to more faithfully replicate at least some aspects of the pathophysiology of a specified disease and are responsive to a broader range of therapeutic agents for that disease, independent of the MOA (Saporito et al. 2012b). It is important to note that the *in vivo* models utilized in phenotypic screening typically do not fit into a selected/specified treatment category; rather, and most importantly, they are

---

TABLE 7.1
Benefits of High-Content Screening

Image-based parametric readout can

- Provide rich phenotypic information that can support holistic identification of hits with a desired MOA
- Provide a "signature fingerprint"
- Lead to a reduced rate of false positives
- Lead to a reduction in attrition rate at the hit verification stage
- Support the characterization of MOA

Sources: Reisen et al. (2013); Bickle (2010); Swinney and Anthony (2011); Kümmel et al. (2012); Dürr et al. (2007); Perlman et al. (2004); Towne et al. (2012); Adams et al. (2006).

---

responsive to a range of clinically approved drugs (Saporito et al. 2012b). For example, learned helplessness rodent models such as tail suspension and forced swim test respond to multiple classes of antidepressants, including tricyclics (TCAs), selective serotonin reuptake inhibitors (SSRIs), serotonin norepinephrine reuptake inhibitors (SNRIs), and ketamine, and are predictive of antidepressant activity in the clinic (Porsolt et al. 1977a,b; Cryan et al. 2002, 2005a,b; McGonigle 2014).

Pharmaceutical companies may consider phenotypic screening to be entirely too unpredictable and unable to pair well with medicinal chemistry optimization (i.e., if a target is not identified) as reflected earlier by Michael Varney's quote (see Section 7.2.1; Kotz 2012). However, phenotypic screening provides a 60% greater chance of discovering first-in-class drugs compared to target-based screening (Swinney and Anthony 2011; Swinney 2013a,b) and most of the targets associated with initial discovery by phenotypic screening have gone on to yield second-generation drug products through target-based medicinal chemistry. The odds of success in the integration of a phenotypic screening component to complement a drug discovery approach and to help replenish depleted drug pipelines depend somewhat on the therapeutic area. In cancer, the impact of phenotypic screening has been modest (Moffat et al. 2014). By contrast, in antibacterial drug discovery, it appears that phenotypic screening is the only option as exemplified in the discovery of bedaquiline, the only new drug with a novel mechanism discovered to treat drug-resistant tuberculosis in many decades (Barry 2009).

7.4.3 *In Vivo* Screening for Neurotherapeutics

The need for translatable animal models may be greater for CNS disorders than other therapeutic areas because the human brain is relatively inaccessible compared to organs of the periphery that can be biopsied and tested using various measurements of function. Access to the human brain is restricted to imaging and electroencephalography (EEG). In addition, and as discussed earlier, the particular complexity associated with CNS diseases, compared to other disease areas, has established a trend of especially high failure rates for target-based approaches in this therapeutic area. Unfortunately, human CNS disease also presents special challenges to phenotypic screening, with the human brain's vastly expanded prefrontal cortex compared to that of rodents, thereby presenting important physiological distinctions when trying to model psychiatric disorders or neurodegenerative diseases, such as Alzheimer's disease. These features of the human brain largely preclude the use of lower model organisms, such as fruit flies and zebra fish (see Table 7.2). Although closely physiologically matched, nonhuman primates are not a practical organism for screening early-stage compounds, thereby leaving rodents as generally the most suitable compromise. Criticism is often raised that it is challenging to evaluate mood in rodents and that models of depression for example, such as models of learned helplessness, (forced swim test, tail-suspension) do not have good face validity (i.e., similar manifestation to the clinical condition). Nonetheless, it is worthy to note that these models of depression can be validated across a broad range of antidepressants, including TCAs, SSRIs, SNRIs, and most recently, ketamine. Certainly much of the basic neuronal circuitry, such as that responsible for basic emotions, like fear and reward, is conserved through the expanse of evolution. Despite the challenges, there is a broad range of animal models that have been established for an array of CNS diseases and

TABLE 7.2
Benefits of *In Vivo* Screening

Model Organism	Benefits
Fruit fly	• Low cost • Small and easy to grow • 10-day generation time • High fecundity (Redei 2001) • Has only 4 pairs of chromosomes (3 autosomes and 1 sex chromosome) • Complete genome sequenced and published in 2000 (Adams et al. 2000)
Zebra fish	• Regenerative abilities (Goldshmit et al. 2012) include fins, skin, heart, lateral line hair cells, and larvae stage brain (Lush and Piotrowski 2014; Wade 2015) • Fully sequenced genome • Rapid embryonic development • Utilized in neurodegenerative and musculoskeletal research
Mouse	• Share >95% of the translatable genome with humans • Genome can be manipulated to mimic human disease (ALS, cancer, obesity, anxiety, etc.) (Spencer 2015; The Jackson Laboratory 2015) • Accelerated life span • Cost-effective • Small • Reproduce quickly

there are many anecdotes, both in psychiatry and neurodegeneration, of these models providing the critical preclinical validation for drug candidates that were subsequently advanced to approval (McGonigle 2014). On a still more positive note, advances in molecular genetics and genetic modification technology promise to continue to contribute to more sophisticated rodent models that better recapitulate the human condition.

7.5 DRUG REPOSITIONING AND PHENOTYPIC SCREENING PLATFORMS

7.5.1 Drug Repositioning, a Strategy for Solving the Drug Discovery Crisis

Pharmaceutical companies will continuously struggle with the demand for drug discovery and its sustainability issues. It is crucial to note that by repurposing drugs already proven to be safe, the failure risk and costs are significantly reduced. One of the biggest reasons drugs fail from Phase 1 to submission is due to safety reasons (Arrowsmith and Harrison 2012). The rising importance of drug repositioning (drug repurposing) can be seen in the impact of phenotypic screening of the National Center for Advancing Translational Sciences (NCATS) pharmaceutical collection (NPC) (Lee et al. 2015). Table 7.3 offers several examples of repurposed drugs, as well as their original indication. Keen observation was critical in identifying side effects or benefits, in addition to the drug's original indication.

TABLE 7.3
Examples of Repurposed Drugs and Their Original Indication

Drug	Innovator	Mechanism	Original Indication	Repurposed Indication
Gemcitabine	Eli Lilly	Inhibition of DNA synthesis	Antiviral	Anticancer
Raloxifene	Eli Lilly	Estrogen agonist/antagonist	Breast cancer	Osteoporosis
Bupropion	GSK	Norepinephrine–dopamine reuptake inhibitor	Depression	Smoking cessation
Dapoxetine	Eli Lilly	SSRI	Analgesia	Premature ejaculation
Fluoxetine	Eli Lilly	SSRI	Depression	Premenstrual dysphoria
Hydroxychloroquine	Sanofi	Lysosomal alkalinization; TLR inhibitor	Antiparasitic	Antiarthritic
Doxepine	Boehringer Mannheim	SNRI	Antidepressant	Antipruritic
Bimatoprost	Allergan	Prostaglandin analog	Glaucoma	Eyelash growth

Source: Reprinted from Arrowsmith, J. and Harrison, R., Drug repositioning: The business case and current strategies to repurpose shelved candidates and marketed drugs, in: *Drug Repositioning: Bringing New Life to Shelved Assets and Existing Drugs*, Eds. M. Barratt and D. Frail, John Wiley & Sons, Hoboken, NJ, 2012, pp. 9–30 (Print).

An excellent example of drug repositioning is *sildenafil*. This drug was originally purposed to treat angina, a symptom in coronary heart disease that results in chest pain caused by reduced blood flow to the heart (Angina 2015; Roberts 2015). Sildenafil is a phosphodiesterase-5-inhibitor and acts as a potent vasodilator by increasing or improving blood flow. During clinical trials, keen observation indicated a secondary serendipitous discovery that this drug improved blood flow not only to the heart but also to the penis, therefore aiding in improving male erection. During these clinical trials, striking side effects helped define a new disorder, male erectile dysfunction (ED). Sildenafil was then marketed for male ED (Viagra®), instead of the original purpose in treating patients with angina (Arrowsmith and Harrison 2012; Roberts 2015).

7.5.2   *thera*TRACE® PHENOTYPIC SCREENING PLATFORM

In vivo phenotypic screening provides the most fertile platforms for drug repositioning, including pre- and postmarketing human clinical trials and preclinical experimental animal models of diseases (Ashburn and Thor 2004; Dimond 2010; Saporito et al. 2012b). Melior Discovery has coined the term *thera*TRACE for its *in vivo* phenotypic screening platform that utilizes a multiplexing strategy

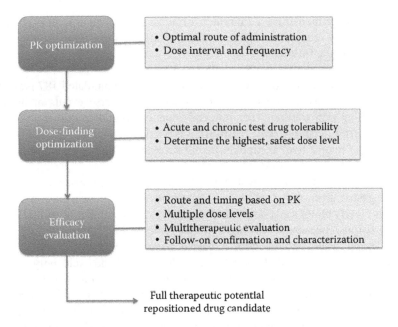

FIGURE 7.3 *thera*TRACE. This multiplexed, therapeutic phenotypic *in vivo* platform is used to identify the therapeutic potential of preclinical drug candidates and to reposition development stage drugs. *thera*TRACE occurs in three steps: (1) PK, (2) dose finding, and (3) efficacy evaluation. (From Saporito, M.S. et al., Phenotypic *in vivo* screening to identify new, unpredicted indications for existing drugs and drug candidates, in: *Drug Repositioning: Bringing New Life to Shelved Assets and Existing Drugs*, Eds. M. Barratt and D. Frail, John Wiley & Sons, Hoboken, NJ, 2012, pp. 253–290 (Print); Saporito, M.S. and Reaume, A.G., *Drug Discov. Today Therap. Strat.*, 8(3–4), 89, 2011.)

capable of evaluating compounds through approximately 40 individual animal models representing 14 therapeutic areas (Melior Discovery 2015). *thera*TRACE (Figure 7.3) integrates three essential steps in the drug discovery process: (1) PK evaluation (PK), (2) dose-level finding, and (3) efficacy at three dose levels (Saporito et al. 2012b).

7.5.3 MLR-1023: A Candidate for Diabetes Repositioned Using *thera*TRACE

MLR-1023 (tolimidone) was originally developed in the late 1970s by Pfizer to treat gastric ulcer. A chemical series of compounds was identified that demonstrated activity in a rat model of gastric ulcer (Lipinski et al. 1980). MLR-1023 was synthesized out of a medicinal chemistry effort around that series using the rat model as a biological screening endpoint for compound activity. The molecular target was not identified and, as was common in the 1970s, the compound advanced through Phase 2 studies. These studies were then discontinued due to a lack of significant improvement on gastric and duodenal ulcer healing rates;

however, it did demonstrate a favorable clinical safety and tolerability profile. Melior Discovery recognized the compound as a good drug repositioning candidate based upon the following:

1. *Good safety and tolerability*: The compound had accumulated 187 patient-years of exposure with only mild-to-moderate adverse events at higher doses that were transient (i.e., relieved upon lowering dose of drug)
2. *Good drug-like properties*: MLR-1023 is a small molecule (MW: 202) that is fully "rule-of-five" compliant (Lipinski et al. 2001) with good oral bioavailability and exposure.
3. *Good biological activity*: Although not effective on the chosen endpoint in the gastric ulcer study (reduction of gastric ulcer lesion size), the compound was clearly biologically active as demonstrated by a number of animal studies but also because it was highly mucogenic in humans even though this did not translate into reduction in gastric ulcer lesion size.
4. The compound provided good "freedom to operate" and exclusivity opportunity. The composition-of-matter that Pfizer had been granted was expired and the structure of the compound was highly differentiated from others in the patent literature, suggesting that a method-of-use patent for this compound and the genus around it, if awarded, may not be easily overrun by the use of a similar compound outside the genus.

Melior Discovery ran this compound through its *thera*TRACE platform comprised of 39 animal models and identified a pattern of activity around metabolic disease (see Figure 7.4).

Continued investigation on the pharmacology and mechanism of MLR-1023 revealed that it is an insulin sensitizer (see Figure 7.5), but unique compared to previously described insulin sensitizers in that it does not interact with peroxisome proliferator-activated receptors (PPARs) (Ochman et al. 2012; Saporito et al. 2012a). Cumulatively, the preclinical pharmacology data supported a projected product profile for an agent that could be novel and effect addition to the armamentarium for type 2 diabetes.

Subsequently, procurement of the clinical dossier for the original ulcer healing trials provided the clinical chemistry data that were collected weekly during the trial as a routine safety endpoint for the study. Retrospective analysis of the data revealed statistically significant blood glucose lowering for those subjects treated with MLR-1023. Moreover and fortuitously, there were five diabetics included in this study, four of which were treated with MLR-1023. Amongst those four subjects, average blood glucose lowering was greater than 100 mg/dL (see Figure 7.6).

In the course of investigating the mechanism of MLR-1023, it was determined that the molecular target of the compound is lyn kinase and that MLR-1023 is a specific and potent activator of lyn kinase (Saporito et al. 2012a). Existing literature shows that a substrate of lyn kinase is insulin receptor substrate 1 (IRS-1) (Müller et al. 2001) and that at least one preexisting antidiabetic compound attributed a portion of its glycemic control activity to a nonspecific and indirect activation of lyn kinase (Müller 2000). Although this literature linkage existed, lyn kinase was not regarded as a diabetes target candidate until the results with MLR-1023 emerged.

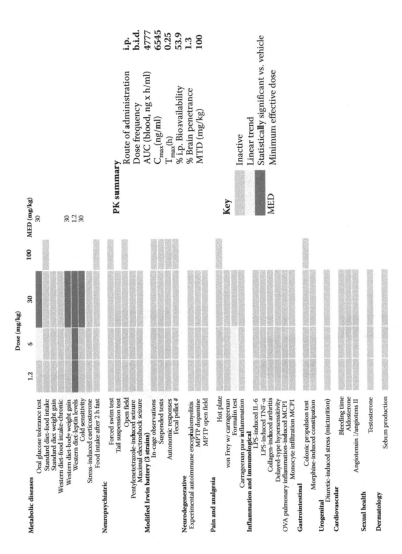

FIGURE 7.4 Dashboard summary of MLR-1023 *thera*TRACE screen. The PK characteristics of MLR-1023 were first determined to help design a dosing paradigm. For the initial screen, MLR-1023 was administered b.i.d. at three dose levels, as indicated. The key shows models and dose levels where statistically significant activity or trends were observed. A gastric ulcer model was not included in this panel. A related set of activities around metabolic disease (oral glucose tolerance test, decreased weight gain on high fat diet, decreased leptin levels on high fat diet, reduced cold sensitivity) was observed. *Note*: i.p., intraperitoneal; b.i.d., *bis in die*.

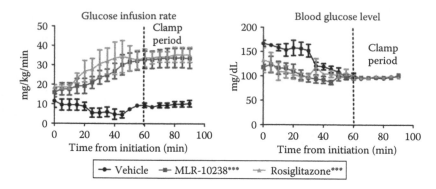

FIGURE 7.5   Hyperinsulinemic euglycemic clamp characterization of MLR-1023. Eight-week-old Zucker (ZDF) rats were infused with a constant rate of insulin (20 mU/min/kg) and a variable rate of glucose in order to establish an equilibrium of 100 mg/dL blood glucose. Animals treated with MLR-1023 or rosiglitazone (a PPAR γ activator) required higher glucose infusion rates, compared to placebo, in order to establish the 100 mg/dL equilibrium (Ochman et al. 2012). ***$p < 0.001$ compared to vehicle.

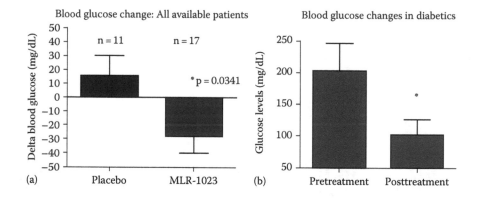

FIGURE 7.6   Retrospective analysis of blood glucose from MLR-1023 gastric ulcer study. The Pfizer MLR-1023 gastric ulcer study was a 6-week protocol with weekly visits involving an endoscopy and clinical chemistry measures. Clinical study reports, but not case report forms, were available for this study. The appendices of the clinical study reports contained summary data for clinical chemistry measures including baseline (prior to treatment) and in-study values. Panel (a) is a summary for all subjects with reported blood glucose and shows the difference between a subject's baseline blood glucose value and their in-treatment blood glucose value. Panel (b) represents the four diabetic subjects that received MLR-1023 treatment and shows the average baseline blood glucose value and their average in-treatment blood glucose value. *$p < 0.0341$ compared to placebo; *$p < 0.05$ compared to pretreatment.

In summary, this account of the discovery of MLR-1023 as a candidate, therapeutic for type 2 diabetes, from a phenotypic screen and the understanding that emerged from this discovery, as regards action on lyn kinase and the role of lyn kinase on influencing insulin sensitivity (Reaume and Saporito 2010), is a good illustration for how phenotypic screening allows for the discovery of new mechanisms without the limitations associated with hypothesis-based approaches where knowledge gaps may confine our ability to make certain discoveries.

7.5.4 MLR-1019: A PARKINSON'S DISEASE CANDIDATE REPOSITIONED USING *thera*TRACE

MLR-1019 is the active L-enantiomer of sydnocarb, a racemic compound that was marketed for a variety of psychiatric disorders in Russia and the former Soviet Union, for over 35 years beginning in 1971 (Anokhina et al. 1974). Mclior Discovery identified sydnocarb as worthy of exploration by phenotypic screening, based in part on its unusual specificity as a dopamine transporter (DAT) inhibitor. Unlike most CNS-active drugs, especially ones developed in the 1970s or before, this compound showed great specificity when screened across a large panel of 66 CNS receptors, channels, and transporters. At a 10 µM screening concentration, the only activity that was observed was at the dopamine transporter (DAT; 100% inhibition) and the norepinephrine transporter (NET; 51% inhibition) (Gruner et al. 2011). Moreover, of any DAT inhibitor that has been described (see Table 7.4; Gruner et al. 2011),

---

TABLE 7.4

Data Showing the Selectivity of Sydnocarb for the Dopamine Transporter (DAT), the Norepinephrine Transporter (NET), and the Serotonin Transporter (SERT) (Expressed as K_i Values)

Compound	RHS	Uptake Inhibition K_i Values (nM)		
		DAT	NET (×Fold DAT)	SERT (×Fold DAT)
MLR-1017	No	8.3	1,500 (**181**)	>10,000 (>**1,205**)
GBR-12909	No	4.3	79.2 (**18.4**)	73.2 (**17.0**)
GBR-12935	No	4.9	277 (**56.5**)	289 (**59.0**)
Mazindol	No	25.9	2.9 (**0.1**)	272 (**10.5**)
Nomifensine	No	93.1	32 (**0.3**)	1,889 (**20.3**)
Cocaine	Yes	478	779 (**1.6**)	304 (**0.6**)
Methylphenidate	Yes	260	170 (**0.7**)	1,143 (**4.4**)
Bupropion	Yes	1534	>10,000 (>**6.5**)	34,707 (**22.6**)
D-Amphetamine	Yes	550	120 (**0.2**)	2,382 (**4.3**)
Tesofensine	Unknown	72	1.7 (**0.02**)	11 (**0.15**)
Modafinil	No	6400	35,600 (**5.56**)	500,000 (**7.8**)

Source: Taken from Gruner, J.A. et al., *J. Pharmacol. Exp. Therap.*, 337(2), 380, 2011.
Sydnocarb is the most selective DAT inhibitor described to date. Rebound hypersomnolence (RHS) is a proposed surrogate for dopamine release activity and potential abuse liability.

sydnocarb showed the highest selectivity toward DAT *versus* other catecholamine transporters. This feature, combined with the compound's known clinical safety and tolerability profile, made it a promising repositioning candidate.

Melior screened sydnocarb in a modified version of its *thera*TRACE platform where it uncovered previously unreported activity in models of Parkinson's disease (PD). In a toxin-based model of PD, 6-OHDA is intracranially injected into the median forebrain bundle, comprised of dopaminergic neurons and produces toxic reactive oxygen species (ROS), which ultimately kill the dopaminergic neurons (McGonigle 2014). In rodents, this unilateral lesion results in asymmetric circling or rotating behavior, dyskinesia, and abnormal involuntary movements (AIMs) that are useful in characterizing the disease (Tolwani et al. 1999; Dauer and Przedborski 2003; Duty and Jenner 2011; McGonigle 2014). Upon more in-depth characterization using different animal models of PD and different aspects of the disease, it was discovered that the compound was especially useful for treating dyskinesia associated with L-DOPA administration. This L-DOPA-induced dyskinesia (LID) presents as abnormal involuntary movements (AIMs) in a rat 6-hydroxydopamine (6-OHDA) unilateral lesion model of PD. As an added benefit, sydnocarb also potentiates the anti-parkinsonian effects of L-DOPA. Finally, the compound promotes wakefulness and therefore is anticipated to address the excessive daytime sleepiness (EDS) associated with PD (Mitler et al. 2000; Gjerstad et al. 2002; Larsen 2003; Arnulf 2005; Lökk 2010), which is a significant unmet medical need in this patient population. Melior's further investigations showed that the therapeutic activity described here was greatest when administering optimal dose levels of the active L-enantiomer (MLR-1019) compared to optimal dose levels of the racemic mixture (sydnocarb).

PD is characterized by the specific loss of dopaminergic neurons of the substantia nigra, which project to the striatum. Therefore, the pharmacological effects of MLR-1019 as a DAT inhibitor and the resulting modulation of synaptic dopamine levels that it would produce make mechanistic sense. Given the longstanding understanding of PD etiology and the central role of dopamine in the disease various dopamine-influencing compounds including a number of DAT inhibitors have been previously evaluated in the clinic. However, none of the DAT inhibitors that have been studied in PD patients (methylphenidate (Nutt et al. 2007), tesofensine (Hauser et al. 2007), and modafinil (Lökk 2010)) exhibited notable benefit to the disease state, including the treatment of PD-LID. To further explore this paradox, Melior Discovery has characterized a range of DAT inhibitors in a toxin-induced rat 6-OHDA unilateral lesion model of PD including an evaluation of AIMs (i.e., PD-LID). A clear differentiation of MLR-1019 from other DAT inhibitors was observed with MLR-1019, showing a distinct superiority toward alleviating PD-LID compared to other DAT inhibitors (see Figure 7.7).

In contrast, a search revealed evidence that elevated synaptic norepinephrine levels exacerbate or potentiate PD-LID. Therefore, a DAT inhibitor with significant NET activity may present opposing effects toward PD-LID. This indicates that the DAT activity that it possesses would ameliorate PD-LID while the NET activity of the compound would potentiate PD-LID. Arguably, all DAT inhibitors except MLR-1019, and certainly all DAT inhibitors that have been clinically tested against PD, have NET inhibitor activity that is potent enough to be of pharmacological relevance *in vivo*.

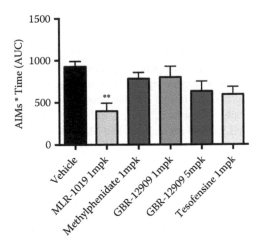

FIGURE 7.7 Modified Abnormal Involuntary Movements (AIMs) and rotational scoring after 7 days of treatment in a 6-OHDA unilateral lesion rat model of PD. Two weeks after 6-OHDA lesion and 7 days of treatment, rats treated with optimal dose level of test article, previously determined in separate pilot experiments. MLR-1019 exhibited significantly reduced overall AIMs and translatable rotational scores (AIMs*Time [AUC]) compared to vehicle-treated rats. Data are mean ± SEM; **$p < 0.01$ compared to vehicle.

Collectively, this information supports a hypothesis that specific inhibition of DAT and, importantly, without appreciable NET inhibition will serve as a useful L-DOPA cotherapy for PD with simultaneous benefits, including

- Mitigation of PD-LID
- Potentiation of L-DOPA anti-parkinsonian activity
- Mitigation of excessive daytime sleepiness associated with PD

In summary, this account, as was true for the case of MLR-1023 described earlier, is a case where the disease state (Parkinson's disease) was very well studied and a good amount of detail around the etiology and biochemistry had been described. A number of therapeutic strategies have been attempted based on this understanding of the science. Yet, phenotypic screening was able to uncover an aspect of the science that was overlooked and underappreciated.

7.6 CONCLUSION

The pharmaceutical industry has faced significant R&D productivity issues over the last two decades (Munos 2009; Allison 2012; Scannell et al. 2012). While regulatory policy issues may contribute to the challenge in maintaining rates of new drug discovery, much focus has been given to discovery strategy paradigms and the significant shift toward a target-based medicinal chemistry paradigm that the industry, as a whole, turned to in the 1990s. It is especially worthy to note that even well into this era of hypothesis-based drug discovery, from 1999 to 2008, more than half of all newly approved, small molecule, first-in-class therapies were discovered by pheno-typic screening (Swinney and Anthony 2011).

Moreover, on top of these challenges, with the significant attrition that continues to exist in the drug discovery and development pipeline, with only 1 in 10 drugs that enter the clinic, the market, there is increasing attention on drug repositioning as a means of mitigating development risk and further reducing cost by shortcutting opportunities. We have presented illustrations of cost-effective, risk-mitigated, drug repositioning by pheno-typic screening using MLR-1023 for diabetes and MLR-1019 for PD-LID as examples.

REFERENCES

Adams, C.L. et al. Compound classification using image-based cellular phenotypes. *Methods in Enzymology* 414 (2006): 440–468.

Adams, M.D., S.E. Celniker, and R.A. Holt. The genome sequence of *Drosophila melanogaster*. *Science (New York)* 287(5461) (March 24, 2000): 2185–2195.

Allison, M. Reinventing clinical trials. *Nature Biotechnology* 30(1) (2012): 41–49.

Mayo Clinic. Angina. Mayo Foundation for Medical Education and Research, n.p., September 8, 2015. http://www.mayoclinic.org/diseases-conditions/angina/basics/definition/con-20031194.

Anokhina, I.P., G.D. Zabrodin, and Ia.E. Svirinovskiĭ. Characteristics of the central action of Sidnocarb. *Zhurnal nevropatologii i psikhiatrii imeni S.S. Korsakova (Moscow, Russia: 1952)* 74(4) (1974): 594–602.

Arnulf, I. Excessive daytime sleepiness in parkinsonism. *Sleep Medicine Reviews* 9(3) (2005): 185–200.

Arrowsmith, C.H. et al. The promise and peril of chemical probes. *Nature Chemical Biology* 11(8) (2015): 536–541.

Arrowsmith, J. and R. Harrison. Drug repositioning: The business case and current strategies to repurpose shelved candidates and marked drugs. In: *Drug Repositioning: Bringing New Life to Shelved Assets and Existing Drugs*. Eds. M. Barratt and D. Frail, John Wiley & Sons, Hoboken, NJ, 2012, pp. 9–30. Print.

Ashburn, T.T. and K.B. Thor. Drug repositioning: Identifying and developing new uses for existing drugs. *Nature Reviews: Drug Discovery* 3(8) (2004): 673–683.

Ban, T.A. Academic psychiatry and the pharmaceutical industry. *Progress in Neuro-Psychopharmacology & Biological Psychiatry* 30(3) (May 2006): 429–441.

Barry, C.E. Unorthodox approach to the development of a new antituberculosis therapy. *The New England Journal of Medicine* 360(23) (2009): 2466–2467.

Bayer HealthCare LLC. Aspirin history, n.p., 2015. http://www.aspirin.com/scripts/pages/en/home.php. Accessed March 29, 2017.

Bickle, M. The beautiful cell: High-content screening in drug discovery. *Analytical and Bioanalytical Chemistry* 398(1) (2010): 219–226.

Boczek, T. et al. Region-specific effects of repeated ketamine administration on the presynaptic GABAergic neurochemistry in rat brain. *Neurochemistry International* 91 (2015): 13–25.

Caddy, C. et al. Ketamine and other glutamate receptor modulators for depression in adults. *The Cochrane Database of Systematic Reviews* 9 (September 23, 2015): n.p.

Cryan, J.F., A. Markou, and I. Lucki. Assessing antidepressant activity in rodents: Recent developments and future needs. *Trends in Pharmacological Sciences* 23(5) (2002): 238–245.

Cryan, J.F., C. Mombereau, and A. Vassout. The tail suspension test as a model for assessing antidepressant activity: Review of pharmacological and genetic studies in mice. *Neuroscience and Biobehavioral Reviews* 29(4–5) (2005a): 571–625.

Cryan, J.F., R.J. Valentino, and I. Lucki. Assessing substrates underlying the behavioral effects of antidepressants using the modified rat forced swimming test. *Neuroscience and Biobehavioral Reviews* 29(4–5) (2005b): 547–569.

Dauer, W. and S. Przedborski. Parkinson's disease: Mechanisms and models. *Neuron* 39(6) (2003): 889–909.

Dimond, P.F. Drug repositioning gains in popularity. *Genetic Engineering and Biotechnology News* 30 (2010): 3–5. Print.

Dürr, O. et al. Robust hit identification by quality assurance and multivariate data analysis of a high-content, cell-based assay. *Journal of Biomolecular Screening* 12(8) (2007): 1042–1049.

Duty, S. and P. Jenner. Animal models of Parkinson's disease: A source of novel treatments and clues to the cause of the disease. *British Journal of Pharmacology* 164(4) (2011): 1357–1391.

Fleming, A. Penicillin: Nobel lecture, pp. 83–93, 1964. http://www.nobelprize.org/nobel_prizes/medicine/laureates/1945/fleming-lecture.pdf. Accessed March 29, 2017

Gjerstad, M.D., D. Aarsland, and J.P. Larsen. Development of daytime somnolence over time in Parkinson's disease. *Neurology* 58(10) (2002): 1544–1546.

Gleeson, M.P. et al. Probing the links between in vitro potency, ADMET and physiochemical parameters. *Nature Reviews: Drug Discovery* 10 (2011): 197–208.

Goldshmit, Y. et al. Fgf-dependent glial cell bridges facilitate spinal cord regeneration in zebrafish. *The Journal of Neuroscience: The Official Journal of the Society for Neuroscience* 32(22) (2012): 7477–7492.

Gruner, J.A. et al. Characterization of pharmacological and wake-promoting properties of the dopaminergic stimulant sydnocarb in rats. *The Journal of Pharmacology and Experimental Therapeutics* 337(2) (2011): 380–390.

Haney, S.A., ed. *High Content Screening: Science, Techniques and Applications*. Hoboken, NJ: John Wiley & Sons, 2008. Print.

Hauptmann, A. Luminal bei epilepsi. *Münch Med Wochenschr* 59(1912): 1907–1912.

Hauser, R.A. et al. Randomized trial of the triple monoamine reuptake inhibitor NS 2330 (tesofensine) in early Parkinson's disease. *Movement Disorders: Official Journal of the Movement Disorder Society* 22(3) (2007): 359–365.

Hughes, J.P. et al. Principles of early drug discovery. *British Journal of Pharmacology* 162(6) (2011): 1239–1249.

Hyman, S.E. Revolution stalled. *Science Translational Medicine* 4(155) (2012): n.p.

Jeffreys, D. *Aspirin: The Remarkable Story of a Wonder Drug*. New York: Bloomsbury, 2004. Print.

Kenakin, T. and M. Williams. Defining and characterizing drug/compound function. *Biochemical Pharmacology* 87(1) (2014): 40–63.

Kotz, J. Phenotypic screening, take two. *SciBx* 5(15) (2012). doi:10.1038/scibx.2012.380. Accessed March 29, 2017.

Kümmel, A. et al. Differentiation and visualization of diverse cellular phenotypic responses in primary high-content screening. *Journal of Biomolecular Screening* 17(6) (2012): 843–849.

Larsen, J.P. Sleep disorders in Parkinson's disease. *Advances in Neurology* 91 (2003): 329–334.

Lee, J.A. et al. Novel phenotypic outcomes identified for a public collection of approved drugs from a publicly accessible panel of assays. *PLoS ONE* 10(7) (July 15, 2015): n.p.

Lee, J.A. et al. Open innovation for phenotypic drug discovery: The PD2 assay panel. *Journal of Biomolecular Screening* 16(6) (2011): 588–602.

Lipinski, C.A. et al. Bronchodilator and antiulcer phenoxypyrimidinones. *Journal of Medicinal Chemistry* 23(9) (1980): 1026–1031.

Lipinski, C.A. et al. Experimental and computational approaches to estimate solubility and permeability in drug discovery and development settings. *Advanced Drug Delivery Reviews* 46 (March 1, 2001): 3–26.

Liu, W.-X. et al. Regulation of glutamate transporter 1 via BDNF-TrkB signaling plays a role in the anti-apoptotic and antidepressant effects of ketamine in chronic unpredictable stress model of depression. *Psychopharmacology (Berl)* 233(3) (2015): 405–415.

Lökk, J. Daytime sleepiness in elderly Parkinson's disease patients and treatment with the psychostimulant modafinil: A preliminary study. *Neuropsychiatric Disease and Treatment* 6 (April 7, 2010): 93–97.

López-Muñoz, F. et al. The role of serendipity in the discovery of the clinical effects of psychotropic drugs: Beyond of the myth. *Actas españolas de psiquiatría* 40(1) (2012): 34–42.

Lounnas, V. et al. Current progress in structure-based rational drug design marks a new mind-set in drug discovery. *Computational and Structural Biotechnology Journal* 5(6) (2013): 1–14.

Lush, M.E. and T. Piotrowski. Sensory hair cell regeneration in the zebrafish lateral line. *Developmental Dynamics: An Official Publication of the American Association of Anatomists* 243(10) (2014): 1187–1202.

McGonigle, P. Animal models of CNS disorders. *Biochemical Pharmacology* 87(1) (2014): 140–149.

Melior Discovery. Melior discovery: In vivo efficacy and drug safety models, n.p., 2015. http://www.meliordiscovery.com/home-models.html. Accessed March 29, 2017.

Miner, J. and A. Hoffhines. The discovery of aspirin's antithrombotic effects. *Texas Heart Institute Journal* 34(2) (2007): 179–186.

Mitler, M.M. et al. Long-term efficacy and safety of modafinil (PROVIGIL ((R))) for the treatment of excessive daytime sleepiness associated with narcolepsy. *Sleep Medicine* 1(3) (2000): 231–243.

Moffat, J.G., J. Rudolph, and D. Bailey. Phenotypic screening in cancer drug discovery—Past, present and future. *Nature Reviews: Drug Discovery* 13(8) (2014): 588–602.

Müller, G. The molecular mechanism of the insulin-mimetic/sensitizing activity of the antidiabetic sulfonylurea drug amaryl. *Molecular Medicine (Cambridge, Mass.)* 6(11) (2000): 907–933.

Müller, G. et al. Redistribution of glycolipid raft domain components induces insulin-mimetic signaling in rat adipocytes. *Molecular and Cellular Biology* 21(14) (2001): 4553–4567.

Munos, B. Lessons from 60 years of pharmaceutical innovation. *Nature Reviews: Drug Discovery* 8(12) (2009): 959–968.

Munos, B. A forensic analysis of drug targets from 2000 through 2012. *Clinical Pharmacology & Therapeutics* 94(3) (2013): 407–411.

Munos, B. Accelerating innovation in the bioscience revolution. In: *Collaborative Innovation in Drug Discovery: Strategies for Public and Private Partnerships*. Ed. R. Chaguturu, John Wiley & Sons, Hoboken, NJ, 2014, pp. 195–212. Print.

Nagy, D. et al. Differential effects of an NR2B NAM and ketamine on synaptic potentiation and gamma synchrony: Relevance to rapid-onset antidepressant efficacy. *Neuropsychopharmacology* 41(6) (2016): 1486–1494.

Newport, D.J. et al. Ketamine and other NMDA antagonists: Early clinical trials and possible mechanisms in depression. *The American Journal of Psychiatry* 172(10) (2015): 950–966.

Nguyen, L. et al. Off-label use of transmucosal ketamine as a rapid-acting antidepressant: A retrospective chart review. *Neuropsychiatric Disease and Treatment* 11 (October 14, 2015): 2667–2673.

Nutt, J.G., J.H. Carter, and N.E. Carlson. Effects of methylphenidate on response to oral levodopa: A double-blind clinical trial. *Archives of Neurology* 64(3) (March 2007): 319–323.

Ochman, A.R. et al. The Lyn kinase activator MLR-1023 is a novel insulin receptor potentiator that elicits a rapid-onset and durable improvement in glucose homeostasis in animal models of type 2 diabetes. *The Journal of Pharmacology and Experimental Therapeutics* 342(1) (2012): 23–32.

Overington, J.P., B. Al-Lazikani, and A.L. Hopkins. How many drug targets are there? *Nature Reviews: Drug Discovery* 5 (2006): 993–996.

Paul, S.M. et al. How to improve R&D productivity: The pharmaceutical industry's grand challenge. *Nature Reviews: Drug Discovery* 9(3) (2010): 203–214.

Perlman, Z.E. et al. Multidimensional drug profiling by automated microscopy. *Science (New York)* 306(5699) (2004): 1194–1198.

Petit-Demouliere, B., F. Chenu, and M. Bourin. Forced swimming test in mice: A review of antidepressant activity. *Psychopharmacology* 177(3) (2005): 245–255.

Pharmaceutical Research and Manufacturers of America. 2016 Biopharmaceutical Research Industry Profile. Washington, DC: PhRMA (April 2016). http://phrma-docs.phrma.org/sites/default/files/pdf/biopharmaceutical-industry-profile.pdf. Accessed on March 29, 2017.

Porsolt, R.D., A. Bertin, and M. Jalfre. Behavioral despair in mice: A primary screening test for antidepressants. *Archives internationales de pharmacodynamie et de thérapie* 229(2) (1977a): 327–336.

Porsolt, R.D., M. Le Pichon, and M. Jalfre. Depression: A new animal model sensitive to antidepressant treatments. *Nature* 266(5604) (1977b): 730–732.

Reaume, A.G. and M.S. Saporito. Methods and formulation for modulating lyn kinase activity and treating related disorders. USPTO, US Patent 8,835,448, September 16, 2014.

Redei, G.P. *Drosophila melanogaster*: The fruit fly. In: *Encyclopedia of Genetics*, 2001, pp. 157–162. Springer Science & Business Media.

Reisen, F. et al. Benchmarking of multivariate similarity measures for high-content screening fingerprints in phenotypic drug discovery. *Journal of Biomolecular Screening* 18(10) (2013): 1284–1297.

Roberts, M. Original heart hope of viagra realised, n.p., September 8, 2015. http://news.bbc.co.uk/2/hi/health/6367643.stm. Accessed March 29, 2017.

Sams-Dodd, F. Is poor research the cause of the declining productivity of the pharmaceutical industry? An industry in need of a paradigm shift. *Drug Discovery Today* 18(5–6) (2013): 211–217.

Saporito, M.S. et al. MLR-1023 is a potent and selective allosteric activator of Lyn kinase in vitro that improves glucose tolerance in vivo. *The Journal of Pharmacology and Experimental Therapeutics* 342(1) (2012a): 15–22.

Saporito, M.S., C.A. Lipinski, and A.G. Reaume. Phenotypic in vivo screening to identify new, unpredicted indications for existing drugs and drug candidates. In: *Drug Repositioning: Bringing New Life to Shelved Assets and Existing Drugs*. Eds. M. Barratt and D. Frail, John Wiley & Sons, Hoboken, NJ, 2012b, pp. 9–30. Print.

Saporito, M.S. and A.G. Reaume. theraTRACE(R): A mechanism unbiased in vivo platform for phenotypic screening and drug repositioning. *Drug Discovery Today: Therapeutic Strategies* 8(3–4) (2011): 89–95.

Scannell, J.W. et al. Diagnosing the decline in pharmaceutical R&D efficiency. *Nature Reviews: Drug Discovery* 11(3) (2012): 191–200.

Shorter, E. *A History of Psychiatry: From the Era of the Asylum to the Age of Prozac*. Hoboken, NJ: John Wiley & Sons, 1997. Print.

Spencer, G. Background on mouse as a model organism. National Human Genome Research Institute, NIH News Advisory 2002 Release, The mouse genome and the measure of man (2002): https://www.genome.gov/10005834/background-on-mouse-as-a-model-organism/. Accessed March 29, 2017.

Swinney, D.C. Phenotypic vs. target-based drug discovery for first-in-class medicines. *Clinical Pharmacology and Therapeutics* 93(4) (2013a): 299–301.

Swinney, D.C. The contribution of mechanistic understanding to phenotypic screening for first-in-class medicines. *Journal of Biomolecular Screening* 18(10) (2013b): 1186–1192.

Swinney, D.C. and J. Anthony. How were new medicines discovered? *Nature Reviews: Drug Discovery* 10(7) (2011): 507–519.

Swinney, D.C. and S. Xia. The discovery of medicines for rare diseases. *Future Medicinal Chemistry* 6(9) (2014): 987–1002.

Taylor, D.L., K.A. Giuliano, and J.R. Haskins, eds. *High Content Screening: A Powerful Approach to Systems Cell Biology and Drug Discovery*. Totowa, NJ: Humana Press, 2010. Print.

The Jackson Laboratory. Why mouse genetics? The Jackson Laboratory, Bar Harbor, ME. https://www.jax.org/research-in-action/why-mouse-genetics. Accessed March 29, 2017.

Tizabi, Y. Duality of antidepressants and neuroprotectants. *Neurotoxicity Research* 30(1) (2016): 1–13.

Tolwani, R.J. et al. Experimental models of Parkinson's disease: Insights from many models. *Laboratory Animal Science* 49(4) (1999): 363–371.

Towne, D.L. et al. Development of a high-content screening assay panel to accelerate mechanism of action studies for oncology research. *Journal of Biomolecular Screening* 17(8) (2012): 1005–1017.

Vincent, F. et al. Developing predictive assays: The phenotypic screening 'Rule of 3'. *Science Translational Medicine* 7(293) (2015): n.p.

Wade, N. Research offers clue into how hearts can be regenerated in some species. *New York Times*, n.p., 2015. http://www.nytimes.com/2010/03/25/science/25heart.html?_r=0. Accessed March 29, 2017.

Xu, Y. et al. Effects of low-dose and very low-dose dose ketamine among patients with major depression: A systematic review and meta-analysis. *The International Journal of Neuropsychopharmacology/Official Scientific Journal of the Collegium Internationale Neuropsychopharmacologicum (CINP)* (2015): n.p., http://www.ncbi.nlm.nih.gov/pubmed/26578082.

Section III

Drug Repositioning for
Nervous System Diseases

8 A Case Study
Chlorpromazine

Francisco López-Muñoz, Cecilio Álamo,
and Silvia E. García-Ramos

CONTENTS

8.1 INTRODUCTION

The discovery of the antipsychotic properties of chlorpromazine has been presented by many authors as an example of serendipitous scientific discovery in which chance would have played a prominent role. This would explain the case of chlorpromazine as a paradigm of drug repositioning for nervous system diseases: a drug intended to be an antihistamine, which was studied as an adjuvant anesthetic, and eventually became an antipsychotic (Dronsfield and Ellis 2006). However, our research team has shown, using a new operational definition of serendipity, that in the area of psychopharmacology, purely serendipitous discoveries, in contrast to what has been postulated by some, are rather rare (Baumeister et al. 2010; López-Muñoz et al. 2012). The majority presents a mixed pattern, a combination of serendipitous and nonserendipitous finds. Most, as the case of chlorpromazine, follow a pattern that begins with an initial serendipitous observation followed by investigations that led to a nonserendipitous discovery of clinical application. In this chapter, we will analyze this historical development.

The treatment of psychotic disorders, until the middle of the twentieth century, was based on the application of a series of remedies with limited clinical effectiveness, such as the so-called biological therapies (paludization techniques, application of tuberculin or trementine, insulin or cardiozolic comas, electroconvulsive therapy, etc.) or certain highly unspecific pharmacological agents (opium, morphine, cocaine, hashish, codeine, digitalis, chloral hydrate, bromide, etc.) (López-Muñoz et al. 1998, 2014). In this inhospitable therapeutic framework, at the beginning of the 1950s, was the appearance in the psychiatric therapy of a chemically synthesized molecule, chlorpromazine (Delay et al. 1952a; Laborit et al. 1952a), whose introduction into clinical practice contributed to mark the beginning of what came to be called the "psychopharmacological revolution" (Caldwell 1970; Jacobsen 1986; Deniker 1989; Lehmann 1989; Ayd 1991; Frankenburg 1994; Shepherd 1994; Healy 1996, 1999, 2000; Lehmann and Ban 1997; López-Muñoz et al. 1998, 2000, 2002, 2005, 2014; Shen 1999; Cancro 2000; Ban 2001, 2007; Preskorn 2007; Carpenter and Davis 2012; Jašović-Gašic et al. 2012). On August 9, 1955, just 3 years after the introduction of chlorpromazine, Mark D. Altschule, a Harvard lecturer and director of the Laboratory of Clinical Physiology at McLean Hospital (Boston), addressing *the Gordon Conference on Medicinal Chemistry* at Colby Junior College in New London, affirmed that this drug with reserpine had already "totally changed psychiatric practice" (Altschule 1956).

The advent of chlorpromazine, derided by some of the great figures of psychiatry at the time, such as Henri Ey—who referred to it as "psychiatric aspirin" (Ey and Faure 1956)—represented not only the first selective and effective approach to the treatment of schizophrenic patients, but also opened the way for the synthesis of numerous drugs for treating mental disorders, thus heralding the psychopharmacological era (López-Muñoz et al. 1998, 2004, 2005, 2014). The introduction into clinical practice of chlorpromazine can also be considered as the first of the three milestones in the history of antipsychotic drugs that would mark great advance in the treatment of schizophrenia, the others being the synthesis and subsequent use of haloperidol (López-Muñoz and Álamo 2009) and, finally, the discovery of the atypical characteristics of clozapine, which permitted the development of the second-generation (atypical) antipsychotic agents (risperidone, olanzapine, quetiapine, ziprasidone, etc.) (Shen 1999; López-Muñoz et al. 2015), with a new pharmacodynamic profile and improved neurological tolerance (Álamo et al. 2000).

However, the application of chlorpromazine to patients with mental illnesses was never directly sought; rather, as Lickey and Gordon so rightly put it, "their introduction in therapeutic use is more like the story of a drug in search of an illness" (Lickey and Gordon 1986).

8.2   DISCOVERY OF PHENOTHIAZINES: FROM THE CHEMICAL DYEING INDUSTRY TO ANTI-INFECTIOUS THERAPY

The discovery of the first family of antipsychotic agents was made within the context of widespread research on antihistaminic substances in France after World War II, and more specifically in that of the work being carried out on phenothiazines (see López-Muñoz et al. 2005). These substances had been known since the late nineteenth century, having been used by the dyeing industry. Later, in the early 1930s, they were employed as antiseptics and antihelminthics. Finally, in the second half of the 1940s, their antihistaminic properties were studied, though their toxicity made clinical use impossible.

The first phenothiazinic substances were developed in Germany at the end of the nineteenth century, within the framework of the burgeoning German textile industry (Swazey 1974). The history of these substances began with the work of Carl Graebe and Carl Liebermann, who in 1868 synthesized alizarin, a dye derived from coal tar. The Badische Anilin und Soda Fabrik (BASF) company undertook its manufacture and commercialization, and further research by the same company resulted in their obtaining a large number of new dyes, including methylene blue, synthesized by Caro in 1876. It was precisely while working on the development of dyes derived from this aniline that the organic chemist August Bernthsen synthesized the first molecule of this family in 1883 (Zircle 1973; Shen 1999).

The introduction of phenothiazines in medicine coincides with the development of microscopy, and with the need to obtain tinctures that would permit the visualization of histological preparations. It was in this context that the aniline dyes developed in England by William H. Perkin were used. Among the pioneers in this field was Paul Ehrlich, who observed that some of these substances had bactericide capacities and began studying them with the aim of finding a product capable of destroying pathogenic agents while respecting human cells (the famous "magic bullet"). Thus, in 1907, he discovered trypan red, a lithic substance for parasites of the genus *Trypanosoma*, responsible for sleeping sickness, and subsequently arsphenamine (Salvarsan®), a lethal agent for *Treponema pallidum*, the microorganism that induces syphilis (Zircle 1973).

An indirect but decisive role in the story of the clinical use of phenothiazines was played by the needs and strategies involved in the two world wars (Shen 1999). During World War I, the supplements of quinine, the only remedy for malaria at the time, and obtained from the tropical tree *quina cinchona*, were affected by military blockades that made them inaccessible to the German army, so that their researchers undertook to find synthetic derivatives of the substance. Thus, Werner Schulemann and his team decided to continue studying the antimalarial effect of methylene blue, a phenothiazine derivative used as a dye in histological dyeing techniques, with which Ehrlich and Guttman had made considerable research progress in 1891. The results of this work led to the synthesis of several derivatives of methylene blue, such as a diethyl-amino-ethyl derivative, with greater antimalarial activity but high toxicity, and finally quinacrine, which became as commonly used against malaria as quinine itself (Zircle 1973). This antimalarial action of phenothiazines continued to be studied until the end of the 1930s, since these substances were found to have a toxic effect on the mosquito larvae as well as on porcine parasites, and research increased throughout World War II. During that conflict, the Japanese expansion in Southeast Asia affected the supply of quinine, in this case to the Allied forces, and this obliged scientists to seek new therapeutic alternatives, so that they turned once more to phenothiazines. Thus, Henry Gilman and colleagues (Gilman et al. 1944) synthesized a series of compounds, through the addition of aminoalkilate chains to the central nitrogen atom of the phenothiazine ring, although these agents showed a complete absence of antimalarial activity.

The compounds synthesized by Gilman's team continued to be studied by French researchers at the Société des Usines Cliniques of Rhône-Poulenc Laboratories (Vitry-sur-Seine, France), who also confirmed that the amino-alkylate derivatives of the phenothiazines had no effect on the symptoms of malaria but decided to investigate, following the classic research lines, their antihistaminic properties. Thus, the team led by Paul Charpentier at Rhône-Poulenc developed phenothiazine derivatives with an aminate chain, similar to that found in molecules with antimalarial activity. The result of this development process was the synthesis, between 1946 and 1948, of promethazine (RP-3277) and diethazine, subsequently commercialized as Diparcol®.

8.3 PHENOTHIAZINES AS ANTIHISTAMINE AND ANTISHOCK AGENTS

Concurrently with the developments and events mentioned here, other groups of scientists were researching the antihistamine properties of different substances in relation to the study of shock and stress reactions. Notable among them was the group led by Daniel Bovet (Figure 8.1), a Swiss-born Italian pharmacologist at the Institut Pasteur, which in 1937 was working on the first substance capable of exercising a histaminergic blocking action, 2-isopropyl-5-methylphenoxiethyldiethylamine, derived from aniline and developed as a dye by Ernest Fourneau in 1910, under the name F-929.

However, this substance could not be used in clinical practice, in the treatment of allergies, due to its potential toxicity. Following this line of research, in 1944 Bovet's team described the antihistamine properties of pyrilamine maleate, and subsequently, working by now at Société Rhône-Poulenc, Bovet studied (with others, such as Halpern and Ducrot) the antihistamine effects of the phenothiazines synthesized by Fourneau. The result of this research was the clinical introduction, within the field of allergies, of phenbenzamine (RP-2339; Antergan®), diphenhydramine (Benadryl®), and, finally, in 1947, of promethazine (RP-3277), whose

FIGURE 8.1   Italian pharmacologist Daniel Bovet (1907–1992). Between 1929 and 1947 he worked at the Pasteur Institute in Paris on antihistamines. In 1957, he won the Nobel Prize in Physiology or Medicine for his discovery of drugs that block the actions of specific neurotransmitters.

commercial name was Fenergan®, and which was also used in the treatment of Parkinson's disease. Its sedative effects were also later discovered (Swazey 1974; Frankenburg 1994).

Some of these antihistamines were even tested in the field of psychiatry. Phenbenzamine was studied by Georges Daumezon, in 1942, in patients with manic-depressive disorder, with the aim of reducing the number of relapses and limiting the use of electroshock, the only therapeutic alternative at the time for this type of patients (Daumezon and Cassan 1943). Although the preliminary results were encouraging, research did not continue. Promethazine was also tested in psychiatry. In July 1950, Paul Guiraud reported his experience with this antihistamine-hypnotic agent in 24 patients with manic-depressive psychosis, though his conclusions (inducement of drowsiness and sedation in agitated psychotic patients or reduction of the duration of manic episodes) were questioned and made little impact (Guiraud and David 1950).

The early use of phenothiazine compounds as neuroleptic agents resulted from the research of Henri-Marie Laborit (Figure 8.2; López-Muñoz et al. 2004, 2005). This French Army surgeon, working in 1949 at the Hôpital Maritime in Bizerte (Tunisia), was interested in finding a pharmacological method for preventing surgical shock.

According to one of the prevailing hypotheses at the time, proposed by Canadian endocrinologist Hans Selye and defended by French surgeon René Leriche, surgical shock was due to an excessive defensive reaction of the organism to stress, so that a peripheral and/or central inhibition of the autonomic nervous system would be a highly advantageous alternative antishock therapy. Thus, from 1947 Laborit studied the ganglionic blocking effect of curare, with the aim of achieving chemical sympathectomy.

FIGURE 8.2 Henri Laborit (1914–1995) in his office at the Val-de-Grâce Military Hospital in Paris.

His idea was received with scepticism by the scientific community at the time, though it did prove successful later on, with the incorporation into the anesthetic techniques of another ganglioplegic substance, tetraethylammonia. Subsequently, Laborit continued to test different substances endowed with inhibitory effects of the visceral vasomotor reactions of the vegetative system—substances that included the antihistamines then available. This "Laborit's idea" was described by Leriche, in 1952, as "revolutionary, fascinating and extremely promising" (Laborit 1952).

Among the antihistamine drugs of the era under study, Laborit found that promethazine, whose capacity for prolonging the sleep induced by barbiturates had been demonstrated in rodents, had acceptable antishock activity, so that he added it to another morphine-type substance, dolantine (Dolosal®), creating the so-called "lytic cocktail," a landmark in the history of anesthesia in that it constituted the origin of neuroleptoanalgesia. This early cocktail was widely used in Tunisian women affected by eclampsia. Laborit himself actually predicted the potential psychiatric implications of these agents and recalls in an interview recounted by Judith Swazey that "I asked an army psychiatrist to watch me operate on some of my tense, anxious, Mediterranean-type patients. After surgery, he agreed with me that the patients were remarkably calm and relaxed. But I guess he didn't think any more about his observations, as they might apply to psychiatric patients" (Swazey 1974).

Subsequently, Laborit's cocktail would undergo numerous modifications, including the addition of diethazine (Dip-Dol cocktail, Diparcol-Dolosal), or even, later, chlorpromazine. The Dip-Dol cocktail was introduced by a colleague of Laborit, Pierre Huguenard, anesthetist at the Hôpital de Vaugirard in Paris, who in a nostril operation on a highly agitated patient, to whom he was unable to apply the ether or chloroform mask, administered diethazine mixed with dolantine. The patient underwent general relaxation while remaining conscious, even being capable of answering questions from the hospital staff (Thuillier 1994)—a result that some authors described as "pharmacological lobotomy" (Courvoisier et al. 1953). However, despite the success of the intervention, this cocktail was not applied in psychiatric practice, possibly due to fears that the opiate nature of its formula would create dependence.

8.4 SYNTHESIS OF CHLORPROMAZINE

In the light of these discoveries, Specia Laboratories at Rhône-Poulenc (Vitry-sur-Seine, France), the company that synthesized and commercialized promethazine, undertook to continue the line of research opened up by Laborit and, in 1950, attempted to find a lytic agent that would prevent surgical shock through depressant actions on the central nervous system. Thus, Simone Courvoisier (Figure 8.3a) analyzed all the phenothiazines synthesized by Paul Charpentier (Figure 8.3b) since 1944 as antihistaminic agents.

Of these, promazine appeared to be the best option, despite its low antihistaminic activity, and so Charpentier synthesized various derivatives of it. A chlorinated derivative (RP-4560) (Figure 8.4), produced in December 1950, displayed, according to Courvoisier's test, extraordinary activity, not only of an antihistaminic nature, but also of a parasympathetic and adrenolytic character, capable of canceling out (at intravenous doses of 1–3 mg/kg), and even inverting (at higher doses), the effect of adrenalin on blood pressure (Courvoisier et al. 1953).

(a)   (b)

FIGURE 8.3 The two pioneers of chlorpromazine drug development, scientists from Rhone-Poulenc, Simone Courvoisier (a) and Paul Charpentier (b).

FIGURE 8.4 Chemical structure of chlorpromazine.

Furthermore, it was demonstrated in experiments with rats, such as tests of conditioned avoidance (also carried out by Leonard Cook's group at SmithKline & French Corporation, Philadelphia, who had designed them), that RP-4560 was capable of extinguishing conditioned reflexes (animals would climb a rope after an auditory stimulus, when this was previously associated with an electrical discharge) without modifying the animal's strength.

Similarly, RP-4560 was capable of prolonging the sleep induced by barbiturates in rodents and preventing the emesis induced by apomorphine in dogs (Courvoisier 1956). Although the pharmacology of the new product was studied by Courvoisier and Pierre Koetschet in 1951, the first data were not published until 1953, after the publication of the first clinical experience with the substance (Courvoisier et al. 1953).

The following year, between April and August, RP-4560 was tested by numerous doctors, both French and from other countries. Among those who received samples was Laborit, now working at the Physiology Laboratory of the Val-de-Grâce Military Hospital in Paris, who confirmed that this could be the lytic agent he had been seeking for so long. After the statutory studies with experimental animals, Laborit tried the new drug on patients undergoing surgery, at endovenous doses of 50–100 mg. The results as an anesthetic booster were striking. However, Laborit observed that not only did these patients feel much better during and after the operation, due to the antishock action, but they also felt much more relaxed and calm (*désintéressement*) in the preoperative period, a time associated with intense stress and high levels of anxiety (Laborit et al. 1952a). Another interesting property of the product was its hypothermic effect, which allowed reduction of

ustunm

Specia

L'ANNÉE

1 9 5 3

qui vient de s'achever,

marque une étape importante dans l'évolution de la

Chimiothérapie, par la confirmation, sur le plan

clinique, des activités expérimentales multiples du

4560 R. P.

LARGACTIL

CHLORPROMAZINE

CHLORHYDRATE DE CHLORO-3 (DIMÉTHYLAMINO-3' PROPYL)-10 PHÉNOTHIAZINE

découverte originale

des Laboratoires de Recherches

RHÔNE-POULENC

FIGURE 8.5 Publicity advertisement of Largactil® (chlorpromazine) of the pharmaceutical company Rhône-Poulenc.

the body temperature to 28°C–30°C. This effect, attributed by Laborit to a fall in basal metabolism and oxygen consumption, together with the hypnotic properties of the new drug, allowed Laborit and Huguenard to propose, in 1951, the concept of "artificial hibernation" (Laborit and Huguenard 1951), a technique that would make possible greater efficacy of certain types of operation, such as cardiac surgery. Indeed, as Jacobsen (1986) relates, the "artificial hibernation" technique was applied on a large scale by Laborit and Huguenard in 1953 in Vietnam, during the French campaign in Indo-China, and permitted them to save the lives of hundreds of soldiers.

In relation to Laborit's work, it is interesting to note the comment of René Leriche, in 1952, in the preface to a work by the naval surgeon *Réaction organique à l'agression et choc* that what is most original in Henri Laborit's work is the conception he has of therapy for shock. It is frankly revolutionary. While until now we have tried to reanimate the elements of a life that was dying, he has the idea of putting them into a vegetative sleep, of slowing down all the changes, since it is the vegetative reactions that give rise to and maintain shock (Laborit 1952).

The new drug, described by numerous authors at the time as "Laborit's drug," was called chlorpromazine and was commercialized in France by Rhône-Poulenc in 1952. Its commercial name, Largactil® ("large" = broad; "acti*" = activity) (Figure 8.5), was designed to reflect its wide spectrum of pharmacological activities: gangliolytic, adrenolytic, antifibrillatory, antiedema, antipyretic, antishock, anticonvulsant, antiemetic, and so on (Courvoisier 1956).

8.5 CLINICAL PSYCHIATRIC INTRODUCTION OF CHLORPROMAZINE

Laborit's observations allowed him to hypothesize other therapeutic uses for the new drug, which he called a "vegetative stabilizer" (Laborit et al. 1952a), including, in addition to the boosting of anesthesia, the management of surgical stress, serious burns, cardiovascular disorders (such as Raynaud's disease), and psychiatric disorders. Thus, in November 1951, Laborit and Marcel Montassut administered a dose of chlorpromazine intravenously to Cornelia Quarti, a fellow psychiatrist acting as a healthy volunteer at the Villejuif mental hospital. Although there were no effects worthy of mention, save a certain sensation of indifference, on getting up to go to the toilet, Quarti fainted; as a result, the head of the hospital's Psychiatric Service decided to discontinue experimentation with the substance (Chertock 1982; Deniker 1989).

In spite of these events, in one of his first publications on the surgical results obtained with RP-4560, in early February of 1952, Laborit argued that the observations made "may anticipate certain indications for the use of this compound in psychiatry, possibly related to sleep cures with barbiturates" (Laborit et al. 1952b).

Thus, during a meal in the canteen at the Hôpital Val-de-Grâce, he persuaded his colleagues from the Neuropsychiatry Service, headed by Joseph Hamon, to test the drug in psychotic patients, though, as Swazey (1974) recounts, the psychiatrists were not initially enthusiastic about Laborit's proposal. On January 19, 1952, it was administered for the first time, as an adjunct to an opiate (petidine), a barbiturate (pentotal), and an electroconvulsive therapy, to Lh. Jacques, an extremely agitated manic patient aged 24, who rapidly began to calm down, maintaining a state of calm for several hours. By February 7, Jacques had calmed down sufficiently to be able to play bridge and carry out normal activities, though he maintained certain hypomanic attitudes.

Finally, after a 3-week treatment, with a total quantity of 855 mg of RP-4560 administered, the patient was discharged from hospital. Colonel Jean Paraire presented these data on February 25, at a meeting of the Société Médico-Psychologique in Paris, and they were published in March of that same year (1952). In prophetic tone, he said: "We have quite probably introduced a series of products that will enrich psychiatric therapy" (Hamon et al. 1952). This event marked the culmination of what may constitute one of the most important landmarks in the history of psychopharmacology, since this was the first time chlorpromazine had been administered in the field of psychiatry, even though, as Shen and Giesler (1998) point out, reference to this contribution has been omitted by many researchers, due possibly to the multiple therapeutic drugs used.

There soon began to appear in the literature scientific works on clinical experience in psychiatry with chlorpromazine, notable among which are the pioneering studies by Jean Delay (professor of psychiatry at the Sorbonne and director of the Hôpital Sainte-Anne in Paris) and Pierre Deniker (men's service chief at the same hospital) (Figure 8.6).

Deniker heard about Laborit's hibernation experiments from his brother-in-law, who was a surgeon, and ordered from Specia Rhône-Poulenc some samples of the substance RP-4560 for administration to psychiatric patients. Doctor Beal, head of clinical research at Rhône-Poulenc, sent him some of the product, together with a brief note on its pharmacological characteristics and instructions for the hibernation technique. Thus, Deniker and Delay, several weeks after Paraire's presentation, administered chlorpromazine alone, with no other drug in combination, for the first time, and confirmed its great efficacy as a tranquillizing agent in psychotic or agitated patients (Delay et al. 1952a). Furthermore, they observed that the dosage of chlorpromazine employed by Laborit in his hibernation techniques was insufficient when the drug was used alone, and that dosages 4–6 times higher were necessary for an antipsychotic effect (75–100 mg/day).

In 1952, Delay and Deniker described the clinical condition caused by the administration of an injection of 15–100 mg of chlorpromazine, which was characterized by a slowing down of motor activity, affective indifference, and emotional neutrality, a condition they referred to as "neuroleptic syndrome" (Delay and Deniker 1952a). According to Ginestet (1991), in January 1955, Jean Delay proposed to the French Académie Nationale de Médicine the term neuroleptic (from the Greek: "that take the nerve") to designate chlorpromazine and all the drugs producing a similar motor side effect. The term neuroleptic was widely accepted in Europe, but not in America, where it was considered inappropriate to define a family of drugs by their adverse effects, rather than by their therapeutic qualities. Thus, the preferred term in the United States was initially "tranquillizer," and this was replaced by the expression "major tranquilliser" before the introduction of the current term "antipsychotic" drug (King and Voruganti 2002). In this period, the concept of "rapid tranquillisation" was also outlined, referring to the practice

FIGURE 8.6 The French psychiatrists Jean Delay (1907–1987) (left) and Pierre Deniker (1917–1998) (right) in the courtyard of the Sainte-Anne Hospital in Paris in the years following the clinical introduction of chlorpromazine.

of emergency sedation for behavioral disturbance in psychiatry, and where chlorpromazine played a significant role (Allison and Moncrieff 2014).

Between May and July 1952, Delay and Deniker, together with the interns Jean-Marie Harl and André Grasset, presented six clinical reports containing the results of chlorpromazine use in 38 patients in states of agitation and excitation, mania, or mental confusion, or undergoing acute psychotic processes. They confirmed therapeutic effectiveness in these patients, as well as poor response in cases of depression and to the negative symptoms of schizophrenia. Case 1 is a good illustration (Delay et al. 1952b), and referred to A. Giovanni, a 57-year-old manual worker with a long history of mental pathology, admitted for "giving improvised political speeches, getting into fights with strangers, and walking along the street with a plant pot on his head proclaiming his love of liberty." After a 9-day treatment with chlorpromazine, he was able to maintain a normal conversation, and within 3 weeks he was in such a calm state that he could be discharged.

The first of the reports, presented on May 22 at the centenary meeting of the Société Médico-Psychologique and dealing with "shock and reactions of alarm," was published a month later in the prestigious French journal *Annales Médico-Psychologiques* (Delay et al. 1952a). Curiously, the article made no reference whatsoever to the research and previous experience of Laborit, nor to the work of Hamon, Paraire, and Velluz, suggesting that there was some degree of conflict between the two groups. On June 26, the group presented its second report at a meeting of the same society (Delay et al. 1952b), and the third was presented on July 7 (Delay et al. 1952c). Both were published in the same review as the first. The end of July saw the presentation of the three remaining studies within the framework of *the 50th French Congress of Psychiatry and Neurology*, held in Luxembourg (Delay and Deniker 1952a,b,c).

In the story of chlorpromazine, the year 1955 marks a point of no return. In addition to the publication of the first randomized and controlled clinical trial with the drug, by Elkes and Elkes (1954), that year saw the celebration of a series of important scientific events. Between March 29 and April 1, it took place in Barcelona the first international conference on this neuroleptic (*I Coloquio Internacional sobre la Terapéutica Narcobiótica*), organized by Professor Ramón Sarró. In June, a symposium set up by SmithKline & French in Philadelphia assembled 117 psychiatrists under the title *Chlorpromazine and Mental Health*. In September and October, there were plenary conferences in Italy on chlorpromazine and reserpine, respectively (*Convegno Nazionale su Sonno prolungato, Ibernazione artificale, Neuroplegici in Neuropsichiatria*, Vercelli, and *Symposium Nazionale sulla Reserpina e la Chlorpromazina in Neuropsichiatria*, Milan). Finally, also in October, Delay and his assistant Deniker organized, at the Hôpital Sainte-Anne in Paris, the *I Colloque International sur la Chlorpromazine et les Médicaments Neuroleptiques en Thérapeutique Psychiatrique* (October 20–22, 1955). This last meeting, considered by many authors as the first event of a new era in the field of psychiatry and psychopharmacology, was attended by over 400 specialists from 22 countries, who debated at length on the new chemical tools (chlorpromazine and reserpine) in the treatment of psychoses. The scientific result of the Colloquium amounted to more than 150 papers, all published in a special issue of almost 1000 pages of the journal *L'Encéphale* in 1956 (Figure 8.7).

This Colloquium provided an opportunity for the presentation and discussion of the possible therapeutic indications of the new drug. Finally, the Colloquium provided the occasion for numerous contributions on the possible action mechanism of chlorpromazine, and on its profile of adverse effects (for more information, see López-Muñoz et al. 2005).

However, it would not be until the beginning of the 1960s that the first trials were carried out with an adequate methodological design and a substantial sample, in order to assess the antipsychotic effectiveness of chlorpromazine. Among such studies were that of the project designed by Jonathan Cole and his colleagues at the Psychopharmacology Service of the U.S. National Institute of Mental Health (NIMH), begun in April 1961 and published in 1964 (Cole and the NIMH Psychopharmacology Service Center Collaborative Study Group 1964). This was

COLLOQUE
INTERNATIONAL

sur la

CHLORPROMAZINE

ET LES MÉDICAMENTS NEUROLEPTIQUES

EN THÉRAPEUTIQUE PSYCHIATRIQUE

PARIS

20, 21, 22 octobre 1955

Clinique des Maladies mentales et de l'Encéphale
de la Faculté de Médecine
— Professeur JEAN DELAY —

Secrétaire du Colloque : Pierre DENIKER

Recueil des textes parus dans l'Encéphale
•
NUMÉRO SPÉCIAL 1956

G. DOIN et Cⁱᵉ

FIGURE 8.7 Cover of the journal *L'Encéphale* (1956), containing the scientific contributions to the *I Colloque International sur la Chlorpromazine et les Médicaments Neuroleptiques en Thérapeutique Psychiatrique* (Paris, October 20–22, 1955).

a multicenter study (nine hospitals), randomized, double-blind, and controlled with placebo, which assessed the efficacy of three antipsychotics (chlorpromazine, fluphenazine, and thioridazine) in 344 patients recently admitted to hospital and diagnosed with schizophrenia, after 6 weeks of treatment. The results of the trial, shown in Figure 8.8, indicated the clear effectiveness of the new drugs.

Indeed, in 1957, the American Public Health Association awarded the prestigious Lasker Prize for Medicine to Laborit and Deniker, together with Heinz Lehmann, a Canadian psychiatrist, for the discovery of the antipsychotic effect of chlorpromazine. The plinth of Deniker's award bore the inscription: "Prize awarded for the introduction of chlorpromazine in psychiatry and the demonstration that a medication can influence the clinical course of the major psychoses." Nevertheless, the disputes between the groups from Val-de-Grâce and Sainte-Anne and the subsequent controversy over the discovery of the antipsychotic properties of chlorpromazine deprived these researchers,

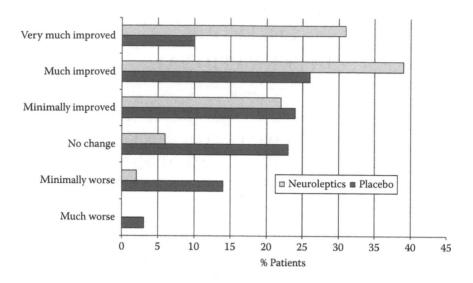

FIGURE 8.8 NIMH (1964) study results: effect of neuroleptics administration in schizo-phrenic patients, evaluated by the researcher by means of the *Global Rating of Improvement*. (Modified from Cole, J.O. and the NIMH Psychopharmacology Service Center Collaborative Study Group, *Arch. Gen. Psych.*, 10, 246, 1964.)

as noted by Pichot (1996), of winning the Nobel Prize, for which they were nomi-nated in view of the great clinical significance of their contribution, since the Swedish Academy preferred not to give the award to either so as to avoid problems within the French scientific community; this was indeed an even more obvious outcome if we take into account that Delay was at the time a foreign member of the Academy.

8.6 USE OF CHLORPROMAZINE IN OTHER DISORDERS: ANOTHER WAY OF DRUG REPOSITIONING

Although the main therapeutic uses of chlorpromazine in its 65 years of existence have been in psychiatric disorders, particularly psychotic disorders, the clinical use of this drug has also been extended to other diseases outside the psychiatric sphere, usu-ally *off-label* indications (Di Pietro 2015), although in many cases under protocols in good clinical practice guidelines. Such is the case of intractable hiccups, migraines, or the use of this agent as a coadjuvant analgesic, as antiemetic, or for sedation in palliative care. We can mention that the neuroleptics (and chlorpromazine), outside the psychiatric framework, have also been used, but their use is far less documented, in the treatment of tics present in neurological disorders, such as Tourette's syndrome and Huntington's chorea, as well as in managing symptoms of agitation, aggression, or nightmares in subjects with senile dementia. These extra-psychiatric uses could be considered as a new form of drug repositioning.

Nausea and vomiting are very common eventualities in different clinical fields. Drugs that block the dopamine receptors have an antiemetic ability, which have been used since the 1960s to this end. However, the ability to cause extrapyramidal effects has seriously limited their use, in favor of agents acting through serotonergic mechanisms. Chlorpromazine possess antiemetic effects demonstrated in preclinical studies on dogs, according to a pioneering work by Janssen et al. (1960).

The role of the neuroleptics as analgesic coadjuvants has been a source of debate since John W. Dundee's group, from the Department of Anesthesia at The Queen's University (Belfast), in the late 1950s, studied them for acute experimental pain in healthy volunteers (Dundee and Moore 1960). According to this work, several classical neuroleptics showed moderate analgesic activity and some of them exhibited a biphasic effect of initial hyperalgesia, followed by analgesia. Another of the postulated effects, the potentiation of opioid analgesia, was also erratic. Thus, promazine slightly enhanced meperidine's analgesia, but promethazine was antianalgesic and diminished meperidine's effect. Subsequent studies, comparative with a placebo, showed that various phenothiazines, including chlorpromazine, were not analgesics *per se* nor did they potentiate meperidine's analgesic effect (Siker et al. 1966). Apparently, methotrimeprazine could be an exception, showing acute analgesic properties (Lasagna and DeKornfeld 1961), similar to meperidine, although it did not show any differential advantage over the opioid (Atkinson et al. 1999). In conclusion, chlorpromazine, like other classic neuroleptics, for their analgesic effect and opioid potentiating effects, as well as their severe adverse effects (sedation, extrapyramidal effects, orthostasis) or chronic adverse effects, especially tardive dyskinesia, does not appear to be a good analgesic adjuvant (Goldstein 2002).

Short spells of hiccups are relatively common, and despite being annoying, are not grave, but hiccups that last more than 48 hours, called persistent hiccups, or that last more than a month, called incoercible hiccups, are more serious, although fortunately very rare, so there are few data to treat them according to evidence-based protocols. However, chlorpromazine has been used for the treatment of hiccups almost since its discovery, because Friegood and Ripstein (1955), on the one hand, and Davignon et al. (1955), on the other hand, published the first cases of treating incoercible hiccups. More recently, Martínez-Rey and Villamil-Cajoto (2007) have published the results obtained after the administration of chlorpromazine to 23 patients diagnosed with intractable hiccups, obtaining control of symptoms in 65.2% of patients (n = 15). Currently, chlorpromazine has been approved by the U.S. Food and Drug Administration regulatory authorities, but not for the European Medicines Agency, so use in the European Union countries is an off-label use for this disorder (Friedman 1996).

Severe migraine attacks and migraine status are medical conditions that interfere with the patient's daily life, often require hospitalization, and usually do not respond to standard drug therapy. In these cases, chlorpromazine is positioned as an alternative therapy administered by slow intravenous injection (Herman 2003). Furthermore, in these particular cases, it is also very frequent the occurrence of nausea and vomiting, which improve the administration of chlorpromazine. Dopamine antagonists, in general, appear to be equivalent to the migraine "specific" medications sumatriptan and dihydroergotamine for migraine pain relief (Kelley and Tepper 2012). Numerous clinical practice guidelines support the use of this agent in cases of severe migraine resistant to standard treatment (Pryse-Phillips et al. 1997; Silberstein and for the U.S. Headache Consortium 2000; British Association for the Study of

Headache 2009). More recently, a Neurologist's Guide to Acute Migraine Therapy in the Emergency Room (Gelfand and Goadsby 2012) proposes the use of chlorpromazine, as dopamine antagonist, for the treatment of acute migraine and migraine status, given that the results of clinical trials show a therapeutic efficacy over 80% (Bell et al. 1990; Kelly et al. 1997).

Chlorpromazine is included in various protocols used to induce palliative sedation in terminally ill patients. In this sense, the Palliative Sedation Guide (2011) prepared by the Spanish Society for Palliative Care puts chlorpromazine as an agent of second choice, after midazolam, to induce sedation in patients with terminal illness (Sociedad Española de Cuidados Paliativos, 2011). In refractory cases, the recommended setting is the association of benzodiazepines and neuroleptics (Nogueira and Sakata 2012), such as chlorpromazine (Kohara et al. 2005). Chlorpromazine has also been used to induce sedation in pediatric patients who have to undergo imaging tests. Heng Vong et al. (2012) have published a review to assess the efficacy and safety of chlorpromazine intravenously to induce sedation in children less than 6 years with cancer in imaging tests. The authors conclude that chlorpromazine is an effective and safe alternative for these patients.

8.7 CONCLUSIONS

The clinical introduction of chlorpromazine in the treatment of psychotic patients is one of the best examples of repositioning drug in the history of pharmacology. All drug development of this phenothiazine was intended to achieve an antihistamine drug and finally an antipsychotic agent was obtained (Dronsfield and Ellis 2006). Thus, the case of chlorpromazine represents another instance of an unforeseen observation (*désintéressement*) resulting in the programmatic development of an application; in this instance, one of the most important applications in the history of psychopharmacology. As mentioned, the ability to recognize the phenomenon of ataraxia by Laborit, an event totally unexpected and unforeseeable, was fundamental to the development of chlorpromazine, which led to the treatment of psychosis by Deniker and Delay with the expectation (or hope) that this drug could calm their excited patients. Thus, the finding of the antipsychotic effect of chlorpromazine cannot be categorized as purely serendipitous, but of mixed character (Baumeister et al. 2010; López-Muñoz et al. 2012).

The discovery of the antipsychotic properties of chlorpromazine in the 1950s was a fundamental event for the practice of psychiatry and for the genesis of the so-called "psychopharmacological revolution" (López-Muñoz et al. 2003, 2005, 2014; Ban 2007). Arriving as it did in a desert landscape as far as therapy was concerned, chlorpromazine made it clear that mental illness could be treated effectively by chemical means. It also paved the way for the clinical use of new psychoactive drugs, such as lithium salts, imipramine, or chlordiazepoxide, which continue, at the dawn of the twenty-first century, to be of great therapeutic, conceptual, and practical importance.

The introduction of chlorpromazine made possible a series of great clinical, healthcare, and scientific advances (Shen 1999; López-Muñoz et al. 2003; Ban 2007). Clear proof of the significance of the introduction to clinical practice of chlorpromazine is the enormous number of patients that have benefited from its use, which rose, just in the decade 1955–1965, to more than 50 million, or more than 10,000 publications on chlorpromazine that appeared in the same period (Jarvik 1970). Chlorpromazine has also played a fundamental role in the progressive phenomenon of "deinstitution-alization" of psychiatry and in the implication of primary care in matters of mental health, two developments that have helped to reduce the stigmatization associated with psychiatric assistance. Between 1954 and 1996, the official figure for inpatients at public psychiatric hospitals in the United States fell by 89%, while the number of these institutions decreased by 34% between 1954 and 1988, according to Geller (2000). As a European example, at the University Psychiatric Hospital in Basel (Switzerland), the mean number of days' stay per patient fell from 150 in 1950 to 95 in 1960 (Battegay 2001). Moreover, the development of chlorpromazine attracted interest from researchers and from the pharmaceuticals industry in the development of new psychoactive drugs in general and antipsychotic agents in particular (López-Muñoz and Álamo 2011; Jašović-Gašic et al. 2012). Other consequences of this "rev-olution in psychiatry" must be placed within a more strictly scientific framework, such as the postulate of the first biological hypotheses on the genesis of mental disor-ders, the introduction of changes in nosological conceptualizations, and in the design of a new set of diagnostic criteria, and the development of a modern methodology in clinical research within the psychiatric field (López-Muñoz et al. 2003). The role of chlorpromazine in the history of psychopharmacology and in the consolidation of modern psychiatry has been, in sum, not only fundamental, but truly indispensable.

Thus, a century and a half after Philippe Pinel physically freed the inmates of the Parisian Hôpital de la Salpêtrière from their chains, French psychiatrists once more released psychiatric patients from the torment of confinement, this time thanks in part to the "power of Serendip" and by means of a pharmacological tool, chlorpromazine. In the words of Edward Shorter, "chlorpromazine initiated a revolution in psychiatry, comparable to the introduction of penicillin in general medicine" (Shorter 1997).

REFERENCES

Álamo C, López-Muñoz F, and Cuenca E. El desarrollo de la clozapina y su papel en la con-ceptualización de la atipicidad antipsicótica. Psiquiatria.COM (electronic journal), Vol. 4(3), September 2000. https://www.psiquiatria.com/revistas/index.php/psiquiatriacom/article/viewFile/562/540. Accessed April 10, 2017.

Allison L and Moncrieff J. 'Rapid tranquillisation': An historical perspective on its emergence in the context of the development of antipsychotic medications. *History of Psychiatry* 25 (2014): 57–69.

Altschule MD. Use of chlorpromazine and reserpine in mental disorders. *New England Journal of Medicine* 254 (1956): 515–519.

Atkinson JH, Slater MA, Wahlgren DR et al. Effects of noradrenergic and serotonergic antide-pressants in chronic low back pain intensity. *Pain* 83 (1999): 137–145.

Ayd FJ. The early history of modern psychopharmacology. *Neuropsychopharmacology* 5 (1991): 71–84.

Ban TA. Pharmacotherapy of mental illness. A historical analysis. *Progress in Neuro-Psychopharmacology and Biological Psychiatry* 25 (2001): 709–727.

Ban TA. Fifty years chlorpromazine: A historical perspective. *Neuropsychiatric Disease and Treatment* 3 (2007): 495–500.

Battegay R. Forty-four years of psychiatry and psychopharmacology. In: Healy D, ed., *The Psychopharmacologists III*. London, U.K.: Arnold, 2001, pp. 371–394.

Baumeister AA, Hawkins MF, and López-Muñoz F. Toward standardized usage of the word serendipity in the historiography of psychopharmacology. *Journal of the History of Neuroscience* 19 (2010): 254–271.

Bell R, Montoya D, Shuaib A, and Lee MA. A comparative trial of three agents in the treatment of acute migraine headache. *Annals of Emergency Medicine* 19 (1990): 1079–1082.

British Association for the Study of Headache. Guidelines for all doctors in the diagnosis and management of migraine and tension-type headache, 2009. http://www.bash.org. uk/wp-content/uploads/2012/07/10102-BASH-Guidelines-update-2_v5-1-indd.pdf. Acceded July 20, 2015.

Caldwell AE. History of psychopharmacology. In: Clark WG and Del Giudice J, eds., *Principles of Psychopharmacology*. New York: Academic Press, 1970, pp. 9–30.

Cancro R. The introduction of neuroleptics: A psychiatric revolution. *Psychiatric Services* 51 (2000): 333–335.

Carpenter WT and Davis JM. Another view of the history of antipsychotic drug discovery and development. *Molecular Psychiatry* 17 (2012): 1168–1173.

Chertock L. 30 ans après: la petite histoire de la découverte des neuroleptiques. *Annales Médico-Psychologiques* 140 (1982): 971–974.

Cole JO and the NIMH Psychopharmacology Service Center Collaborative Study Group. Phenothiazine treatment in acute schizophrenia. *Archives of General Psychiatry* 10 (1964): 246–261.

Courvoisier S. Pharmacodynamic basis for the use of chlorpromazine in psychiatry. *Journal of Clinical and Experimental Psychopathology* 17 (1956): 25–37.

Courvoisier S, Fournel J, Ducrot R, Kolsky M, and Koetschet P. Propiétés pharmacodynamiques du chlorhydrate de chloro-3 (dimethypamine 3′propyl)-10 phénotiazine (4560RP). *Archives Internationales de Pharmacodynamie et de Thérapie* 92 (1953): 305–361.

Daumezon G and Cassan L. Essai de thérapeutique abortive d'accès maniaco-dépressifs par le 2339 RP. *Annales Médico-Psychologiques* 101 (1943): 432–435.

Davignon A, Lemieux G, and Genest J. Chlorpromazine in the treatment of stubborn hiccup. *L'Union Médicale du Canada* 84 (1955): 282–284.

Delay J and Deniker P. 38 cas de psychoses traitées par la cure prolongée et continué de 4560RP. *Comptes Rendus du 50 Congrès des Médecins Aliénistes et Neurologistes de Langue Française*, Luxembourg, 1952a, pp. 503–513.

Delay J and Deniker P. Le traitement des psychoses par une méthode neurolytique dérivée de l'hibernothérapie (le 4560 RP utilisé seul en cure prolongée et continue). *Comptes Rendus du 50 Congrès des Médecins Aliénistes et Neurologistes de Langue Française*, Luxembourg, 1952b, pp. 495–502.

Delay J and Deniker P. Réactions biologiques observées au cours du traitement par le chlorte de deméthylaminoprppyl-N-chlorophénothi chlorophénothiazine. *Comptes Rendus du 50 Congrès des Médecins Aliénistes et Neurologistes de Langue Française*, Luxembourg, 1952c, pp. 514–518.

Delay J, Deniker P, and Harl JM. Utilisation en thérapeutique d'une phénothiazine d'action centrale selective (4560 RP). *Annales Médico-Psychologiques* 110 (1952a): 112–117.

Delay J, Deniker P, and Harl JM. Traitement des états d'excitation et d'agitation par une méthode médicamenteuse dérivée de l'hibernothérapie. *Annales Médico-Psychologiques* 110 (1952b): 267–273.

Delay J, Deniker P, Harl JM, and Grasset A. Traitements des états confusionnels par le chlorte de deméthylaminoprppyl-N-chlorophénothiazine (4560 RP). *Annales Médico-Psychologiques* 110 (1952c): 398–403.

Deniker P. From chlorpromazine to tardive dyskinesia (brief history of the neuroleptics). *Psychiatric Journal of the University of Ottawa* 14 (1989): 253–259.

Di Pietro N. A brief history of the science and ethics of antipsychotics and off-label prescribing. In: Di Pietro N and Illes J, eds., *The Science and Ethics of Antipsychotic Use in Children*. Oxford, U.K.: Academic Press, 2015, pp. 1–12.

Dronsfield AT and Ellis PM. Chlorpromazine—Unlocks the asylums. *Education in Chemistry* 43 (2006): 74–76.

Dundee JW and Moore J. Alterations in response to somatic pain associated with anaesthesia. An evaluation of a method of analgesimetry. *British Journal of Anaesthesia* 32 (1960): 396–406.

Elkes J and Elkes C. Effects of chlorpromazine on the behaviour of chronically overactive psychotic patients. *British Medical Journal* 2 (1954): 560–565.

Ey H and Faure H. Les diverses méthodes d'emploi de la chlorpromazine en thérapeutique psychiatrique et leurs indications. *L'Encéphale* Numéro Spécial (1956): 61–71.

Frankenburg FR. History of the development of antipsychotic medication. *Psychiatric Clinics of North America* 17 (1994): 531–540.

Friedman NL. Hiccups: A treatment review. *Pharmacotherapy* 16 (1996): 986–995.

Friegood CE and Ripstein CB. Chlorpromazine (Thorazine) in the treatment of intractable hiccups. *Journal of American Medical Association* 157 (1955): 309–310.

Gelfand AA and Goadsby PJ. A neurologist's guide to acute migraine therapy in the emergency room. *Neurohospitalist* 2 (2012): 51–59.

Geller JL. The last half-century of psychiatric services as reflected in Psychiatric Services. *Psychiatric Services* 51 (2000): 41–67.

Gilman H, van Ess PR, and Shirley DA. The metalation of 10-phenylphenothiazine and of 10-ethyl-phenothiazine. *Journal of the American Chemical Society* 66 (1944): 1214–1216.

Ginestet D. Les neuroleptiques. Développement et situation actuelle. *L'Encéphale* XVII (1991): 149–152.

Goldstein FJ. Adjuncts to opioid therapy. *Journal of the American Osteopathic Association* 102(Suppl. 3) (2002): S15–S20.

Guiraud P and David C. Traitement de l'agitation motrice par un antihistaminique (3277 RP). *Comptes Rendus du 48 Congrès des Médecins Aliénistes et des Neurologistes de Langue Française*, Bcsançon, 1950, pp. 599–602.

Hamon J, Paraire J, and Velluz J. Remarques sur l'action du 4560RP sur l'agitation maniaque. *Annales Médico-Psychologiques* 110 (1952): 332–335.

Healy D. *The Psychopharmacologists*. New York: Chapman & Hall, 1996.

Healy D. *The Psychopharmacologists II*. London, U.K.: Arnold, 1999.

Healy D. *The Psychopharmacologists III*. London, U.K.: Arnold, 2000.

Heng Vong C, Bajard A, Thiesse P, Bouffet E, Seban H, and Marec Bérard P. Deep sedation in pediatric imaging: Efficacy and safety of intravenous chlorpromazine. *Pediatric Radiology* 42 (2012): 552–561.

Herman LY. The efficacy of a safe, effective non-narcotic treatment of migraine headache. *Journal of Emergency Medicine* 25 (2003): 322–323.

Jacobsen E. The early history of psychotherapeutic drugs. *Psychopharmacology* 89 (1986): 138–144.

Janssen PA, Niemegeers CJ, and Schellekens KH. Chemistry and pharmacology of compounds related to 4-(4-hydroxy-4-phenyl-piperidino)-butyrophenone. III. Duration of antiemetic action and oral effectiveness of Haloperidol (R 1625) and of chlorpromazine in dogs. *Arzneimittelforschung* 10 (1960): 955.

Jarvik ME. Drugs used in the treatment of psychiatric disorders. In: Goodman LS and Gilman A, eds., *The Pharmacological Basis of Therapeutics*, 4th edn. New York: The McMillan Company, 1970, pp. 151–203.

Jašović-Gašic M, Vuković O, Pantović M, Cvetić T, and Marić-Bojović N. Antipsychotics—History of development and field of indication, new wine—Old glasses. *Psychiatria Danubina* 24(Suppl. 3) (2012): 342–344.

Kelley NE and Tepper DE. Rescue therapy for acute migraine, part 2: Neuroleptics, antihistamines, and others. *Headache* 52 (2012): 292–306.

Kelly AM, Ardagh M, Curry C, D'Antonio J, and Zebic S. Intravenous chlorpromazine versus intramuscular sumatriptan for acute migraine. *Journal of Accident and Emergency Medicine* 14 (1997): 209–211.

King C and Voruganti LNP. What's in a name? The evolution of the nomenclature of antipsychotic drugs. *Journal of Psychiatry and Neuroscience* 27 (2002): 168–175.

Kohara H, Ueoka H, Takeyama H, Murakami T, and Morita T. Sedation for terminally ill patients with cancer with uncontrollable physical distress. *Journal of Palliative Medicine* 8 (2005): 20–25.

Laborit H. *Réaction organique à l'agression et choc.* Paris, France: Masson & Cie, 1952.

Laborit H and Huguenard P. L'hibernation artificielle par moyens pharmacodynamiques of physiques. *La Presse Médicale* 59 (1951): 1329.

Laborit H, Huguenard P, and Alluaume R. Un nouveau stabilisateur végétatif (le 4560 RP). *La Presse Médicale* 60 (1952a): 206–208.

Laborit H, Jaulmes C, and Bénitte A. Certain experimental aspects of artificial hibernation. *Anesthesia and Analgesia* 9 (1952b): 232.

Lasagna L and DeKornfeld TJ. Methotrimeprazine a new phenothiazine derivative with analgesic properties. *Journal of American Medical Association* 178 (1961): 887–890.

Lehmann HE. The introduction of chlorpromazine in North America. *Psychiatric Journal of the University of Ottawa* 14 (1989): 263–265.

Lehmann HE and Ban TA. The history of the psychopharmacology of schizophrenia. *Canadian Journal of Psychiatry* 42 (1997): 152–163.

Lickey ME and Gordon B. *Medicamentos para las enfermedades mentales.* Barcelona, Spain: Labor, 1986, p. 78.

López-Muñoz F and Álamo C. The consolidation of neuroleptic therapy: Janssen, the discovery of haloperidol and its introduction into clinical practice. *Brain Research Bulletin* 79 (2009): 130–141.

López-Muñoz F and Álamo C. Neurobiological background for the development of new drugs in schizophrenia. *Clinical Neuropharmacology* 34 (2011): 111–126.

López-Muñoz F, Álamo C, and Cuenca E. Farmacos antipsicóticos. In: López-Muñoz F and Álamo C, eds., *Historia de la Neuropsicofarmacología. Una nueva aportación a la terapéutica farmacológica de los trastornos del Sistema Nervioso Central.* Madrid, Spain: Ediciones Eurobook, S.L. and Servicio de Publicaciones de la Universidad de Alcalá, 1998, pp. 207–243.

López-Muñoz F, Alamo C, and Cuenca E. La "Década de Oro" de la Psicofarmacología (1950–1960): Trascendencia histórica de la introducción clínica de los psicofármacos clásicos. Psiquiatria.COM (electronic journal), Vol. 4(3), September 2000. https://www.psiquiatria.com/revistas/index.php/psiquiatriacom/article/viewFile/561/539. Accessed April 10, 2017.

López-Muñoz F, Álamo C, and Cuenca E. Aspectos históricos del descubrimiento y de la introducción clínica de la clorpromazina: medio siglo de psicofarmacología. *Frenia Revista de Historia de la Psiquiatría* 2 (2002): 77–107.

López-Muñoz F, Álamo C, and Cuenca E. Aportación de la clorpromazina al desarrollo de la psiquiatría. *Archivos de Psiquiatria* 66 (2003): 16–34.

López-Muñoz F, Álamo C, Rubio G, and Cuenca E. Half a century since the clinical introduction of chlorpromazine and the birth of modern psychopharmacology. *Progress in Neuro-Psychopharmacology and Biological Psychiatry* 28 (2004): 205–208.

López-Muñoz F, Álamo C, Cuenca E, Shen WW, Clervoy P, and Rubio G. History of the discovery and clinical introduction of chlorpromazine. *Annals of Clinical Psychiatry* 17 (2005): 113–135.

López-Muñoz F, Álamo C, and Domino E, eds. *History of Psychopharmacology*, 4 vols. Arlington, TX: NPP Books, 2014.

López-Muñoz F, Baumeister AA, Hawkins MF, and Álamo C. El papel de la serendipia en el descubrimiento de los efectos clínicos de los psicofármacos: más allá del mito. *Actas Españolas de Psiquiatría* 40 (2012): 34–42.

López-Muñoz F, Sanz-Fuentenebro FJ, Rubio G, García-García P, and Álamo C. Quo vadis clozapine? A bibliometric study of 45 years of research in international context. *International Journal of Molecular Sciences* 16 (2015): 23012–23034.

Martínez-Rey C and Villamil-Cajoto L. Hipo (singultus): revisión de 24 casos. *Revista Médica de Chile* 135 (2007): 1132–1138.

Nogueira FL and Sakata RK. Sedación paliativa del paciente terminal. *Revista Brasileira de Anestesiologia* 62 (2012): 1–7.

Pichot P. The discovery of chlorpromazine and the place of psychopharmacology in the history of the psychiatry. In: Healy D, ed., *The Psychopharmacologists*. New York: Chapman & Hall, 1996, pp. 1–27.

Preskorn SH. The evolution of antipsychotic drug therapy: Reserpine, chlorpromazine, and haloperidol. *Journal of Psychiatric Practice* 13 (2007): 253–257.

Pryse-Phillips WEM, Dodick DW, Edmeads JG et al. Guidelines for the diagnosis and management of migraine in clinical practice. *Canadian Medical Association Journal* 156 (1997): 1273–1287.

Shen WW. A history of antipsychotic drug development. *Comprehensive Psychiatry* 40 (1999): 407–414.

Shen WW and Giesler MC. The discoverers of the therapeutic effect of chlorpromazine in psychiatry: Qui étaient les vrais premiers practiciens? *Canadian Journal of Psychiatry* 43 (1998): 423–424.

Shepherd M. Neurolepsis and the psychopharmacological revolution: Myth and reality. *History of Psychiatry* 5 (1994): 89–96.

Shorter E. A history of psychiatry. *From the Era of Asylum to the Age of Prozac.* New York: John Wiley & Sons, Inc., 1997.

Siker ES, Wolfson B, Stewart WD, and Schaner PJ. The earlobe algesimeter. *Anesthesiology* 2 (1966): 497–500.

Silberstein SD and for the US Headache Consortium. Practice parameter: Evidence-based guidelines for migraine headache (an evidence-based review). *Neurology* 55 (2000): 754–763.

Sociedad Española de Cuidados Paliativos. Guía de Sedación Paliativa, 2011. www.cgcom.es/sites/default/files/guia_sedaccion_paliativa.pdf.

Swazey JP. *Chlorpromazine in Psychiatry: A Study of Therapeutic Innovation.* Cambridge, MA: MIT Press, 1974.

Thuillier J. Naissance de la psychopharmacologie ou "une histoire de nez". *Urgences* 1–2 (1994): 21–22.

Zircle CL. To tranquilizers and antidepressants: From antimalarials and antihistamines. In: Clarke FH, ed., *How Modern Medicines Are Discovered*. Mt. Kisco, New York: Futura, 1973, pp. 55–77.

9 Drug Repurposing for Central Nervous System Disorders
A Pillar of New Drug Discovery

Mondher Toumi, Aleksandra Caban,
Anna Kapuśniak, Szymon Jarosławski,
and Cecile Rémuzat

CONTENTS

9.1 INTRODUCTION

Traditionally, drug repurposing has been considered as a strategy for developing new drugs, adopted in order to increase cost-efficiency and reduce risks related to novel molecule discovery or to extend the patent protection period for existing products (PriceWaterhouseCoopers 2007, Hemphill and Sampat 2012). However, upgradation of regulatory frameworks that discourages the development of the so-called "me too" drugs by isolating enantiomers or introducing minor chemical modifications to known molecules is changing the face of repurposing nowadays. Furthermore, information-sharing platforms about molecules that are suitable for repurposing have been created in order to encourage collaboration between the industry and academia (Allarakhia 2013, Murteira et al. 2014b). Poorly addressed therapeutic areas, such as the central nervous system (CNS), oncology, pediatric, or orphan diseases, are expected to most benefit from drug repurposing.

While the underlying physiopathology of many CNS disorders is still poorly understood, drug repurposing may prove to be a critical approach in this area (Hardman et al. 2008, Murteira et al. 2013). To date, the mechanism of action of many approved CNS drugs is not fully understood. Therefore, disease characterization within the CNS is often based on clinical aspects rather than the underlying physiopathology (Aminoff et al. 2014). Consequently, the discovery and development of drugs for CNS diseases has one of the lowest success rates (Pangalos et al. 2007, World Health Organization 2014). Several studies concluded that not only is the number of drugs available for clinical development in the CNS lower than in other areas, but also the regulatory approval times are longer (Pardridge 2002, Kola and Landis 2004, Riordan and Cutler 2011).

Serendipity, trial and error, and systematic screening of approved products remain critical sources of new CNS candidate drugs. Well-known examples of repurposing by serendipity in CNS are amantadine, repositioned from the treatment of influenza to the treatment of Parkinson's disease in the late 1960s, and propranolol, initially approved for the treatment of angina and hypertension and repurposed for migraine prophylaxis in the early 1980s (Padhy and Gupta 2011, Sekhon 2013). More recently, dimethyl fumarate repositioned from the treatment of psoriasis to a disease-modifying product for multiple sclerosis is considered as the most promising oral drug for this CNS disorder (Murteira et al. 2014a).

High unmet needs as well as high prevalence of CNS disorders in the overall population make repurposing a suitable approach for drug development in this specific area.

In this chapter, we present a systematic review of repurposing of drugs in the CNS area and adopt the following terminology: *repurposing* of a molecule consists in finding new therapeutic uses for an already known drug; *reformulation* consists in developing different formulations for the same drug; a new *combination* consists in developing a single product that combines at least two drugs previously used separately (Murteira et al. 2013).

9.2   EXTENT OF REPURPOSING IN CNS DRUG DEVELOPMENT

A comprehensive review of literature published until January 2016 was recently performed by the authors (Caban et al. 2016). The authors have excluded from the research all products that had been minimally reformulated, unless this resulted in change of the route of administration (e.g., injectable to oral and vice versa). The search revealed that 118 drugs have undergone repurposing 203 times in the CNS area (Table 9.1).

Furthermore, the highest number of source drugs was in the CNS therapeutic area (66 out of 118 drugs, 122 single cases—some drugs were repurposed more than one time), followed by cardiovascular (11 drugs, 22 single cases), endocrine, metabolic, and genetic disorders (10 drugs, 11 single cases), and oncology (8 drugs, 10 single cases) (Figure 9.1). The remaining 23 products (38 single cases) were distributed evenly among the remaining therapeutic areas.

Products sourced from the CNS and cardiovascular areas targeted both neurological and psychiatric disorders, whereas products from endocrine, metabolic and genetic, as well as oncology areas targeted mainly neurological disorders.

Except cilostazol and indobufen, cardiovascular drugs repositioned for the CNS were at source either antihypertensive, antiarrhythmic, or both types of drugs. Despite this common background in terms of cardiovascular effect, there were 13 distinct target CNS indications that ranged from psychiatric to neurological disorders, including neurodegenerative conditions such as Parkinson's or Alzheimer's disease. This heterogeneity can be attributed to different mechanisms of action among the cardiovascular

TABLE 9.1

Number of Source Products and Target Indications Included in the Analysis

Source Products	118
Products repositioned once	80
Products repositioned two times	16
Products repositioned three times or more	22
Target Indications	**203**
Products in development	101
Products approved	102
Products reformulated	16
Products repositioned	171
Products reformulated and repositioned	16

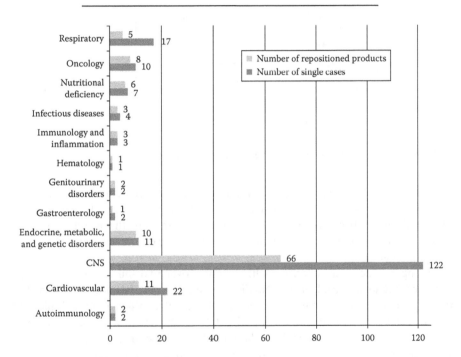

FIGURE 9.1   Drugs repositioned in CNS by initial therapeutic area.

originators, ranging from beta blockers to angiotensin II receptor antagonists among others. Indeed, those receptors may also be involved in CNS disorders and, when repositioned, those drugs can have distinct therapeutic effects.

Among new therapeutic indications, Alzheimer's disease was targeted most often (22 cases), followed by substance dependence (alcohol, drugs/opioids, tobacco), bipolar disorder, depression, neuropathy/neuralgia, multiple sclerosis, and schizophrenia, with 10 and more cases each (Figure 9.2).

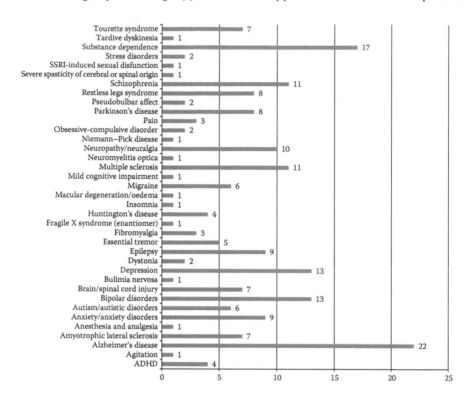

FIGURE 9.2   Drugs repositioned in CNS by target indication.

Within the CNS area, epilepsy, schizophrenia, and depression were the richest sources of repositioned drugs with 10 and more products each, targeting 27, 30, and 20 indications, respectively (Figure 9.3). Indeed, those indications are among the most prevalent and many drug alternatives are available for those source indications.

Among products repositioned multiple times, over two thirds (68%) originated from the CNS area. Indeed, many CNS diseases share a part of a known physiopathology (e.g., partial overlap of receptors or neurotransmitters affected by multiple CNS diseases) and are often comorbid with other CNS diseases. Furthermore, since most CNS drugs target symptoms and not the underlying cause, these symptoms may be common across multiple CNS diseases, such as bipolar disorder, schizophrenia, and depression. Illustratively, deep brain stimulation therapy used to alleviate Parkinson's disease may induce transient acute depression, which resolves after adjustment of the electrode position (Kogan and Haren 2008, Martinez-Ramirez et al. 2015).

A half of the newly developed indications (102 cases) were approved. While the majority of approved cases (80%) originated from the CNS area, the majority (61%) of cases still in development originated from areas outside the CNS. However, it is unclear if this is due to a broader range of source products being used more recently in drug repositioning or to a higher approval rate for drugs sourced from within the CNS compared to those repositioned drugs sourced outside the CNS. Further research will

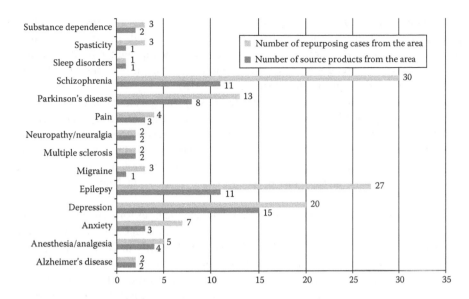

FIGURE 9.3 Initial therapeutic indication of drugs repositioned within CNS.

be necessary to track the success rate in terms of regulatory approval for the products in development retrieved in our research to ascertain this question.

Most of the cases were repositioned (171), whereas only 16 were reformulated and 16 were reformulated and repositioned at the same time. Because the minimally reformulated products were excluded from the search, the number of reformulated products is limited. However, the large number of repositioned products indicates that this approach is of particular interest to drug developers.

9.3 CONCLUSIONS

Drug repurposing in the CNS is a very active approach to drug development as half of the identified cases are drugs still in development. The high prevalence of CNS disorders and the high unmet needs in this area call for novel policies and developments that can remove the inefficiencies of this strategy. For instance, it could be valuable to produce historic databases of drug repurposing cases in the CNS that failed, in order to help avoid duplication of efforts. Furthermore, the role of regulators should be to incentivize the development of value-added drug repurposing in the CNS by drafting suitable guidance for drug approval and health technology assessment.

The importance of repurposing was recently recognized by the European Commission, which constituted the Commission Expert Group on Safe and Timely Access to Medicines for Patients ("STAMP"). STAMP aims at recognizing the importance of comprehensive investigation of different opportunities that a molecule could bring to patients, with faster development times, at reduced costs and risk for pharmaceutical companies (European Commission 2016).

Several initiatives focused on promoting drug repurposing and based on public, private, and academic partnership have also been established. In 2011, the United Kingdom's Medical Research Council signed a partnership with AstraZeneca that gave academic researchers access to the company's clinical and preclinical compounds for potential repurposing (AstraZeneca 2015).

In France, the National Cancer Institute (INCa), in agreement with the French National Agency for Medicines and Health Products Safety (ANSM), launched the AcSé program in 2013. The program aims at securing access to innovative targeted therapies for cancer patients who failed treatment with approved therapies and whose cancer has the same genetic profile but different organ location as compared to the cancer the innovative therapy has been approved for (Institut National du Cancer (INCa) 2016).

In the United States, the Discovering New Therapeutic Uses for Existing Molecules (New Therapeutic Uses) program was initiated by the National Center for Advancing Translational Sciences and Pfizer, AstraZeneca, and Eli Lilly in 2012. The program is matching researchers with a selection of pharmaceutical industry compounds to help scientists explore new treatment options (National Institutes of Health 2016a,b).

Academic researchers and drug manufacturers should carefully consider repositioning as an option for drug development as it is a lower-risk, faster, and more efficient development strategy than new drug discovery.

REFERENCES

Allarakhia M. 2013. Open-source approaches for the repurposing of existing or failed candidate drugs: Learning from and applying the lessons across diseases. *Drug Des Develop Ther* 7:753–766.

Aminoff MJ, Boller F, and Swaab DF. 2014. *Handbook of Clinical Neurology: Neurobiology of Psychiatric Disorders: Foreword*, Vol. 106. Elsevier B.V., Amsterdam, Netherlands.

AstraZeneca. 2015. Novel collaboration with Medical Research Council gives UK academia access to AstraZeneca compounds. https://www.astrazeneca.com/content/dam/az/Our-Science/Partnering/case-studies/Feb15-Medical_Research_Council_Collaboration_Case_Study_UK-External-Final.pdf. Accessed September 26, 2016.

Caban A, Pisarczyk K, Kopacz K, Kapuśniak A, Rémuzat C, and Toumi M. 2016. Filling the gap in CNS drug development: Evaluation of the role of drug repurposing. *J Mark Access Health Policy*, submitted.

European Commission. 2016. Commission Expert Group on Safe and Timely Access to Medicines for Patients ("STAMP"). Repurposing of established medicines/active substances. Agenda item. http://ec.europa.eu/health/files/committee/stamp/2016-03_stamp4/3_repurposing_of_established_medicines_background_paper.pdf. Accessed September 26, 2016.

Hardman JG, Limbird LE, Molinoff PB, Ruddon RW, and Gilman AG. 2008. Neurotransmission and the central nervous system, Section III, Chapter 12. In: *Goodman & Gilman's—The Pharmacological Basis of Therapeutics* (Ed. LL Brunton). McGraw-Hill, New York City.

Hemphill CS and Sampat BN. 2012. Evergreening, patent challenges, and effective market life in pharmaceuticals. *J Health Econ* 31:327–339.

Institut National du Cancer (INCa). 2016. Le programme AcSé. http://www.e-cancer.fr/Professionnels-de-la-recherche/Recherche-clinique/Le-programme-AcSe. Accessed September 26, 2016.

Kogan AJ and Haren M. 2008. Translating cancer trial endpoints into the language of managed care. *Biotechnol Healthcare* 5(1):22–35.

Kola I and Landis J. 2004. Can the pharmaceutical industry reduce attrition rates? *Nat Rev Drug Discov* 3(8):711–715.

Martinez-Ramirez D, Hu W, Bona AR, Okun MS, and Wagle Shukla A. 2015. Update on deep brain stimulation in Parkinson's disease. *Transl Neurodegener* 4:12.

Murteira S, El Hammi E, and Toumi M. 2014a. Fixing the price of the orphan drug Siklos®: The Council of State takes over the decision. *Eur J Health Law* 21(5):505–515.

Murteira S, Ghezaiel Z, Karray S, and Lamure M. 2013. Drug reformulations and repositioning in pharmaceutical industry and its impact on market access: reassessment of nomenclature. *J Mark Access Health Policy* 1. doi:10.3402/jmahp.v1i0.21131.

Murteira S, Millier A, Ghezaiel Z, and Lamure M. 2014b. Drug reformulations and repositioning in the pharmaceutical industry and their impact on market access: Regulatory implications. *J Mark Access Health Policy* 2. doi:10.3402/jmahp.v2.22813.

National Institutes of Health. 2016a. National Center for Advancing Translational Sciences (NCATS). Mission. NIH website: https://www.nih.gov/about-nih/what-we-do/nih-almanac/national-center-advancing-translational-sciences-ncats. Accessed September 26, 2016.

National Institutes of Health. 2016b. News releases: NIH launches collaborative program with industry and researchers to spur therapeutic development, May 3, 2012. NIH website: https://www.nih.gov/news-events/news-releases/nih-launches-collaborative-program-industry-researchers-spur-therapeutic-development. Accessed September 26, 2016.

Padhy BM and Gupta YK. 2011. Drug repositioning: Re-investigating existing drugs for new therapeutic indications. *J Postgrad Med* 57(2):153–160.

Pangalos MN, Schechter LE, and Hurko O. 2007. Drug development for CNS disorders: Strategies for balancing risk and reducing attrition. *Nat Rev Drug Discov* 6(7):521–532.

Pardridge WM. 2002. Why is the global CNS pharmaceutical market so under-penetrated? *Drug Discov Today* 7(1):5–7.

PriceWaterhouseCoopers. 2007. *Pharma 2020: The Vision*. New York: PriceWaterhouseCoopers.

Riordan HJ and Cutler HR. 2011. The death of CNS drug development: Overstatement or Omen. *J Clin Stud* 3:12–15.

Sekhon BS. 2013. Repositioning drugs and biologics: Retargeting old/existing drugs for potential new therapeutic applications. *J Pharm Educat Res* 4(1):1–15.

World Health Organization. 2014. Regional estimates for 2000–2011: Disease burden. WHO health statistics and information systems. World Health Organization, Geneva [Online]. http://www.who.int/healthinfo/global_burden_disease/estimates_regional/en/index1.html. Accessed September 26, 2016.

10 Repurposing Opportunities for Parkinson's Disease Therapies

Giulia Ambrosi, Silvia Cerri, and Fabio Blandini

CONTENTS

10.1   PARKINSON'S DISEASE: PATHOLOGICAL FEATURES, EPIDEMIOLOGY, AND AVAILABLE THERAPIES

Parkinson's disease (PD) is an adult neurodegenerative disease affecting approximately 1% of the population over 65 years and 4%–5% over 80 years of age. The rate of disease progression is widely variable among subjects. PD is primarily a sporadic disease (about 90% of cases) (Rodriguez et al. 2015) while about 10% correspond to purely familial forms and have monogenic basis (Lesage and Brice 2009; Spataro et al. 2015). However, even within the sporadic form, some patients report genetic susceptibility factors, which are responsible for increasing the risk of developing PD. For instance, heterozygous mutations in *GBA1* gene, encoding for the lysosomal enzyme glucocerebrosidase, have been detected in 5%–10% of sporadic PD patients (Sidransky et al. 2009; Schapira and Gegg 2013), thereby representing the most critical genetic risk factor in PD identified to date. PD is caused by the slow and progressive degeneration of dopaminergic neurons within the *substantia nigra pars compacta* (SNc); this leads to dopamine (DA) depletion of the *corpus striatum* followed by profound functional alterations of the basal ganglia circuitry, which controls the correct execution of voluntary movements (Blandini et al. 2000). In fact, PD is typically associated with motor symptoms, such as resting tremor, bradykinesia, slowness of movements, and postural instability. However, nonmotor symptoms are also present in PD patients and often precede the onset of classical motor manifestations (Olanow and Obeso 2012). These symptoms include olfactory impairment, sleep disorders, urogenital and gastrointestinal dysfunctions, and cognitive and psychiatric disturbances (Ferrer et al. 2012). Diagnosis of PD is performed on a clinical basis and can be confirmed only postmortem in patient brains, with the evidence of massive neuronal loss in the SNc and the detection of Lewy bodies (LBs), the main hallmark of PD. LBs are insoluble aggregates mainly containing α-synuclein, the protein encoded by *SNCA* or *PARK1/PARK4* gene, a genetic locus linked to familial PD (Xu et al. 2015), as well as phosphorylated and poly-ubiquitinated proteins. Several lines of evidence clearly suggest that overexpression and oligomerization of α-synuclein are directly related to the toxicity in the nigrostriatal areas (Masliah et al. 2000; Hayashita-Kinoh et al. 2006). The presence of LBs is not restricted to SNc and other cerebral regions (Von Bohlen und Halbach et al. 2004); they also spread in the spinal cord, in the vagus nerve, and in peripheral organs such as the gastrointestinal tract (Beach et al. 2010), confirming that PD is a complex and diffuse disease, affecting not only the brain.

10.1.1   PD PATHOBIOLOGY

The presence of LBs, the identification of genes associated with the disease (i.e., *PARK* genes), epidemiological data, and serendipitous exposure to molecules leading to the development of PD have provided cues on the molecular deficits that characterize the pathology.

Susceptibility to PD is primarily based on the selective vulnerability of dopaminergic neurons in the SNc, which is due to DA metabolism and its tendency to self-oxidation, as well as their high energy requirement and calcium buffering rate (Sulzer 2007).

The identification of mutations in genes such as *PARK2*, encoding for the E3 ubiquitin ligase parkin, suggested that dysfunctions in intracellular protein and organelle metabolism are involved in PD pathogenesis. Indeed, loss of parkin function impairs ubiquitination, leading to altered functioning of the ubiquitin proteasome system (UPS) (Dawson 2006). Impairment of UPS, together with the autophagy-lysosomal pathway, critically impacts neuron viability (Cook et al. 2012). These systems are in charge of the degradation of damaged or misfolded proteins, intracellular protein aggregates (i.e., α-synuclein bulks), as well as nonfunctional organelles such as defective mitochondria. Dysfunctions in these clearance systems can lead to accumulation of α-synuclein aggregates and dysfunctional organelles, thereby affecting cell survival (Osellame and Duchen 2014).

Alterations in mitochondrial function are also strongly involved in PD pathogenesis (Chan et al. 2009; Franco-Iborra et al. 2016). Mitochondria are intracellular organelles responsible for regulating energy production, reactive oxygen species (ROS) formation, calcium buffering, and activation of pro-apoptotic factors. Mutations in PD-linked proteins impact mitochondrial function by affecting the maintenance of mitochondrial turnover (e.g., PINK1 and parkin, Pilsl and Winklhofer 2012) and ROS detoxification (e.g., DJ-1, Canet-Avilés et al. 2004). In parallel, mitochondria, particularly complex I-driven respiration, are the targets of molecules that have been associated with a higher risk of PD and adopted for developing experimental animal models of the disease, such as rotenone, paraquat, and 1-methyl-4-phenyl-1,2,3,6-tetrahydropyridine or MPTP (Blandini and Armentero 2012; Moretto and Colosio 2013).

Overall, the intrinsic vulnerability of dopaminergic neurons, associated with deranged proteolytic machinery, mitochondrial impairment, and inefficient antioxidant defenses, represents the perfect background for the activation of excitotoxic phenomena, which further exacerbate the neurodegenerative process in PD (Ambrosi et al. 2014).

Finally, loss of dopaminergic neurons in PD can be sustained by neuroinflammatory processes, involving chronic activation of resident glial cells, astrocytes, and most critically microglia, as well as infiltration of immune cells and pro-inflammatory cytokines in the brain (González et al. 2014; Macchi et al. 2015). Inflammation, which normally contributes to tissue protection and repair, might become deleterious and amplify cell death in a stressed microenvironment, such as PD brain. In particular, microglia activation triggered by α-synuclein aggregation may sustain oxidative damage to the nigrostriatal system by massive release of nitric oxide and subsequent formation of ROS (Blandini 2013). Interestingly, in PD, among the main promoters of oxidative stress and microglia activation is also the accumulation of metal ions, especially of iron, which further accelerates neuronal loss and therefore PD severity (Ward et al. 2014).

10.1.2 CURRENT TREATMENT

Current therapies for PD are directed at specific symptoms of the disease but do not halt the neurodegenerative process. The pharmacological treatment is mainly based on restoration of the dopaminergic tone in the brain and typically consists in the administration of the DA precursor 3,4-hydroxyphenylalanine (L-dopa) in combination with decarboxylase inhibitors (benserazide or carbidopa) to prevent peripheral metabolism of the molecule (Yuan et al. 2010). L-Dopa efficacy is remarkable but tends to fade with time. The motor response to the drug becomes progressively shorter ("wearing off") or starts to fluctuate, with "ON" periods of full symptom control alternating, during the day, with "OFF" periods of highly impaired mobility. Involuntary movements, known as dyskinesias, are also frequently associated with long-term L-dopa treatment. DA agonists are also in use as an adjunctive therapy or a monotherapy, before or after L-dopa-induced motor complications have appeared (Blandini and Armentero 2014). Nondopaminergic drugs, such as anticholinergics and amantadine, may also be adopted.

Research in this field is focused on the investigation of new targets for neuroprotection, disease-modification, regeneration, as well as therapies to improve motor and nonmotor symptoms or to counteract L-dopa side effects. Among the different approaches adopted to investigate such therapeutic strategies for PD, drug repositioning is obtaining increasing interest and support from researchers, government entities, as well as nonprofit organization, with the aim of optimizing the effort and accelerating the process that may lead to the identification of new therapeutic strategies (Brundin et al. 2013).

10.2 DRUG REPOSITIONING IN PD: CHALLENGES, AND STRATEGIES TO OVERCOME THEM

Repurposing or repositioning drugs is considered a very interesting strategy for opening novel therapeutic perspectives in neurodegenerative diseases. Although serendipitous clinical observations have often contributed to the identification of molecules for repositioning, the choice of transferring a drug developed for one indication to another mainly arises from a combination of improved information on the target, mechanisms of action and pharmacology of a drug, and evolving knowledge about the molecular basis of the disease. In effect, similar risk factors and biological pathways underlying apparently unrelated diseases have opened the door for novel hypothesis-driven repurposing strategies (Figure 10.1).

Consistent with this, the increased findings on PD pathophysiology have expanded opportunities to develop new treatments as well as to repurpose drugs for PD. However, the actual possibility to reposition an approved drug for neurodegenerative diseases collides with relevant commercial and regulatory challenges. First, the difficulty for developers of repurposed drugs to protect new intellectual property and differentiate the repurposed product from the one already present on the market. The lack of obvious commercial incentive represents another obstacle leading to the deprioritization of repurposing projects with limited return on investment. In addition, although repurposed drugs can bypass early-stage development, they need expensive clinical trials to establish efficacy that meets the specific requirements for neurodegenerative diseases. Due to the slow, progressive nature of these diseases, clinical trials must last a long time and, in most cases, the ability of a drug to penetrate the brain must be proven. Moreover, safety represents one of the major concerns for an elderly population with neurodegenerative disease who frequently have

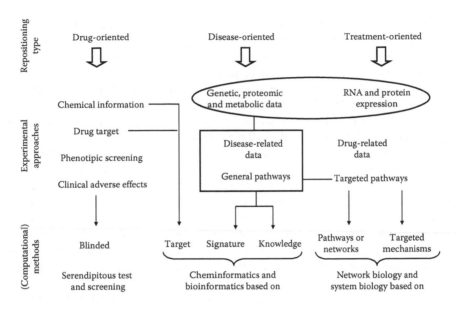

FIGURE 10.1 Schematic representation of strategies driving drug-repurposing.

one or more additional diseases co-occurring with the primary disease and may take medications that interact with the repurposed drug (Shineman et al. 2014). Together with these challenges, the search for new methodological approaches for improving the identification of molecules for repositioning in PD is another major issue.

In order to address financial hurdles in the PD repurposing field, different foundations (i.e., The Michael J. Fox Foundation or MJFF, Cure Parkinson's Trust) have supported small proof-of-concept clinical trials of repurposed agents, whose positive results might catalyze the interest of government and industry, and encourage funding of larger, multicenter trials. In parallel, some of these organizations have devoted resources for promoting the identification of biomarkers with the aim of reducing time and costs of clinical trials by providing reliable and measurable end points of treatment effectiveness. Another strategy adopted to overcome the obstacles in PD repositioning is the creation of scientific collaborative networks and cooperation between academia, industry, and government, to share available experimental data and suggestions on drugs to speed the recognition of new repurposing opportunities.

Within this context, in 2001, the National Institute of Neurological Disorders and Stroke (NINDS) established a committee (CINAPS, Committee to Identify Neuroprotective Agents for Parkinson's) composed by experts in PD, in clinical trials, and in pharmacology with the aim of identifying repurposed agents to slow PD progression. The CINAPS group collected suggestions also from industry, academia, clinicians, and lay society (Tilley and Galpern 2007). After systematic selection with respect to specific criteria (i.e., scientific rational, blood–brain barrier penetration, and availability of safety and efficacy data), they found 12 compounds to be attractive candidates for testing by the NIH Exploratory Trials in PD (NET-PD) program (Ravina et al. 2003). One of the strong points of the NET-PD program was the development of a 1-year futility trial designed to rapidly and efficiently test novel agents, with the benefit of reducing the likelihood to advance ineffective agents into large, long-term phase III trials and to minimize the number of subjects exposed to futile treatments. Based on futility analysis, only creatine monohydrate was not found to be futile, thereby recommending the evaluation of this agent in a large, long-term trial of individuals with early, stable PD (ClinicalTrials.gov Identifier: NCT00449865, Kieburtz et al. 2015).

In the wake of the CINAPS initiative, another international committee of experts was recently involved in the Linked Clinical Trial (LCT) initiative, which described a structured approach to identify which putative treatments were to be prioritized and moved into clinical trials to accelerate new cost-effective treatments for PD. Starting from all available information about brain permeability, safety and experimental/ preclinical data, the LCT committee selected seven available candidate molecules for prioritization in PD clinical trials. The main selection included (1) simvastatin, a cholesterol-lowering drug; (2) deferiprone (DFP), an iron chelator; (3) trehalose, an autophagy inducer; and (4) drugs adopted for treatment of type 2 diabetes, such as exenatide, liraglutide, and lixisenatide (Brundin et al. 2013). Some of these molecules are already undergoing phase II clinical trials, as will be described later in this chapter.

Parallel with scientific networks, the development and *implementation of computational tools* is necessary to prioritize small molecules for drug repositioning. These methods permit the analysis of abundant information on drugs, targets, mechanisms of action, as well as disease pathways and are classified according to the type of information that is processed (i.e., target-based, knowledge-based, signature-based, pathway- or network-based, or targeted-mechanism-based) (Shameer et al. 2015). The use of these technologies to identify drugs for repositioning in PD has been reported in the literature. For example, Gao and colleagues developed a workflow with the aim of optimizing gene expression signatures reported in public databases (general and PD-specific) for subsequent use with the connectivity map, a computational tool for drug repositioning (Gao et al. 2014). Using this approach, the authors selected alvespimycin, one heat shock protein 90 inhibitor typically adopted as an antineoplastic drug, to assess its neuroprotective potential *in vitro* against rotenone-induced toxicity. The results showed the ability of this drug to inhibit cell death and to ameliorate mitochondrial respiratory dysfunction and oxidative stress triggered by 24 h rotenone exposure, suggesting it is worth further exploring its potential.

Similarly, Rakshit and collaborators recently described a bidirectional drug repositioning approach, which takes into account the significance of candidate drugs in PD-associated drug-target networks, as well as the significance of their targets in the PD-specific protein–protein interaction network (Rakshit et al. 2015). Through this method, they identified nine molecules (i.e., diethylstilbestrol, erlotinib, lidocaine, dasatinib, nifedipine, melatonin, nicardipine, sorafenib, and testosterone) as potential repositioning candidates for PD. These molecules are currently adopted for different indications, such as anesthesia, treatment of angina and hypertension, cancer (breast, prostate, and lung cancer; liver and kidney carcinoma; and leukemia), or other disorders of the central nervous system. Preclinical evidence suggests the promising role of some of these drugs (i.e., nicardipine, sorafenib) as agents for the treatment of PD further supporting the reliability of the developed computational method (Liu et al. 2011; Huang et al. 2014).

10.3 REPURPOSED DRUGS IN PD

Drug repositioning concept goes back a long way in PD. The most notable example of successful repositioning in PD dates back to the 1960s and is the case of amantadine. Initially developed as an antiviral medication to treat influenza for its ability to block virus replication, amantadine has been approved as anti-PD treatment since 1969, after demonstrating its efficacy on disease symptoms (Hubsher et al. 2012). This drug is currently used to treat patients during the earlier stages of PD and as an anti-dyskinetic agent. Its effectiveness in PD has been attributed to its DA-releasing and reuptake-inhibitory effects and weak N-methyl-D-aspartate (NMDA) antagonism.

According to the current therapeutic needs, there are *two main investigational streams* for repurposing drugs in PD: one for *ameliorating symptoms and side effects* due to pharmacological treatment, and the other for *disease modification*. In this paragraph, we will discuss examples of repurposed drugs under investigation for PD more extensively. For each drug, we will describe the biological basis and the rationale underlying repositioning, the preclinical studies, and the ongoing and/or concluded clinical trials. A summary of the listed drugs is available in Table 10.1.

10.3.1 ANTIDIABETIC DRUGS

Type 2 diabetes has a complex and heterogeneous nature. Increasing evidence shows that the disease shares pathogenic features and common molecular denominators with PD, such as impaired insulin signaling, chronic inflammation, apoptotic mechanisms, perturbed redox status, and mitochondrial dysfunctions. It is described that insulin and DA may exert reciprocal regulation and that peroxisome proliferator-activated receptor-γ (PPAR-γ), ATP-sensitive potassium channels, AMP-activated protein kinase, glucagon-like peptide-1, and dipeptidyl peptidase-4 receptors might be important therapeutic targets for PD (Lima et al. 2014). Therefore, therapeutic interventions focused on shared molecular pathogenesis, along with effective glycemic control, may provide protection from associated neurodegenerative disorders (Khan et al. 2014). Examples of drugs adopted in the treatment of type 2 diabetes and repurposed for the treatment of PD are reported as follows.

10.3.1.1 Pioglitazone

Pioglitazone is used for the treatment of type 2 diabetes either alone or in combination with sulfonylurea, metformin, or insulin and has been shown to be a specific and reversible inhibitor of human monoamine oxidase B (MAO B) (Binda et al. 2011). Furthermore, pioglitazone selectively stimulates the nuclear receptor PPAR-γ, modulates transcription of insulin-sensitive genes involved in the control of glucose and lipid metabolism as well as of antioxidant genes through nuclear factor (erythroid-derived 2)-like 2 or Nrf2, and is involved in the inhibition of a neuroinflammatory process involving microglia activation (Carta and Simuni 2015).

Neuroprotective action of pioglitazone was reported in the MPTP mouse and monkey model of PD (Breidert et al. 2002; Quinn et al. 2008; Swanson et al. 2011).

The MJFF and the NINDS are supporting a multicenter, double-blind placebo-controlled clinical trial sponsored by the University of Rochester to assess efficacy of pioglitazone treatment in patients with early PD. In the study, subjects on a stable dose

TABLE 10.1

List of Repurposed Drugs in PD

Drug Class	Molecule	Mechanism of Action	Preclinical/Clinical Evidence
Antidiabetic	Pioglitazone	Reversible inhibitor of MAO-B; stimulator of PPAR-γ, modulator of insulin-sensitive genes; anti-inflammatory	PMID:22282722 Binda PMID:26116315 Simuni PMID:25227476 Carta ClinicalTrials Identifier: NCT01280123
	Exenatide	GLP1-receptor agonist	ClinicalTrials Identifier: NCT01174810
	Metformin	Promoter of neurogenesis belonging to the biguanide family	PMID:22498320 Wahlgvist
Cholesterol-lowering	Simvastatin	Statin with anti-inflammatory and anti-aggregation properties	Kumar PMID22789904 Tison PMID23283428
Antineoplastic	Nilotinib	Tyrosine kinase inhibitor for the treatment of leukemia	ClinicalTrials Identifier: NCT02281474 PMID:27434297 Pagan
	Epothilone D	Microtubules and cytoskeleton structure destabilizer, similar to taxanes	PMID: 23670541 Cartelli
	Geldanamycin (SNX-0723 or SNX-9114) Tanespimycin (17-AAG) Alvespimycin (17-DMAG)	Hsp90 inhibitors, blockers of several oncogenic pathways and protein aggregation	Shen PMID16210323 Auluck PMID15556931 McFarland PMID24465863
Antihypertensive	Isradipine	Calcium channel blocker of the dihydropyridine class	ClinicalTrials.gov Identifier: NCT00909545 and NCT02168842
	Candesartan	Selective agonist of angiotensin II type I receptor (AT1)	Wu PMID:23774475
Psychotropic	Duloxetine	Serotonin–norepinephrine reuptake inhibitor	https://www.michaeljfox.org/foundation/grant-detail.php?grant_id=847

(Continued)

TABLE 10.1 (*Continued*)
List of Repurposed Drugs in PD

Drug Class	Molecule	Mechanism of Action	Preclinical/Clinical Evidence
Mitochondrial disorders	EPI-743	Molecule supporting cellular energy metabolism	ClinicalTrials.gov Identifier: NCT01923584
Iron chelator	DFP	Iron clearance promoter	ClinicalTrials Identifier: NCT01539837 and NCT00943748
Immunotherapy	Inosine	Nucleoside producing uric acid, natural antioxidant	ClinicalTrials Identifier: NCT00833690
Proteostasis regulators	Isofagomine Ambroxol	Chemical chaperone	PMID:25037721 Richter ClinicalTrials.gov Identifier: NCT02941822 and NCT02914366

of rasagiline (1 mg/day) or selegiline (10 mg/day) for at least 8 weeks but less than 8 months were randomized to one of the two dosages of oral pioglitazone (15 and 45 mg) or matching placebo. The primary end point adopted in the study was the change in the total Unified Parkinson's Disease Rating Scale (UPDRS) score between the baseline visit and 44 weeks. The study results showed that both doses of pioglitazone are not effective in slowing disease progression in early Parkinson's disease, suggesting that pioglitazone is not recommended for further testing for this indication (Simuni et al. 2015).

10.3.1.2 Exenatide

Exenatide is a glucagon-like peptide-1 (GLP-1) receptor agonist commonly used for treatment of type 2 diabetes. A single-blind pilot study with exenatide (ClinicalTrials. gov Identifier: NCT01174810) was started on PD patients in 2010 at University College London. The results are reported in a publication from Aviles-Olmos et al. (2013) who evaluated the progress of 45 patients (with moderate PD: 45–70 years of age and more than 5 years of disease duration), randomly assigned to receive exenatide for 12 months (two injections per day, starting with 5 μg the first month and 10 μg the following 11 months) or placebo. Their PD severity was compared after overnight withdrawal of conventional PD medication using the blinded video assessment of the Movement Disorders Society UPDRS (MDS-UPDRS), together with several nonmotor tests, at baseline, 6 months, and 12 months and after a further 2-month washout period (14 months). The authors showed that exenatide was well tolerated, although weight loss was common. Single-blinded rating of the exenatide group suggested clinically relevant improvements in PD across motor and cognitive measures compared with the control group. The authors recently reported a follow-up study showing that 24 months after their original baseline visit, that is, 12 months after cessation of exenatide, the patients treated with the antidiabetic drug were still manifesting improved motor and cognitive skills as compared to the control group (Aviles-Olmos et al. 2014).

10.3.1.3   Metformin

Metformin is in use to treat type 2 diabetes and has been investigated and proposed to promote neurogenesis and enhance the spatial memory formation (Potts and Lim 2012; Wang et al. 2012). The neuroprotective potential and positive effects on the motor performance have been recently tested in a preclinical model of PD. MPTP-injected mice were treated with oral 500 mg/kg metformin for 21 days. Dopaminergic degeneration and other markers of oxidative stress were evaluated in the midbrain, and the motor coordination and locomotor activities were assessed. Long-term metformin administration ameliorated motor performance and increased antioxidant activity and nigral neurons' survival (Patil et al. 2014).

As for human studies, indirect evidence of a protective effect of metformin in PD has been obtained on a Taiwanese population cohort affected by type 2 diabetes. The study shows that the administration of antidiabetic drug sulfonylurea alone increased the risk of developing PD, while this negative effect was normalized by the administration of metformin in combination with sulfonylurea (Wahlqvist et al. 2012). Following these data, the Cure Parkinson's Trust is working in association with the Parkinson's center in Nottingham to design a clinical study using this molecule in the treatment of PD.

10.3.2   CHOLESTEROL-LOWERING DRUGS

Epidemiological studies have shown that the use of statins or cholesterol-lowering drugs is associated with lower PD risk (Wahner et al. 2008; Friedman et al. 2013; Lee et al. 2013), suggesting a potential neuroprotective role for these drugs in PD. Moreover, *in vitro* and *in vivo* evidence of the anti-inflammatory and antioxidant properties of this drug class (Malfitano et al. 2014), as well as its ability to reduce aggregation-prone α-synuclein accumulation (Bar-On et al. 2008; Koob et al. 2010), further supported the concept that statins might have a therapeutic use in PD. Preclinical evidence has confirmed the ability of statins to confer neuroprotection against the parkinsonian neurotoxin MPTP (Ghosh et al. 2009; Aguirre-Vidal et al. 2015). Studies performed in animal models of PD have also demonstrated that statins, in particular simvastatin treatment, improved PD-like symptoms (Kumar et al. 2012) and long-term memory performance (Wang et al. 2014), and reduced dyskinesia (Tison et al. 2013). However, the beneficial role of statin use in PD is still a matter of debate as highlighted in a very recent report suggesting that higher total or low-density lipoprotein cholesterol levels, rather than statin use, might be associated with risk reduction of developing PD (Huang et al. 2015). Therefore, although statins have interesting possibilities for the prevention and treatment of PD, further evidence from clinical studies is necessary before drawing any definitive conclusion about their efficacy.

10.3.3   ANTINEOPLASTIC AGENTS

The association between PD and the risk of developing cancer has been discussed and explored in large cohort studies. Overall, these studies highlighted a lower incidence of developing cancer in PD patients as compared to the general population (Devine et al. 2011). However, some types of tumors are reported as being moderately frequent in PD. One study conducted on the Utah population (Kareus et al. 2012), as well as analyses based on English and Swedish health registries (Wirdefeldt et al. 2013; Ong et al. 2014), showed that PD patients and their relatives have a higher risk of developing melanoma. In these studies, the authors also identified a higher risk of prostate, breast, uterine, and renal cancer. The evolving progress in understanding the genetic basis for PD showed that the pathogenic mechanisms involved in this disease share striking similarities to those underlying many cancers. The considerable overlap between cancer and PD suggests that studies in the cancer context might provide a novel insight into the role of PD-associated genes, and vice versa, permitting the identification of new therapeutic agents or supporting the repurposing of antineoplastic drugs in PD.

10.3.3.1   Nilotinib

Nilotinib is a molecule adopted for the treatment of leukemia. Preclinical *in vivo* studies showed that nilotinib decreases brain and peripheral α-synuclein levels reducing, at the same time, neuroinflammation in an animal model of α-synucleinopathy (Hebron et al. 2013). In addition, nilotinib ameliorates motor deficits and counteracts loss of nigral dopaminergic neurons in the MPTP mouse model of PD (Karuppagounder et al. 2014; Tanabe et al. 2014). In line with this evidence, a pilot clinical study to test nilotinib's ability to alter the abnormal protein buildup in in patients with PD and LB diseases has been recently completed (ClinicalTrials.gov Identifier: NCT02281474). In this study, nilotinib efficacy was also determined in terms of improvement of motor and nonmotor symptoms and changes in neuroinflammatory markers. Collectively, data from this study suggested that low doses of Nilotinib are relatively safe and tolerated in subjects with advanced PD or dementia with Lewy bodies supporting that further clinical trials are performed to determine an appropriate therapeutic dose and evaluate Nilotinib effects in a larger cohort of patients (Pagan et al. 2016).

10.3.3.2   Epothilone D

Drugs targeting microtubules represent one of the most commonly prescribed therapies for cancer. Epothilones are a novel class of microtubule-stabilizing agents with a broad range of antitumor activity associated with tolerable side effects (Goodin et al. 2004). In PD, the concept that microtubules' dysfunction can participate in, and perhaps lead to disease progression, has been suggested by the observation that some PD-linked proteins, such as parkin, LRRK2, and α-synuclein, are able to modulate the stability of microtubules (Cartelli et al. 2012; Cappelletti et al. 2015). In addition, evidence from both the cellular and animal models of PD showed that neural functions might be affected as a consequence of microtubule and axonal transport deficits (Brunden et al. 2014). In line with these findings, in 2012, the MJFF funded a project for target validation under Prof. Brunden coordination at the University of Pennsylvania in Philadelphia to evaluate whether epothilone D provides benefit in cell culture and transgenic preclinical models of PD. The results from this study indicate that epothilone D treatment did not affect the changes in microtubule structure observed in the cell culture model, or improve outcomes in the model with LB-like aggregates. Conversely, another study evaluating the effect of epothilone D in the MPTP mouse model of PD demonstrated that repeated daily administration of this agent reduces the number of fibers with altered mitochondria distribution and cytoskeleton organization and attenuates nigrostriatal degeneration following MPTP administration (Cartelli et al. 2013). Consistent with the lack of strong preclinical evidence, no clinical trials in PD patients are currently ongoing for this agent.

10.3.4   HEAT SHOCK PROTEIN 90 INHIBITORS: GELDANAMYCIN, TANESPIMYCIN, AND ALVESPIMYCIN

Heat shock protein 90 (Hsp90) is a ubiquitous molecular chaperone critical for maintaining the functional stability and viability of the cells in pathogenic conditions. Based on the biological functions played by Hsp90, the inhibition of Hsp90 activity can have a dual potential for treatment of neurodegenerative diseases. First, the inhibition of Hsp90 activates the heat shock factor-1 to induce the production of other chaperones that promote protein degradation and disaggregation. In addition, Hsp90 inhibition results in a reduction in the expression and activity of aberrant neuronal protein, namely, "clients" that support the growth of tumor cells as well as the progression of several neurological disorders (Luo et al. 2010). For these reasons, several Hsp90 inhibitors are currently under investigation for neurodegenerative disorders, including PD (Adachi et al. 2009). *Geldanamycin*, a benzoquinone ansamycin antibiotic, is a selective inhibitor of Hsp90 that has been shown to protect against neurotoxicity induced by MPTP in mice (Shen et al. 2005) and by α-synuclein in flies (Auluck et al. 2005). However, due to poor aqueous solubility, inadequate blood–brain barrier permeability, and liver toxicity, derivatives and analogs of geldanamycin with fewer side effects have been adopted. Examples are represented by *tanespimycin* or 17-AAG and *alvespimycin* or 17-DMAG (Ebrahimi-Fakhari et al. 2011). A recent preclinical study based on the chronic administration of the novel synthetic analogs of geldanamycin (i.e., SNX-0723 or the more potent SNX-9114) showed that, despite being ineffective on nigral neurotoxicity, these molecules were well tolerated and exerted a positive neuromodulatory effect on striatal DA levels in rats overexpressing α-synuclein in the SNc (McFarland et al. 2014).

10.3.5   ANTIHYPERTENSIVE DRUGS

The role of arterial hypertension as risk factor for PD is still debated (Mazza et al. 2013). Different evidence has suggested a role for L-type calcium channels and the central renin–angiotensin system (RAS) in PD. L-type calcium channels not only contribute to the electrical activity of dopaminergic neurons, but also seem to render these neurons particularly vulnerable to degeneration by increasing intracellular oxidative stress (Surmeier et al. 2010). In parallel, angiotensin II, the major product of RAS, elicits pro-inflammatory actions potentially resulting in neuronal death. This evidence suggested that antihypertensive agents, especially *angiotensin receptor blockers*, *inhibitors of angiotensin-converting enzyme*, and *calcium channel blockers* (CCBs), may have possible neuroprotective effects in PD (Lee et al. 2014).

10.3.5.1   Isradipine

Isradipine is a CCB of the dihydropyridine class currently approved to treat high blood pressure and to reduce the risk of stroke and heart attack. *In vivo* preclinical studies showed the neuroprotective properties of the systemic administration of israpidine (Meredith et al. 2008; Ilijic et al. 2011). In 2008, a phase II clinical study in PD was supported by the MJFF, with the goal to establish dosage and tolerability of isradipine controlled-release (CR) and to demonstrate preliminary efficacy for use in a future pivotal efficacy study (Safety, Tolerability and Efficacy Assessment of DynaCirc CR in Parkinson Disease—STEADY-PD). This study was completed in 2012 (ClinicalTrials.gov Identifier: NCT00909545). Briefly, a randomized, double-blind parallel group trial was undertaken in subjects with early PD not taking dopaminergic therapy (DA agonists or L-dopa). Patients were randomized equally in four treatment groups and exposed to a daily oral administration of 5, 10, or 20 mg of isradipine CR or matching placebo. The tolerability of isradipine was dose dependent, and 10 mg was the maximal tolerable dosage in this study of early PD. There was no significant difference between treatments and placebo, although a trend toward increasing benefit at higher dosages was observed (Parkinson's Study Group 2013).

More recently, the NINDS in collaboration with the MJFF funded a phase III study to evaluate the disease-modifying potential of isradipine in subjects with early PD (ClinicalTrials.gov Identifier: NCT02168842). The study is still sponsored by the Parkinson Study Group and is co-led by the University of Rochester Medical Center and Northwestern University.

10.3.5.2 Candesartan

Candesartan is a selective antagonist of the angiotensin II type 1 receptor (AT1). The pro-drug (candesartan cilexetil) is adopted for the treatment of hypertension, chronic heart failure, and left ventricular systolic dysfunction. At the preclinical level, candesartan treatment has been shown to decrease rotenone-induced, dopaminergic neuronal death by blocking endothelial reticulum stress (Muñoz et al. 2014) and to significantly reduce L-dopa induced-dyskinesia (Wu et al. 2013). No clinical studies were found in which candesartan was evaluated in PD patients probably because the market has recently turned its attention on the generic, cheaper, and equally efficacious losartan (Grosso et al. 2011). Losartan is protective against α-synuclein toxicity and reduces protein aggregation *in vitro*, supporting its potential in the prevention or treatment of PD (Grammatopoulos et al. 2007).

10.3.6 Psychotropic Agents

Psychotropic agents are a drug class that have effects on psychological function and include antipsychotics, mood stabilizers, antianxiety agents, and antidepressants. Psychotropic agents are generally adopted in PD for the treatment of psychotic disorders that possibly arise in the patients as a consequence of the DA-based medication, and of emotional changes such as anxiety and depression that affect many parkinsonian subjects. Repositioning psychotropic drugs as disease-modifying agents in PD is based on growing evidence showing that these drugs have neurotrophic and neuroprotective properties, likely based on their ability of promoting neurogenesis, increasing the production of neurotrophic factors such as the brain-derived neurotrophic factor, and inducing changes in signaling pathways involved in cell survival and proliferation, both *in vitro* and in animal models of neurodegeneration (Hunsberger et al. 2009).

Duloxetine is a serotonin–norepinephrine reuptake inhibitor prescribed for major depressive disorder and generalized anxiety disorder. It is also approved for use in osteoarthritis and musculoskeletal pain (Bellingham and Peng 2015). In 2011, the MJFF funded a preclinical study to test whether duloxetine hydrochloride has potential to be preventive against the midbrain dopaminergic cell loss typical of PD. The study results showed that the daily systemic administration of 20 mg/kg for 2 weeks upregulated the expression of transcription factors FoxA2 and En-1, which are important for dopaminergic neuron survival, in the SNc of preclinical models of PD. However, duloxetine hydrochloride treatment did not result in a significant increase in nigral neurons' survival and improvement in behavior and motor functions. The results are briefly described on the MJFF website, but no publication is available yet.

10.3.7   EPI-743 AND MITOCHONDRIAL DYSFUNCTION

EPI-743 is a small molecule developed for the treatment of inherited respiratory chain diseases, such as Leigh syndrome. This molecule is a para-benzoquinone increasing endogenous glutathione biosynthesis, necessary for the control of oxidative stress (Enns et al. 2012; Martinelli et al. 2012). Since mitochondrial dysfunctions are a feature of both genetic and nongenetic forms of PD, EPI-73 has been suggested as a potential drug for repurposing. Cure Parkinson's Trust is working in partnership with Edison Pharma to support phase II clinical trials with EPI-743 on PD patients (ClinicalTrials. gov Identifier: NCT01923584). This double-blind, placebo-controlled clinical trial is currently underway to primarily assess the effects of the molecule on visual and neurological function. The study has been completed but results are not yet available.

10.3.8   DEFERIPRONE AND IRON ACCUMULATION

Deferiprone (DFP) is an iron chelator used for the treatment of iron overload in thalassemia or other iron imbalance conditions. DFP works by attaching to iron and promoting its clearance and excretion from the body (Kontoghiorghes et al. 2003). This agent has been selected by the LCT committee as a high-priority candidate for repositioning in PD. Indeed, preclinical studies have shown its neuroprotective action against MPP+ *in vitro* and in animal models of PD in which systemic chelator administration attenuated the loss of dopaminergic neurons and striatal DA content, and oxidative stress response (Dexter et al. 2011; Devos et al. 2014). The effectiveness of DFP on early-stage PD patients stabilized on DA regimens has been assessed in a pilot, double-blind placebo-controlled randomized clinical trial (ClinicalTrials.gov Identifier: NCT00943748). Patients were enrolled in a 12-month single-center study and treated with DFP (30 mg/kg/day) by applying an early-start or a 6-month delayed-start paradigm. The results showed that early-start patients compared to delayed-start patients responded significantly earlier and sustainably to treatment both in terms of nigral iron deposit reduction and amelioration of UPDRS scores (Devos et al. 2014). In addition, a pilot phase II clinical trial with DFP was started in 2012 at Imperial College London (ClinicalTrials.gov Identifier: NCT01539837). In this study, three groups of early-stage drug-free PD patients were treated with 20 or 30 mg/kg/day DFP or placebo for 6 months. Over the 6 months, the patients received serial MRI scans, neurological examinations to assess PD symptoms as well as psychological state, and blood test to monitor for potential side effects. No information is currently available from this clinical study; however positive results from this pilot might support larger clinical trials to evaluate DFP as a disease-modifying drug in PD.

10.3.9 INOSINE AND OXIDATIVE STRESS

Inosine is a nucleoside employed for the treatment of certain viral infections and as an immunostimulant. It possesses anti-inflammatory and regenerative properties suggesting a potential therapeutic role in neurodegenerative diseases (Haskó et al. 2004). However, researchers became interested in this agent when it was discovered that high levels of uric acid—the ultimate catabolic end product of inosine—were associated with a decreased risk of developing PD (Shen et al. 2013). According to this evidence, a randomized, double-blind placebo-controlled dose-ranging trial (ClinicalTrials.gov Identifier: NCT00833690) of oral inosine was sponsored by the MJFF and the Parkinson Study Group in collaboration with Massachusetts General Hospital, the Harvard School of Public Health, and the University of Rochester. The aim was to assess inosine safety and ability to elevate urate levels in blood and cerebral spinal fluid in early PD patients. The study concluded that inosine was generally safe, tolerable, and effective in raising urate levels in these patients, supporting a more definitive development of inosine as a potential disease-modifying therapy for PD (Schwarzschild et al. 2014).

10.3.10 ISOFAGOMINE AND PROTEOSTASIS

AT2101 (isofagomine tartrate) was presented in preclinical studies as a monotherapy and in combination with enzyme replacement therapy for Gaucher disease (GD) by Amicus Therapeutics. AT2101 is a small pharmacological chaperone that targets the glucocerebrosidase enzyme deficient in GD, a mechanism with the potential to address both GD and PD. In fact, GD is due to misfolding and loss of function of the lysosomal enzyme glucocerebrosidase, encoded by the *GBA1* gene, which has been identified as the main genetic risk factor in PD (see Paragraph 1; Schapira and Gegg 2013). Recent evidence showed that oral administration of the pharmacological chaperone AT2101 for 4 months to mice overexpressing human wild-type α-synuclein (Thy1-aSyn mice) improved motor and nonmotor functions, abolished microglial inflammatory response in the SNc, reduced α-synuclein immunoreactivity in nigral dopaminergic neurons, and reduced the number of small α-synuclein aggregates while increasing the number of large α-synuclein aggregates (Richter et al. 2014). This study was originally sponsored by a grant from the MJFF and included the evaluation of the anti-aggregation, therapeutic potential of both *isofagomine* and *ambroxol*. The latter one is a known expectorant with anti-inflammatory properties identified also as a molecular chaperone for glucocerebrosidase in GD. Two clinical trials are currently ongoing to evaluate the safety, tolerability, and pharmacodynamics of ambroxol in subjects with PD (ClinicalTrials.gov Identifier: NCT02941822) and its ability to improve cognitive and motor symptoms in patients with Parkinson's disease dementia (PDD) (ClinicalTrials.gov Identifier: NCT02914366). In fact, preclinical evidence provides interesting clues for the investigation of molecular chaperones as disease-modifying drugs in PD. Together with ambroxol (McNeill et al. 2014; Ambrosi et al. 2015), more examples are represented by ursodeoxycholic acid (Mortiboys et al. 2013), celastrol (Choi et al. 2014), and 4-phenylbutyrate (Ono et al. 2009).

10.4   FUTURE OF REPOSITIONING IN PD

The search for innovative technologies to test and provide preclinical proof-of-concept evidence for the use of known drugs in PD is constantly in motion.

Parkure, a company supported by the Royal Society of Edinburgh has refined the use of fruit flies as a platform for high-throughput screening of drugs against PD, with a focus on those previously tested for safety (website: http://parkure.co.uk/). This technology allows to quickly test a large number of chemicals increasing the probability of discovering new effective drugs.

Another interesting approach for the identification of novel therapeutic targets in PD is based on the ability of drugs to counteract the formation of α-synuclein aggregates, the main component of LBs, or to promote their clearance. The approach was developed by Dr. Herva and funded by the National Center for the Replacement, Refinement, and Reduction of Animals in Research in Cambridge (Herva et al. 2014). Dr. Herva adapted a method typically used to amplify prion aggregates, called the protein misfolding cyclic amplification or PMCA, to develop a fast and reproducible system to induce α-synuclein aggregation. The PMCA α-synuclein aggregates are used as a substrate in a high-throughput platform to assess the anti-aggregation potential of different drugs. Interestingly, the PMCA aggregates can be used to "infect" cells and develop cellular models of synucleinopathy, which can also be used as a substrate for drug screening (Herva et al. 2014).

Recently, a microRNA (miRNA)-driven computational model has been proposed to predict associations between drugs and diseases for drug repositioning (Chen and Zhang 2015). This model combines experimentally confirmed drug–miRNA associations and disease–miRNA associations. miRNAs are a class of regulatory small RNAs involved in the post-transcriptional regulation of target gene expression that play a critical role in many biological processes such as tissue development, cell growth, and cellular signaling. The importance of miRNA machinery in PD arises from the observation that specific miRNAs are implicated in the control of pathogenic proteins such as α-synuclein, as well as have a key role in DA neuron biology (Mouradian 2012). Based on this approach, 23 predicted drug–disease associations have been found for PD, including two antineoplastic agents such as doxorubicin and trastuzumab. Despite the encouraging results, the use of this method is currently limited by the lack of complete data on drug–miRNA and miRNA–disease associations.

In conclusion, the search for symptomatic and disease-modifying therapies that best suit PD patients' needs is currently following new routes that include the repurposing of marketed drugs, whose safety and tolerability are ascertained. As described in this chapter, phase II clinical trials testing repurposed drugs in PD are ongoing for a few candidates. However, it will be essential to expand this investigation by implementing computational and biological high-throughput screening systems, as well as through the development of collaborative interaction among government entities, funding companies, academia, researchers, nonprofit organizations, and patient associations, with the aim of accelerating the identification of key therapeutic molecules and reducing the burden of trial costs.

REFERENCES

Adachi, H., M. Katsuno, M. Waza, M. Minamiyama, F. Tanaka, and G. Sobue. Heat shock proteins in neurodegenerative diseases: Pathogenic roles and therapeutic implications. *International Journal of Hyperthermia: The Official Journal of European Society for Hyperthermic Oncology, North American Hyperthermia Group* 25 (8) (2009): 647–654.

Aguirre-Vidal, Y., S. Montes, L. Tristan-López et al. The neuroprotective effect of lovastatin on MPP(+)-induced neurotoxicity is not mediated by PON2. *Neurotoxicology* 48 (2015): 166–170.

Ambrosi, G., S. Cerri, and F. Blandini. A further update on the role of excitotoxicity in the pathogenesis of Parkinson's disease. *Journal of Neural Transmission (Vienna, Austria: 1996)* 121 (8) (2014): 849–859.

Ambrosi, G., C. Ghezzi, R. Zangaglia, G. Levandis, C. Pacchetti, and F. Blandini. Ambroxol-induced rescue of defective glucocerebrosidase is associated with increased LIMP-2 and Saposin C levels in GBA1 mutant Parkinson's disease cells. *Neurobiology of Disease* 82 (2015): 235–242.

Auluck, P.K., M.C. Meulener, and N.M. Bonini. Mechanisms of suppression of {alpha}-synuclein neurotoxicity by geldanamycin in *Drosophila*. *The Journal of Biological Chemistry* 280 (4) (2005): 2873–2878.

Aviles-Olmos, I., J. Dickson, Z. Kefalopoulou et al. Exenatide and the treatment of patients with Parkinson's disease. *The Journal of Clinical Investigation* 123 (6) (2013): 2730–2736.

Aviles-Olmos, I., J. Dickson, Z. Kefalopoulou et al. Motor and cognitive advantages persist 12 months after exenatide exposure in Parkinson's disease. *Journal of Parkinson's Disease* 4 (3) (2014): 337–344.

Bar-On, P., L. Crews, A.O. Koob et al. Statins reduce neuronal alpha-synuclein aggregation in in vitro models of Parkinson's disease. *Journal of Neurochemistry* 105 (5) (2008): 1656–1667.

Beach, T.G., C.H. Adler, L.I. Sue et al. Multi-organ distribution of phosphorylated alpha-synuclein histopathology in subjects with Lewy body disorders. *Acta Neuropathologica* 119 (6) (2010): 689–702.

Bellingham, G.A. and P.W.H. Peng. Duloxetine: A review of its pharmacology and use in chronic pain management. *Regional Anesthesia and Pain Medicine* 35 (3) (2015): 294–303.

Binda, C., M. Aldeco, W.J. Geldenhuys, M. Tortorici, A. Mattevi, and D.E. Edmondson. Molecular insights into human monoamine oxidase B inhibition by the glitazone anti-diabetes drugs. *ACS Medicinal Chemistry Letters* 3 (1) (2011): 39–42.

Blandini, F. Neural and immune mechanisms in the pathogenesis of Parkinson's disease. *Journal of Neuroimmune Pharmacology: The Official Journal of the Society on NeuroImmune Pharmacology* 8 (1) (2013): 189–201.

Blandini, F. and M.-T. Armentero. Animal models of Parkinson's disease. *The FEBS Journal* 279 (7) (2012): 1156–1166.

Blandini, F. and M.-T. Armentero. Dopamine receptor agonists for Parkinson's disease. *Expert Opinion on Investigational Drugs* 23 (3) (2014): 387–410.

Blandini, F., G. Nappi, C. Tassorelli, and E. Martignoni. Functional changes of the Basal Ganglia circuitry in Parkinson's disease. *Progress in Neurobiology* 62 (1) (2000): 63–88.

Breidert, T., J. Callebert, M.T. Heneka, G. Landreth, J.M. Launay, and E.C. Hirsch. Protective action of the peroxisome proliferator-activated receptor-gamma agonist pioglitazone in a mouse model of Parkinson's disease. *Journal of Neurochemistry* 82 (3) (2002): 615–624.

Brunden, K.R., J.Q. Trojanowski, A.B. Smith, V.M.-Y. Lee, and C. Ballatore. Microtubule-stabilizing agents as potential therapeutics for neurodegenerative disease. *Bioorganic & Medicinal Chemistry* 22 (18) (2014): 5040–5049.

Brundin, P., R.A. Barker, P. Jeffrey Conn et al. Linked clinical trials—The development of new clinical learning studies in Parkinson's disease using screening of multiple prospective new treatments. *Journal of Parkinson's Disease* 3 (3) (2013): 231–239.

Canet-Avilés, R.M., M.A. Wilson, D.W. Miller et al. The Parkinson's disease protein DJ-1 is neuroprotective due to cysteine-sulfinic acid-driven mitochondrial localization. *Proceedings of the National Academy of Sciences of the United States of America* 101 (24) (2004): 9103–9108.

Cappelletti, G., F. Casagrande, A. Calogero, C. De Gregorio, G. Pezzoli, and D. Cartelli. Linking microtubules to Parkinson's disease: The case of parkin. *Biochemical Society Transactions* 43 (2) (2015): 292–296.

Carta, A.R. and T. Simuni. Thiazolidinediones under preclinical and early clinical development for the treatment of Parkinson's disease. *Expert Opinion on Investigational Drugs* 24 (2) (2015): 219–227.

Cartelli, D., F. Casagrande, C.L. Busceti et al. Microtubule alterations occur early in experimental parkinsonism and the microtubule stabilizer epothilone D is neuroprotective. *Scientific Reports* 3 (2013): 1837.

Cartelli, D., S. Goldwurm, F. Casagrande, G. Pezzoli, and G. Cappelletti. Microtubule destabilization is shared by genetic and idiopathic Parkinson's disease patient fibroblasts. *PLoS One* 7 (5) (2012): e37467.

Chan, C.S., T.S. Gertler, and D.J. Surmeier. Calcium homeostasis, selective vulnerability and Parkinson's disease. *Trends in Neurosciences* 32 (5) (2009): 249–256.

Chen, H. and Z. Zhang. A miRNA-driven inference model to construct potential drug-disease associations for drug repositioning. *BioMed Research International* 2015 (2015): 406463.

Choi, B.-S., H. Kim, H.J. Lee et al. Celastrol from 'Thunder God Vine' protects SH-SY5Y cells through the preservation of mitochondrial function and inhibition of P38 MAPK in a rotenone model of Parkinson's disease. *Neurochemical Research* 39 (1) (2014): 84–96.

Cook, C., C. Stetler, and L. Petrucelli. Disruption of protein quality control in Parkinson's disease. *Cold Spring Harbor Perspectives in Medicine* 2 (5) (2012): a009423.

Dawson, T.M. Parkin and defective ubiquitination in Parkinson's disease. *Journal of Neural Transmission. Supplementum* 70 (2006): 209–213.

Devine, M.J., H. Plun-Favreau, and N.W. Wood. Parkinson's disease and cancer: Two wars, one front. *Nature Reviews Cancer* 11 (11) (2011): 812–823.

Devos, D., C. Moreau, J.C. Devedjian et al. Targeting chelatable iron as a therapeutic modality in Parkinson's disease. *Antioxidants & Redox Signaling* 21 (2) (2014): 195–210.

Dexter, D.T., S.A. Statton, C. Whitmore et al. Clinically available iron chelators induce neuroprotection in the 6-OHDA model of Parkinson's disease after peripheral administration. *Journal of Neural Transmission (Vienna, Austria: 1996)* 118 (2) (2011): 223–231.

Ebrahimi-Fakhari, D., L. Wahlster, and P.J. McLean. Molecular chaperones in Parkinson's disease—Present and future. *Journal of Parkinson's Disease* 1 (4) (2011): 299–320.

Enns, G.M., S.L. Kinsman, S.L. Perlman et al. Initial experience in the treatment of inherited mitochondrial disease with EPI-743. *Molecular Genetics and Metabolism* 105 (1) (2012): 91–102.

Ferrer, I., I. López-Gonzalez, M. Carmona, E. Dalfó, A. Pujol, and A. Martínez. Neurochemistry and the non-motor aspects of PD. *Neurobiology of Disease* 46 (3) (2012): 508–526.

Franco-Iborra, S., M. Vila, and C. Perier. The Parkinson disease mitochondrial hypothesis: Where are we at? *The Neuroscientist* 22 (3) (2016): 266–277.

Friedman, B., A. Lahad, Y. Dresner, and S. Vinker. Long-term statin use and the risk of Parkinson's disease. *The American Journal of Managed Care* 19 (8) (2013): 626–632.

Gao, L., G. Zhao, J.-S. Fang, T.-Y. Yuan, A.-L. Liu, and G.-H. Du. Discovery of the neuroprotective effects of alvespimycin by computational prioritization of potential anti-Parkinson agents. *The FEBS Journal* 281 (4) (2014): 1110–1122.

Ghosh, A., A. Roy, J. Matras, S. Brahmachari, H.E. Gendelman, and K. Pahan. Simvastatin inhibits the activation of P21ras and prevents the loss of dopaminergic neurons in a mouse model of Parkinson's disease. *The Journal of Neuroscience: The Official Journal of the Society for Neuroscience* 29 (43) (2009): 13543–13556.

González, H., D. Elgueta, A. Montoya, and R. Pacheco. Neuroimmune regulation of microglial activity involved in neuroinflammation and neurodegenerative diseases. *Journal of Neuroimmunology* 274 (1–2) (2014): 1–13.

Goodin, S., M.P. Kane, and E.H. Rubin. Epothilones: Mechanism of action and biologic activity. *Journal of Clinical Oncology: Official Journal of the American Society of Clinical Oncology* 22 (10) (2004): 2015–2025.

Grammatopoulos, T.N., T.F. Outeiro, B.T. Hyman, and D.G. Standaert. Angiotensin II protects against alpha-synuclein toxicity and reduces protein aggregation in vitro. *Biochemical and Biophysical Research Communications* 363 (3) (2007): 846–851.

Grosso, A.M., P.N. Bodalia, R.J. Macallister, A.D. Hingorani, J.C. Moon, and M.A. Scott. Comparative clinical- and cost-effectiveness of candesartan and losartan in the management of hypertension and heart failure: A systematic review, meta- and cost-utility analysis. *International Journal of Clinical Practice* 65 (3) (2011): 253–263.

Haskó, G., M.V. Sitkovsky, and C. Szabó. Immunomodulatory and neuroprotective effects of inosine. *Trends in Pharmacological Sciences* 25 (3) (2004): 152–157.

Hayashita-Kinoh, H., M. Yamada, T. Yokota, Y. Mizuno, and H. Mochizuki. Down-regulation of alpha-synuclein expression can rescue dopaminergic cells from cell death in the substantia nigra of Parkinson's disease rat model. *Biochemical and Biophysical Research Communications* 341 (4) (2006): 1088–1095.

Hebron, M.L., I. Lonskaya, and C.E.-H. Moussa. Nilotinib reverses loss of dopamine neurons and improves motor behavior via autophagic degradation of A-synuclein in Parkinson's disease models. *Human Molecular Genetics* 22 (16) (2013): 3315–3328.

Herva, M.E., S. Zibaee, G. Fraser, R.A. Barker, M. Goedert, and M.G. Spillantini. Anti-amyloid compounds inhibit A-synuclein aggregation induced by protein misfolding cyclic amplification (PMCA). *The Journal of Biological Chemistry* 289 (17) (2014): 11897–11905.

Huang, B.-R., P.-C. Chang, W.-L. Yeh et al. Anti-neuroinflammatory effects of the calcium channel blocker nicardipine on microglial cells: Implications for neuroprotection. *PLoS One* 9 (3) (2014): e91167.

Huang, X., A. Alonso, X. Guo et al. Statins, plasma cholesterol, and risk of parkinson's disease: A prospective study. *Movement Disorders: Official Journal of the Movement Disorder Society* 30 (4) (2015): 552–559.

Hubsher, G., M. Haider, and M.S. Okun. Amantadine: The journey from fighting flu to treating Parkinson disease. *Neurology* 78 (14) (2012): 1096–1099.

Hunsberger, J., D.R. Austin, I.D. Henter, and G. Chen. The neurotrophic and neuroprotective effects of psychotropic agents. *Dialogues in Clinical Neuroscience* 11 (3) (2009): 333–348.

Ilijic, E., J.N. Guzman, and D.J. Surmeier. The L-type channel antagonist isradipine is neuroprotective in a mouse model of Parkinson's disease. *Neurobiology of Disease* 43 (2) (2011): 364–371.

Kareus, S.A., K.P. Figueroa, L.A. Cannon-Albright, and S.M. Pulst. Shared predispositions of parkinsonism and cancer: A population-based pedigree-linked study. *Archives of Neurology* 69 (12) (2012): 1572–1577.

Karuppagounder, S.S., S. Brahmachari, Y. Lee, V.L. Dawson, T.M. Dawson, and H.S. Ko. The c-Abl inhibitor, nilotinib, protects dopaminergic neurons in a preclinical animal model of Parkinson's disease. *Scientific Reports* 4 (2014): 4874.

Khan, N.M., A. Ahmad, R.K. Tiwari, M.A. Kamal, G. Mushtaq, and G.M. Ashraf. Current challenges to overcome in the management of type 2 diabetes mellitus and associated neurological disorders. *CNS & Neurological Disorders Drug Targets* 13 (8) (2014): 1440–1457.

Kieburtz, K., B.C. Tilley, J.J. Elm et al. Effect of creatine monohydrate on clinical progression in patients with Parkinson disease: A randomized clinical trial. *JAMA* 313 (6) (2015): 584–593.

Kontoghiorghes, G.J., K. Neocleous, and A. Kolnagou. Benefits and risks of deferiprone in iron overload in thalassaemia and other conditions: Comparison of epidemiological and therapeutic aspects with deferoxamine. *Drug Safety* 26 (8) (2003): 553–584.

Koob, A.O., K. Ubhi, J.F. Paulsson et al. Lovastatin ameliorates alpha-synuclein accumulation and oxidation in transgenic mouse models of alpha-synucleinopathies. *Experimental Neurology* 221 (2) (2010): 267–274.

Kumar, A., N. Sharma, A. Gupta, H. Kalonia, and J. Mishra. Neuroprotective potential of atorvastatin and simvastatin (HMG-CoA reductase inhibitors) against 6-hydroxydopamine (6-OHDA) induced Parkinson-like symptoms. *Brain Research* 1471 (2012): 13–22.

Lee, Y.-C., C.-H. Lin, R.-M. Wu et al. Discontinuation of statin therapy associates with Parkinson disease: A population-based study. *Neurology* 81 (5) (2013): 410–416.

Lee, Y.-C., C.-H. Lin, R.-M. Wu, J.-W. Lin, C.-H. Chang, and M.-S. Lai. Antihypertensive agents and risk of Parkinson's disease: A nationwide cohort study. *PLoS One* 9 (6) (2014): e98961.

Lesage, S. and A. Brice. Parkinson's disease: From monogenic forms to genetic susceptibility factors. *Human Molecular Genetics* 18 (R1) (2009): R48–R59.

Lima, M.M.S., A.D.S. Targa, A.C.D. Noseda et al. Does Parkinson's disease and type-2 diabetes mellitus present common pathophysiological mechanisms and treatments? *CNS & Neurological Disorders Drug Targets* 13 (3) (2014): 418–428.

Liu, Z., S. Hamamichi, B.D. Lee et al. Inhibitors of LRRK2 kinase attenuate neurodegeneration and Parkinson-like phenotypes in *Caenorhabditis elegans* and *Drosophila* Parkinson's disease models. *Human Molecular Genetics* 20 (20) (2011): 3933–3942.

Luo, W., W. Sun, T. Taldone, A. Rodina, and G. Chiosis. Heat shock protein 90 in neurodegenerative diseases. *Molecular Neurodegeneration* 5 (2010): 24.

Macchi, B., R. di Paola, F. Marino-Merlo, M.R. Felice, S. Cuzzocrea, and A. Mastino. Inflammatory and cell death pathways in brain and peripheral blood in Parkinson's disease. *CNS & Neurological Disorders Drug Targets* 14 (3) (2015): 313–324.

Malfitano, A.M., G. Marasco, M.C. Proto, C. Laezza, P. Gazzerro, and M. Bifulco. Statins in neurological disorders: An overview and update. *Pharmacological Research: The Official Journal of the Italian Pharmacological Society* 88 (2014): 74–83.

Martinelli, D., M. Catteruccia, F. Piemonte et al. EPI-743 reverses the progression of the pediatric mitochondrial disease—Genetically defined leigh syndrome. *Molecular Genetics and Metabolism* 107 (3) (2012): 383–388.

Masliah, E., E. Rockenstein, I. Veinbergs et al. Dopaminergic loss and inclusion body formation in alpha-synuclein mice: Implications for neurodegenerative disorders. *Science (New York)* 287 (5456) (2000): 1265–1269.

Mazza, A., R. Ravenni, A. Antonini, E. Casiglia, D. Rubello, and P. Pauletto. Arterial hypertension, a tricky side of Parkinson's disease: Physiopathology and therapeutic features. *Neurological Sciences: Official Journal of the Italian Neurological Society and of the Italian Society of Clinical Neurophysiology* 34 (5) (2013): 621–627.

McFarland, N.R., H. Dimant, L. Kibuuka et al. Chronic treatment with novel small molecule Hsp90 inhibitors rescues striatal dopamine levels but not A-synuclein-induced neuronal cell loss. *PLoS One* 9 (1) (2014): e86048.

McNeill, A., J. Magalhaes, C. Shen et al. Ambroxol improves lysosomal biochemistry in glucocerebrosidase mutation-linked Parkinson disease cells. *Brain: A Journal of Neurology* 137 (Pt 5) (2014): 1481–1495.

Meredith, G.E., S. Totterdell, J.A. Potashkin, and D. James Surmeier. Modeling PD pathogenesis in mice: Advantages of a chronic MPTP protocol. *Parkinsonism & Related Disorders* 14 (Suppl. 2) (2008): S112–S115.

Moretto, A. and C. Colosio. The role of pesticide exposure in the genesis of Parkinson's disease: Epidemiological studies and experimental data. *Toxicology* 307 (2013): 24–34.

Mortiboys, H., J. Aasly, and O. Bandmann. Ursocholanic acid rescues mitochondrial function in common forms of familial Parkinson's disease. *Brain: A Journal of Neurology* 136 (Pt 10) (2013): 3038–3050.

Mouradian, M.M. MicroRNAs in Parkinson's disease. *Neurobiology of Disease* 46 (2) (2012): 279–284.

Muñoz, A., P. Garrido-Gil, A. Dominguez-Meijide, and J.L. Labandeira-Garcia. Angiotensin type 1 receptor blockage reduces L-dopa-induced Dyskinesia in the 6-OHDA model of Parkinson's disease. Involvement of vascular endothelial growth factor and interleukin-1β. *Experimental Neurology* 261 (2014): 720–732.

Olanow, C.W. and J.A. Obeso. The significance of defining preclinical or prodromal Parkinson's disease. *Movement Disorders: Official Journal of the Movement Disorder Society* 27 (5) (2012): 666–669.

Ong, E.L.H., R. Goldacre, and M. Goldacre. Differential risks of cancer types in people with Parkinson's disease: A national record-linkage study. *European Journal of Cancer (Oxford, England: 1990)* 50 (14) (2014): 2456–2462.

Ono, K., M. Ikemoto, T. Kawarabayashi et al. A chemical chaperone, sodium 4-phenylbutyric acid, attenuates the pathogenic potency in human alpha-synuclein A30P + A53T transgenic mice. *Parkinsonism & Related Disorders* 15 (9) (2009): 649–654.

Osellame, L.D. and M.R. Duchen. Quality control gone wrong: Mitochondria, lysosomal storage disorders and neurodegeneration. *British Journal of Pharmacology* 171 (8) (2014): 1958–1972.

Pagan, F., M. Hebron, E.H. Valadez et al. Nilotinib effects in Parkinson's disease and dementia with Lewy bodies. *Journal of Parkinson's Disease* 6 (3) (2016): 503–517.

Parkinson's Study Group. Phase II safety, tolerability, and dose selection study of isradipine as a potential disease-modifying intervention in early Parkinson's disease (STEADY-PD). *Movement Disorders* 28 (13) (2013): 1823–1831.

Patil, S.P., P.D. Jain, P.J. Ghumatkar, R. Tambe, and S. Sathaye. Neuroprotective effect of metformin in MPTP-induced Parkinson's disease in mice. *Neuroscience* 277 (2014): 747–754.

Pilsl, A. and K.F. Winklhofer. Parkin, PINK1 and mitochondrial integrity: Emerging concepts of mitochondrial dysfunction in Parkinson's disease. *Acta Neuropathologica* 123 (2) (2012): 173–188.

Potts, M.B. and D.A. Lim. An old drug for new ideas: Metformin promotes adult neurogenesis and spatial memory formation. *Cell Stem Cell* 11 (1) (2012): 5–6.

Quinn, L.P., B. Crook, M.E. Hows et al. The PPARgamma agonist pioglitazone is effective in the MPTP mouse model of Parkinson's disease through inhibition of monoamine oxidase B. *British Journal of Pharmacology* 154 (1) (2008): 226–233.

Rakshit, H., P. Chatterjee, and D. Roy. A bidirectional drug repositioning approach for Parkinson's disease through network-based inference. *Biochemical and Biophysical Research Communications* 457 (3) (2015): 280–287.

Ravina, B.M., S.C. Fagan, R.G. Hart et al. Neuroprotective agents for clinical trials in Parkinson's disease: A systematic assessment. *Neurology* 60 (8) (2003): 1234–1240.

Richter, F., S.M. Fleming, M. Watson et al. A GCase chaperone improves motor function in a mouse model of synucleinopathy. *Neurotherapeutics: The Journal of the American Society for Experimental NeuroTherapeutics* 11 (4) (2014): 840–856.

Rodriguez, M., C. Rodriguez-Sabate, I. Morales, A. Sanchez, and M. Sabate. Parkinson's disease as a result of aging. *Aging Cell* 14(3) (2015): 293–308.

Schapira, A.H.V. and M.E. Gegg. Glucocerebrosidase in the pathogenesis and treatment of Parkinson disease. *Proceedings of the National Academy of Sciences of the United States of America* 110 (9) (2013): 3214–3215.

Schwarzschild, M.A., A. Ascherio, M. Flint Beal et al. Inosine to increase serum and cerebro-spinal fluid urate in Parkinson disease: A randomized clinical trial. *JAMA Neurology* 71 (2) (2014): 141–150.

Shameer, K., B. Readhead, and J.T. Dudley. Computational and experimental advances in drug repositioning for accelerated therapeutic stratification. *Current Topics in Medicinal Chemistry* 15 (1) (2015): 5–20.

Shen, C., Y. Guo, W. Luo, C. Lin, and M. Ding. Serum urate and the risk of Parkinson's disease: Results from a meta-analysis. *The Canadian Journal of Neurological Sciences [Le Journal Canadien Des Sciences Neurologiques]* 40 (1) (2013): 73–79.

Shen, H.-Y., J.-C. He, Y. Wang, Q.-Y. Huang, and J.-F. Chen. Geldanamycin induces heat shock protein 70 and protects against MPTP-induced dopaminergic neurotoxicity in mice. *The Journal of Biological Chemistry* 280 (48) (2005): 39962–39969.

Shineman, D.W., J. Alam, M. Anderson et al. Overcoming obstacles to repurposing for neurodegenerative disease. *Annals of Clinical and Translational Neurology* 1 (7) (2014): 512–518.

Sidransky, E., M.A. Nalls, J.O. Aasly et al. Multicenter analysis of glucocerebrosidase mutations in Parkinson's disease. *The New England Journal of Medicine* 361 (17) (2009): 1651–1661.

Simuni, T., K. Kieburtz, B. Tilley et al. Pioglitazone in early Parkinson's disease: A phase 2, multicentre, double-blind, randomised trial. *Lancet Neurology* 14 (8) (2015): 795–803.

Spataro, N., F. Calafell, L. Cervera-Carles et al. Mendelian genes for Parkinson's disease contribute to the sporadic forms of the disease. *Human Molecular Genetics* 24 (7) (2015): 2023–2034.

Sulzer, D. Multiple hit hypotheses for dopamine neuron loss in Parkinson's disease. *Trends in Neurosciences* 30 (5) (2007): 244–250.

Surmeier, D.J., J.N. Guzman, J. Sanchez-Padilla, and J.A. Goldberg. What causes the death of dopaminergic neurons in Parkinson's disease? *Progress in Brain Research* 183 (2010): 59–77.

Swanson, C.R., V. Joers, V. Bondarenko et al. The PPAR-γ agonist pioglitazone modulates inflammation and induces neuroprotection in parkinsonian monkeys. *Journal of Neuroinflammation* 8 (2011): 91.

Tanabe, A., Y. Yamamura, J. Kasahara, R. Morigaki, R. Kaji, and S. Goto. A novel tyrosine kinase inhibitor AMN107 (nilotinib) normalizes striatal motor behaviors in a mouse model of Parkinson's disease. *Frontiers in Cellular Neuroscience* 8 (2014): 50.

Tilley, B.C. and W.R. Galpern. Screening potential therapies: Lessons learned from new paradigms used in Parkinson disease. *Stroke: A Journal of Cerebral Circulation* 38 (2 Suppl.) (2007): 800–803.

Tison, F., L. Nègre-Pagès, W.G. Meissner et al. Simvastatin decreases levodopa-induced dyskinesia in monkeys, but not in a randomized, placebo-controlled, multiple crossover ('n-of-1') exploratory trial of simvastatin against levodopa-induced dyskinesia in Parkinson's disease patients. *Parkinsonism & Related Disorders* 19 (4) (2013): 416–421.

Von Bohlen und Halbach, O., A. Schober, and K. Krieglstein. Genes, proteins, and neurotoxins involved in Parkinson's disease. *Progress in Neurobiology* 73 (3) (2004): 151–177.

Wahlqvist, M.L., M.-S. Lee, C.-C. Hsu, S.-Y. Chuang, J.-T. Lee, and H.-N. Tsai. Metformin-inclusive sulfonylurea therapy reduces the risk of Parkinson's disease occurring with type 2 diabetes in a Taiwanese population cohort. *Parkinsonism & Related Disorders* 18 (6) (2012): 753–758.

Wahner, A.D., J.M. Bronstein, Y.M. Bordelon, and B. Ritz. Statin use and the risk of Parkinson disease. *Neurology* 70 (16 Pt 2) (2008): 1418–1422.

Wang, J., D. Gallagher, L.M. DeVito et al. Metformin activates an atypical PKC-CBP pathway to promote neurogenesis and enhance spatial memory formation. *Cell Stem Cell* 11 (1) (2012): 23–35.

Wang, Q., X. Wei, H. Gao et al. Simvastatin reverses the downregulation of M1/4 receptor binding in 6-hydroxydopamine-induced Parkinsonian rats: The association with improvements in long-term memory. *Neuroscience* 267 (2014): 57–66.

Ward, R.J., F.A. Zucca, J.H. Duyn, R.R. Crichton, and L. Zecca. The role of iron in brain ageing and neurodegenerative disorders. *The Lancet Neurology* 13 (10) (2014): 1045–1060.

Wirdefeldt, K., C.E. Weibull, H. Chen et al. Parkinson's disease and cancer: A register-based family study. *American Journal of Epidemiology* 179 (1) (2013): 85–94.

Wu, L., Y.-Y. Tian, J.-P. Shi et al. Inhibition of endoplasmic reticulum stress is involved in the neuroprotective effects of candesartan cilexitil in the rotenone rat model of Parkinson's disease. *Neuroscience Letters* 548 (2013): 50–55.

Xu, W., L. Tan, and J.-T. Yu. The link between the SNCA gene and parkinsonism. *Neurobiology of Aging* 36 (3) (2015): 1505–1518.

Yuan, H., Z.-W. Zhang, L.-W. Liang et al. Treatment strategies for Parkinson's disease. *Neuroscience Bulletin* 26 (1) (2010): 66–76.

11 Drug Candidates for Repositioning in Alzheimer's Disease

Maria P. del Castillo-Frias and Andrew J. Doig

CONTENTS

11.1   ALZHEIMER'S DISEASE AS A WORLD HEALTH ISSUE

Alzheimer's disease (AD) is the most common form of dementia among elderly people. AD is a progressive neurodegenerative disorder that damages neurons, causing cognitive decline, memory deterioration, personality changes, and language impairment, and gradually prevents patients from performing their daily activities independently (Burns et al., 1995; World Health Organisation, 1992).

The increasing prevalence of AD, high treatment costs, and the lack of disease-modifying therapies (preventing onset, slowing progression, or curing AD) make AD a major global health concern. It has been estimated that 35.6 million people lived with dementia worldwide in 2010, with numbers expected to almost double every 20 years, to 65.7 million in 2030 and 115.4 million in 2050 (Prince et al., 2013). The global costs associated with dementia in 2015 are approximately $600 billion, which is equivalent to 1% of the entire world's GDP (http://www.alzheimers.net/resources/alzheimers-statistics/). The rising costs of AD are leading to a huge burden that national health institutions, insurance companies, and patients are struggling to afford.

Five drugs have been FDA approved for the treatment of AD, but they only ameliorate the symptoms of the disease for 6–12 months and do not target the underlying pathology. Research groups and pharmaceutical companies are keen to develop disease-modifying drugs that change the progression/outcome of AD. Nevertheless, most novel drugs have failed to demonstrate efficacy or suitable safety profiles in clinical trials, possibly because the etiology of AD has not been fully elucidated, drugs are administered too late, and assignment of patients into AD or non-AD groups may be in error. The last AD drug approved was memantine in 2004. From 2002 to 2012, 413 AD drug trials were performed including 83 at Phase III, with a 99.6% failure rate (Cummings et al., 2014). New strategies are therefore needed to generate efficient, safe, and economic drugs to treat AD, notably drug repositioning.

11.2   BIOLOGY OF ALZHEIMER'S DISEASE

11.2.1   ALZHEIMER'S DISEASE PATHOLOGICAL HALLMARKS

The main pathological hallmarks of AD are high levels of extracellular amyloid plaques, mainly formed from the aggregated β-amyloid peptide (Aβ), and extracellular neurofibrillary tangles of hyperphosphorylated tau protein. These hallmarks are located in regions controlling memory and learning (hippocampus, entorhinal cortex, and basal forebrain), and behavior and emotions (amygdala).

11.2.2   β-AMYLOID PEPTIDE

The Aβ peptide is 38–43 amino acids long. It is a proteolytic product of a larger transmembrane protein known as amyloid precursor protein (APP) coded by a single multiexonic gene located on chromosome 21. Mutations and multiple copies of APP are known to increase Aβ formation causing the familial form of AD (FAD) (Tanzi and Bertram, 2005). For instance, the high occurrence of AD in individuals with Down syndrome (trisomy 21), where APP gene is triplicate, has been paramount evidence of the role of APP and Aβ in AD pathogenesis.

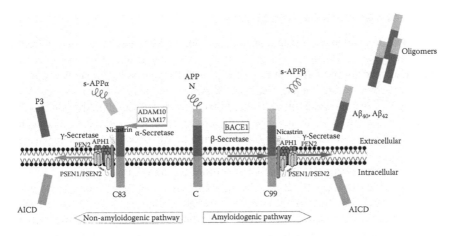

FIGURE 11.1 Amyloidogenic and non-amyloidogenic APP processing. (Modified from Zhang, C., *Discov. Med.*, 76, 189, 2012.)

The APP encodes an integral membrane type I protein consisting of a large N-terminal extracellular domain and a short C-terminal cytoplasmic domain (Hardy and Selkoe, 2002). APP processing involves two sequential proteolytic cleavages: first within the extracellular domain by either α- or β-secretase, and then in the transmembrane region by γ-secretase (Figure 11.1). Three enzymes from the ADAM family (a disintegrin and metalloprotease domain) including ADAM9, ADAM10, and ADAM17 have shown α-secretase activity (Allinson et al., 2003). The β-site APP-cleaving enzyme 1 BACE1 (a type I integral membrane protein) has been identified as β-secretase (Vassar et al., 1999). γ-Secretase is a complex of the enzymes presenilin 1 or 2 (PS1 or PS2), nicastrin, APH-1, and the presenilin enhancer 2 (PEN-2) (Siemers et al., 2006; Yu et al., 2000).

APP can be processed *via* the non-amyloidogenic or amyloidogenic pathways. In the non-amyloidogenic pathway, α-secretase (commonly ADAM10) cleaves the APP within the Aβ domain, producing a soluble s-APPα and a C-terminal fragment of 83 amino-acid residues (C83). γ-Secretase then cleaves the C83 fragment resulting in an N-terminally truncated Aβ (P3), which is non-neurotoxic and an intracellular signalling domain (AICD) (Figure 11.1) (Thinakaran and Koo, 2008). In the amyloidogenic pathway, the β-secretase cleaves APP to soluble sAPPβ leaving intact the Aβ domain in the C-terminal fragment of 99 amino acid residues (C99). Then γ-secretase cleaves C99 between amino acids 38–43 forming Aβ peptides and an intracellular AICD (Figure 11.1) (Thinakaran and Koo, 2008). This cleavage forms two main Aβ species of 40 and 42 amino acids in length ($A\beta_{40}$, $A\beta_{42}$) in a ratio of 10:1. The $A\beta_{42}$ species is more neurotoxic than $A\beta_{40}$ and is the major component of amyloid plaques characteristic of AD (Younkin, 1998). N-terminally truncated forms of Aβ with pyroGlu groups at position 3 are found in AD brains and are more toxic and aggregation prone than $A\beta_{42}$ (Frost et al., 2013). N-terminally extended forms of Aβ also appear to be important in cell culture models of AD, though their relevance *in vivo* is unclear (Welzel et al., 2014).

11.2.3 THE AMYLOID CASCADE HYPOTHESIS

The amyloid cascade hypothesis points to deposition of Aβ as the initiating step leading to AD pathogenesis (Hardy and Higgins, 1992). The deposition of Aβ can occur due to increased processing of APP through the β- and then γ-secretase pathways, imbalance between the production and clearance of Aβ, or the increased ratio of $A\beta_{42}/A\beta_{40}$. According to this hypothesis, the increased levels of Aβ, especially $A\beta_{42}$, are toxic to neurons (Goodman and Mattson, 1994) by disrupting synapses, activating inflammatory responses, increasing oxidative injury, and altering kinase/phosphatase activity leading to tau hyperphosphorylation and tangle formation (Figure 11.2). These alterations result in neuronal dysfunction and ultimately in neuronal death (Selkoe, 2004).

Strong evidence supporting this hypothesis relies on the genetic information obtained from FADs. Autosomal dominant mutations linked to FAD in APP, PS1, and PS2 genes increase the processing of APP through the amyloidogenic pathway, thus elevating the normal production of Aβ (Bertram and Tanzi, 2005; Citron et al., 1992; Scheuner et al., 1996). Both *in vivo* and *in vitro* models, expressing FAD-linked mutations in APP, PS1 and PS2 genes, show elevated levels of $A\beta_{42}$ in comparison with wild type models (Borchelt et al., 1997; Citron et al., 1992; Suzuki et al., 1994).

The hypothesis is challenged by the lack of correlation between the amount of amyloid plaques and the degree of dementia in AD patients or animal models (Dickson et al., 1995). Nevertheless, nearly all potential AD therapies attack part of the amyloid cascade. Soluble Aβ species may be a more toxic form (McLean et al., 1999), such as dimers, trimers, Aβ-derived diffusible ligands (ADDLs), and protofibrils (Walsh and Selkoe, 2007). They have shown synaptotoxic effects in cell cultures and animal models: blocking long-term potentiation (LTP), and causing synaptic loss by decreasing the activity of N-methyl-D-aspartate (NMDA) receptor and calcineurin (Li et al., 2011; Shankar et al., 2008). The blockage of LTP was confirmed *in vivo* after adding dimers extracted from brain tissue of patients with AD to hippocampal slices (Shankar et al., 2008). In addition, dimers trigger hyperphosphorylation of tau resulting in microtubule abnormalities and paired helical formation (Li et al., 2011). Some of these "dimers" may be misidentified extended forms of Aβ such as −40 to +40 (Welzel et al., 2014). Indeed, a whole "peptide soup" of N-terminally extended forms of Aβ has been proposed to be of importance in AD (Kaneko et al., 2014).

11.2.4 Aβ CLEARANCE

In addition to overproduction, accumulation of Aβ in the brain could result from the imbalance between its production and clearance to the bloodstream. Aβ clearance can be performed *via* mediated transport of Aβ, proteolytic degradation, and from efflux from interstitial fluid to bloodstream (Tanzi et al., 2004). The transport of Aβ is mediated by multi-ligand cell surface receptors: LRP and RAGE. LRP is a low-density lipoprotein receptor-related protein that mediates the flux of Aβ from the brain to the periphery. Mice injected with an LRP antagonist and radiolabeled $A\beta_{40}$ showed a reduction up to 90% in the efflux of Aβ from brain to plasma, thus demonstrating the role of LRP in Aβ clearance (Shibata et al., 2000). RAGE is a receptor for advanced glycation products that mediates the influx of Aβ from plasma to brain. In animal models with down-regulated RAGE, the influx of Aβ to the periphery is inhibited (Deane et al., 2003).

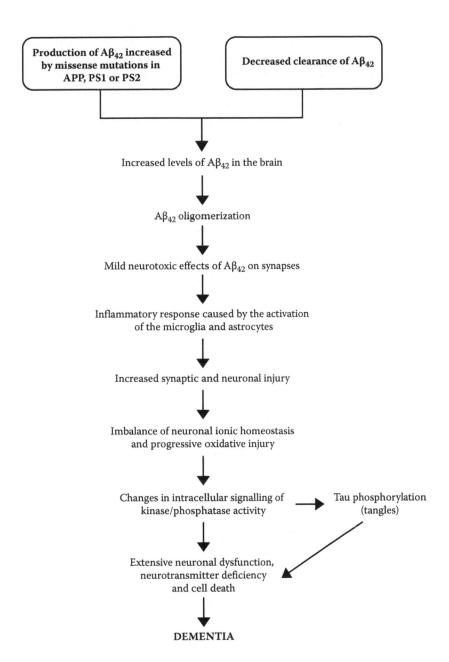

FIGURE 11.2 The amyloid cascade hypothesis points to Aβ$_{42}$ aggregation as the initial event of the AD pathogenesis. Aggregation could be due to overproduction of Aβ$_{42}$ or failure in clearance mechanisms. (Modified from Hardy, J. and Selkoe, D.J., *Science*, 297, 353, 2002.)

Proteolytic removal of Aβ is mediated by zinc metalloendopeptidases: insulin-degrading enzyme (IDE) and neprilysin. IDE is located in the cytosol and hydrolases regulatory peptides including insulin, glucagon, Aβ, and AICD (Bennett et al., 2000). The role of IDE in Aβ degradation was demonstrated in mice where the IDE gene was knocked out resulting in increased levels of Aβ and AICD (Farris et al., 2003). Neprilysin, a type II membrane protein, has been identified as the rate-limiting peptidase that cleaves Aβ. *In vivo* studies have demonstrated that the absence of neprilysin increases Aβ accumulation resulting in amyloid deposition (Madani et al., 2006).

11.2.5 TAU PROTEIN

Tau is a microtubule-associated proteins mainly found in axons. In normal conditions, tau is hydrophilic, highly soluble, and natively unfolded. Tau plays a key role interacting with tubulin and promoting the assembly of microtubules, stabilizing its structure. Tau also helps with intracellular transport of organelles and biomolecules, and prevents apoptosis by stabilizing β-catenin. Tau is subject to different posttranslational modifications, including glycosylation, ubiquitination, and phosphorylation. Phosphorylation is most widely studied because of its relevance to AD. In healthy neurons, two to three residues of tau can be phosphorylated, whereas in patients with AD, phosphorylated sites account to approximately 9 per molecule (Köpke et al., 1993).

Hyperphosphorylation occurs due to an alteration in the balance of tau kinase and tau phosphatase activity (mainly protein phosphatase 2), causing phosphorylation of different serine, threonine, and tyrosine residues. As a result, loss of microtubule binding may contribute to a breakdown of intracellular traffic and consequently neuronal death. Likewise, there is a redistribution of tau from axonal to somatodendritic compartments. Tau then starts aggregating to paired helical filaments , which turn into a bundle to form neurofibrillary tangles (Mandelkow and Mandelkow, 1998; Medeiros et al., 2011). Studies on animal models suggest that the hyperphosphorylation of tau is caused by multiple processes, including accumulation of Aβ plaques, disruption in glucose metabolism, and inflammation (Liu et al., 2009).

11.3 CURRENT PANORAMA OF AD DRUGS

11.3.1 DRUGS CURRENTLY APPROVED TO TREAT AD

Five drugs have been approved by the U.S. FDA for the treatment of AD: cholinesterase inhibitors (ChEIs) for mild-to-moderate AD, and an NMDA receptor antagonist for moderate-to-severe AD.

The four ChEIs that were approved are donepezil, galantamine, rivastigmine, and tacrine. The use of ChEIs for the treatment of AD is supported by the cholinergic hypothesis of AD. According to this hypothesis, the basal forebrain of individuals with AD shows a deficit in choline acetyltransferase resulting in a reduction in the production of acetylcholine and cholinergic dysfunction. This cholinergic dysfunction prompts progressive deterioration in cognitive function. The ChEIs inhibit the acetyl- and butyrylcholinesterases responsible for the degradation of acetylcholine. Hence, increasing the amount of this neurotransmitter at the synaptic cleft improves the cholinergic transmission and reduces the cholinergic deficit (Klafki et al., 2006). Patients treated with ChEIs improve cognitive function for up to 12 months reporting mild adverse events, including nausea, vomiting, and fatigue (Birks, 2006).

Glutamate is the main excitatory neurotransmitter in the central nervous system (CNS) and is regulated by glutamate receptors such as NMDA. Studies suggest that the excessive activation of NMDA receptors increases the accumulation of calcium in the cholinergic cells accelerating the neurodegeneration process (Reisberg et al., 2003). Memantine, a NMDA noncompetitive glutamate receptor antagonist, blocks the NMDA receptor regulating glutamate concentration, thus preventing neuronal damage (Massoud and Gauthier, 2010). Patients with moderate AD treated with memantine showed mild-to-moderate improvement in cognitive function for a limited period of time (about 12 months) (Hellweg et al., 2012).

The drugs available to treat AD just offer temporary benefits to patients with AD, because they only ameliorate the symptoms of AD, but do not target the underlying pathology of the disease (Klafki et al., 2006). Now, research is being focused on the development of disease-modifying drugs able to prevent the onset, slow the progression, or modify the course of the disease.

11.3.2 Drugs in Development for Alzheimer's Disease

Since current treatments are merely symptomatic, new strategies to generate drugs for the treatment of AD are focused on disease modification. The aim of these drugs is to slow the progression of the pathology in the disease; the agents under development target mainly the amyloid cascade and tau hyperphosphorylation. Some drugs can be classified according their targets: to decrease the production of Aβ, to decrease Aβ aggregation, and to promote Aβ clearance (Massoud and Gauthier, 2010). In addition, some research groups have focused their research on the inhibition of tau phosphorylation or alternative hypotheses to explain AD pathology.

Drugs aiming to reduce the production of Aβ most often focus on α-, β-, or γ-secretase because they process APP. Upregulators of α-secretase promote APP processing through the non-amyloidogenic pathway, thus preventing the formation of Aβ. Steroid hormones, protein kinase inhibitors, activators of muscarinic or glutamate γ-aminobutyric acid (GABA) receptor, and statins have been demonstrated to activate α-secretase (Mangialasche et al., 2010; Rockwood, 2006). For instance, in Phase II clinical trials, patients taking etazolate (a GABA receptor modulator) showed cognitive improvement in some subjects and a good safety profile (Vellas et al., 2011).

Drugs targeting β-secretase suppress this enzyme, precluding the formation of Aβ, and are currently being tested in Phase III trials (e.g., rosiglitazone (Gold et al., 2010)). γ-secretase inhibitors prevent the generation of Aβ in the last step of APP processing. However, γ-secretase also cleaves other proteins, including Notch. When Notch is cleaved, an intracellular domain is released to the nucleus activating transcription factors that control important functions such as cell differentiation. Clinical trials of the γ-secretase inhibitor Semagacestat demonstrated it to be effective in reducing Aβ concentrations in plasma and CNS but worsened cognition in patients and caused severe adverse events, including abnormal bleeding, gastrointestinal and skin toxicity (Siemers et al., 2006). These adverse events were attributed to the blockage of Notch proteins, and the trial was halted. After these failures, drug developers attempt to develop Notch-sparing γ-secretase inhibitors, which do not interfere with Notch cleavage (Wolfe, 2012). Tarenflurbil was tested, but in Phase III clinical trials failed to demonstrate efficacy, perhaps because of low penetration through the blood–brain barrier (BBB), and adverse events including anemia and dizziness (Rafii and Aisen, 2009).

Another strategy is to preventing Aβ aggregation, especially into oligomers. Drugs preventing aggregation include antibodies and small molecules. For example, solanezumab is a humanized monoclonal IgG1 antibody that binds to the mid-domain of monomeric Aβ (Crespi et al., 2015; Villemagne et al., 2013). It may be beneficial to mild AD (Doody et al., 2014). Bapineuzumab, a monoclonal antibody, binds to the N-terminus of Aβ and facilitates the clearance of Aβ by crossing the BBB. Bapineuzumab made it to clinical trials, but was discontinued in Phase III due to lack of improvement in cognition in patients with mild and moderate AD (Moreth et al., 2013). Tramiprosate, a modified molecule from taurine, failed to demonstrate improvement of cognition in Phase III clinical trials (Aisen et al., 2011). The drugs focused on reducing Aβ aggregation may prevent the binding of plaques by metallic ions such as zinc or copper.

The tau phosphorylation strategy aims to develop an inhibitor of tau hyperphosphorylation and/or drugs that prevent tau aggregation. For instance, studies on animals showed that valproate inhibits glycogen synthase kinase 3-β (GSK3β) involved in tau phosphorylation. In clinical trials, adverse events and lack of efficacy discouraged further evaluation of the compound (Tariot et al., 2011). There is considerable evidence that oxidative stress is important for the initiation and commencement of AD, causing mitochondrial dysfunction and transition metal accumulation (Cu and Fe) (Zhao and Zhao, 2013). Antioxidants and metal chelators may therefore be beneficial to AD.

AD is associated with an inflammatory response as shown by an increased presence of activated microglia and astrocytes, activated complement proteins, cytokines, and reactive oxygen, nitrogen, and carbonyl species. While inflammation may be beneficial in the short term, prolonged chronic inflammation may be highly damaging. Triggers such as Aβ may cause microglial activation. These release neurotoxic factors such as cytotoxic cytokines and reactive oxygen/nitrogen species and damage neighboring neurons. Damaged or dying neurons release additional microglial activators, resulting in a vicious cycle of neurotoxicity. Anti-inflammatories, such as curcumin, apigenin, and tenilsetam, may therefore be beneficial to AD (Millington et al., 2014).

Despite the efforts to develop a disease-modifying therapy for AD, current strategies targeting amyloid hypothesis have not yielded any marketable drug. Common reasons for failures are that drugs fail to show a suitable safety profile and do not demonstrate significant benefits improving or slowing progression of AD, in some cases because they have low or no permeability through the BBB. It is plausible that drugs have been tested in patients with irreversible advanced AD, where they might have shown benefit at an earlier stage. A good alternative to take advantage of the available models and find successful disease modifying drugs for AD could be the use of alternative strategies to screen drugs, such as drug repositioning.

11.4 EXAMPLES OF REPOSITIONED DRUGS FOR AD

11.4.1 GALANTAMINE

Galantamine was discovered in the 1950s in the bulbs and flowers of wild Caucasian snow drops, *Galanthus woronowii* (Sramek et al., 2000). It was first used to reverse the effect of the alkaloid poison curare, which functions by competitively inhibiting the nicotinic acetylcholine receptor. Galantamine inhibits acetylcholinesterase, raising levels of acetylcholine to compensate for its receptor being blocked (Schuh, 1976). Galantamine was later used for various other diseases of the peripheral and central nervous systems. Since galantamine has the same mode of action as the first Alzheimer's drugs, it was successfully repositioned as an AD drug and approved by the FDA in 2001.

11.4.2 PROTRIPTYLINE

The initial screen to identify new hits can use computational as well as target-based or phenotypic assays. Bansode et al. used virtual screens against acetylcholinesterase, BACE1, and Aβ aggregation to search for new AD drugs (Bansode et al., 2014). Virtual docking was tested with 140 FDA-compounds against crystal structures of all three targets. Remarkably, five antidepressant drugs (protriptyline, amitriptyline, maprotiline, doxepin, and nortriptyline), with related structures, showed strong predicted binding affinity to all three targets, supported by experiment. All five had submillimolar binding against acetylcholinesterase. Protriptyline inhibited the aggregation of $A\beta_{13-22}$, had an IC_{50} of ~0.025 mM for BACE1 inhibition, and was not a general protease inhibitor. Affecting multiple targets simultaneously is a promising strategy for any disease. Protriptyline is an FDA-approved drug for the treatment of depression and narcolepsy, attention deficit hyperactivity disorder, and headaches and is already known to cross the BBB.

11.4.3 ANGIOTENSIN-CONVERTING ENZYME INHIBITORS

Inhibition of the renin–angiotensin system, which regulates blood pressure and fluid balance, has been suggested as a potential therapeutic strategy for AD and other neurodegenerative disorders (Kaur et al., 2015; Kehoc and Wilcock, 2007; Koronyo-Hamaoui et al., 2014; Phillips and de Oliveira, 2008; Wright and Harding, 2010). Angiotensin-converting enzymes (ACEs) produce the AT-II peptide, where elevated levels of AT-II increase Aβ-induced neurotoxicity (Saavedra, 2012). In particular, elevation of ACE activity has been reported in the brains of AD patients (Miners et al., 2009), suggesting that brain-penetrating ACE inhibitors may be beneficial in preventing AD. Several ACE inhibitors have been reported to show promise in AD models.

Dong et al. compared the brain-penetrating ACE inhibitor perindopril with the nonpenetrant ACE inhibitors imidapril and enalapril, using mice that underwent intracerebroventricular injection of $A\beta_{1-40}$ or PS2/APP double-transgenic mice that overexpress Aβ in brain tissue. Perindopril lowered hippocampal ACE activity and prevented cognitive impairment, brain inflammation, and oxidative stress, compared to the non-brain-penetrating ACE inhibitors (Dong et al., 2011). Perindopril was also effective in rodent models for vascular dementia (Yamada et al., 2011).

Intracerebroventricular injection of streptozotocin (STZ) can be used to induce AD-like dementia in mice, causing impaired learning and memory. Administration of lisinopril, an ACE inhibitor, significantly reduced STZ-induced behavioral and biochemical changes (Singh et al., 2013). The ACE inhibitor captopril was tested on the widely used mouse AD model Tg2576. Captopril decreased the excessive hippocampal ACE activity of AD mice, reducing neurodegeneration, by decreasing amyloidogenic processing of APP and hippocampal reactive oxygen species (AbdAlla et al., 2013).

11.4.4 NILVADIPINE

Nilvadipine is an L-type calcium channel (LCC) antagonist used for the treatment of hypertension and chronic major cerebral artery occlusion. It was found to be effective on Aβ-induced vasoconstriction in isolated arteries in Tg2576 transgenic mice (Paris et al., 2004). Nilvadipine is an LCC antagonist with (+)-nilvadipine, the active enantiomer. Both enantiomers inhibited Aβ production and increased its clearance across the BBB, thus revealing that this effect was not LCC related. Further studies of nilvadipine in PS013 mutant human tau transgenic mice revealed that (−)-nilvadipine reduced tau phosphorylation at various AD pertinent epitopes. Elucidation of (−)-nilvadipine's mechanism of action showed that it attained its effects by inhibiting the spleen tyrosine kinase (Syk), resulting in the activation of Protein Kinase A (PKA), which phosphorylates the inhibitory residue Ser9 in GSK3β, resulting in a decrease in tau hyperphosphorylation. Similarly, PKA also phosphorylates CREB, which is essential in neuroprotection and improvement of cognition. Furthermore, inhibition of Syk impedes the stimulation of the NFκ B pathway resulting in a decrease in neuroinflammation, BACE1 expression, and subsequent reduction in the accumulation of Aβ (Paris et al., 2014). These multiple effects mean that the inhibition of Syk by nilvadipine or derivatives represents a very attractive therapeutic target for the treatment of AD. A pilot study of nilvadipine in 55 patients with AD found stabilization of cognition and improvement in executive function in treated individuals (Kennelly et al., 2012). A clinical trial on 500 people with mild-to-moderate AD is currently underway (Lawlor et al., 2014).

11.4.5 CARMUSTINE

Patients with cancer are less likely to have AD (Roe et al., 2010), suggesting that anticancer drugs may be beneficial to AD. Hayes et al. therefore screened FDA-approved oncology drugs using CHO cells stably expressing APP751wt to measure changes in the secretion of Aβ (Hayes et al., 2013). Carmustine was found to decrease secreted Aβ levels and increase sAPPα, though without inhibiting β- and γ-secretases, suggesting that the activity of the drug may arise from altered trafficking and processing of APP. Carmustine decreased plaque burden in mice, suppressed microglial activation and decreased levels of $A\beta_{40}$, and increased sAPPα in mouse brains (Araki, 2013; Hayes et al., 2013).

11.4.6 MINOCYCLINE

Minocycline is a lipid-soluble tetracycline-class antibiotic, often used to treat acne and numerous other bacterial conditions. Tetracycline antibiotics were first shown to inhibit Aβ aggregation (Forloni et al., 2001). Minocycline has anti-inflammatory properties, so it was investigated in microglia. It inhibited neuronal death and glial activation induced by hippocampal injection of Aβ in rat hippocampus (Ryu et al., 2004). In a transgenic mouse, it reduced Aβ accumulation, neuroinflammatory markers, and behavioral deficits (Fan et al., 2007; Parachikova et al., 2010).

11.4.7 LEVETIRACETAM

Levetiracetam is an orally available antiepileptic drug that modulates the synaptic vesicle protein modulator SV2A. A human transcriptome study on how the ApoE4 risk allele affects APP processing implicated SV2A. Even though this work has been partly retracted, levetiracetam was effective in reducing Aβ generation in cells cultured from ApoE4 carriers. Levetiracetam reduced epileptiform activity and reversed cognitive deficits in human APP transgenic mice (Sanchez et al., 2012; Shi et al., 2013). A small trial in 17 people with mild cognitive impairment (MCI) showed that levetiracetam reduced hippocampal activity and improved performance on a hippocampal memory task (Bakker et al., 2012). A one-year study of levetiracetam in AD patients who had seizures reported improved attention, verbal fluency, and controlled seizures (Cumbo and Ligori, 2010). Additional clinical studies of levetiracetam are going ahead, following these encouraging data.

11.4.8 ACITRETIN

Acitretin is a vitamin A retinoid analog. It is orally delivered to treat the skin disease psoriasis. It promotes activity of retinoic acid receptors that are known to be impaired in AD, possibly causing deposition of Aβ (Corcoran et al., 2004; Goodman and Pardee, 2003). ADAM10 activity is regulated by retinoic acid. A test of synthetic retinoic acid derivatives found a strong enhancement of non-amyloidogenic processing of APP by the vitamin A analog acitretin (Tippmann et al., 2009). Acitretin was tested for activation of α-secretase disintegrin and ADAM10 in patients with mild-to-moderate AD. Measurement of CSF levels of sAPPα, the product of α-secretase on APP, showed a significant increase in CSF sAPPα levels (Endres et al., 2014).

11.4.9 TUMOR NECROSIS FACTOR INHIBITORS

Tumor necrosis factor alpha (TNF-α) is a cytokine produced by macrophages. It was discovered and named after its ability to kill cancerous cells in mice. Later it was found that TNF-α plays a key role in mediating inflammatory and immunity processes. In the brain, this cytokine is expressed by neuronal and glial cells in response to brain injury, viral infection, or degenerative disorders (Feuerstein et al., 1998), such as AD. The amyloid cascade hypothesis suggests that inflammation might be activated downstream of Aβ overproduction and deposition. Based on this possible relationship, investigators studied in depth the possible role of inflammation in AD. TNF-α is found to be upregulated in patients with neurodegeneration (Fillit et al., 1991). In addition, there is evidence that cytokines, including TNF-α, upregulate BACE1 expression, thus increasing Aβ load (Hickman and El Khoury, 2014; Yamamoto et al., 2007).

As TNF-α seems to affect APP processing and Aβ plaque generation, it has been suggested as a drug target for AD. The inhibition of TNF-α could improve or slow down the cognitive decline in patients with AD. TNF-α inhibitors available in the market are monoclonal antibodies such as infliximab, fusion proteins such as etanercept, or small molecules like thalidomide. The use of these inhibitors includes the treatment of rheumatoid arthritis, leprosy, Crohn's disease, and multiple myeloma. Despite being successful treatments, TNF-α inhibitors' adverse effects are considerable. In this section, we describe current efforts to use these types of drugs to treat AD.

11.4.9.1   Infliximab (Remicade)

Infliximab is a chimeric (mouse–human) immunoglobulin type 1 (IgG1) monoclonal antibody that is not able to cross the BBB. It specifically binds to both monomers and trimers and membrane-bound and soluble TNF-α (Weinberg et al., 2005; Winterfield et al., 2005). Infliximab thus neutralizes the biological activity of TNF-α by blocking its binding to natural receptors. This drug is approved by the FDA for the treatment of autoimmune diseases, such as rheumatoid arthritis, psoriasis, and Crohn's disease. It has been further investigated for the treatment of other diseases, including inflammatory bowel disease and AD.

In preclinical trials, infliximab reduced the number of amyloid plaques, the levels of TNF-α and tau phosphorylation in APP/PS1 double transgenic mice (Shi et al., 2011). In addition, it improved memory and cognition in rats with induced dementia (Elcioglu et al., 2015).

11.4.9.2   Etanercept (Enbrel)

Etanercept (Enbrel) is a soluble dimeric fusion protein receptor for TNF-α. It is a fusion between human TNFR2-α receptor and human IgG1 Fc domain. This artificial receptor has greater affinity for TNF-α than its natural receptors, thereby decreasing the inflammatory response triggered by TNF-α (Tristano, 2010). Similar to infliximab, etanercept is approved for rheumatoid arthritis, psoriasis, and ankylosing spondylitis.

Given the evidence of the possible role of TNF-α in AD, investigators were keen to use etaneracept to treat AD, even though etanercept poorly penetrates the BBB. An alternative route of administration, consisting of injection into the tissues close to the spinal column (perispinally), was proposed to facilitate etanercept delivery to the brain through the cerebrospinal venous system (Tobinick et al., 2006). In a 6-month open-labeled study, etanercept was administered weekly by perispinal administration in 15 patients with mild-to-severe AD. Patients showed significant cognitive improvement, thus providing sufficient clinical evidence to perform further trials (Tobinick and Gross, 2008a; Tobinick et al., 2006). One patient with late-onset AD reported a rapid cognitive improvement (just 2 h) after the perispinal administration of etanercept. The authors' hypothesis is that this rapid improvement could be related to the effect of the drug inhibiting TNF-α (Tobinick and Gross, 2008b).

These studies have prompted further study of etanercept as an AD drug. An ongoing Phase I clinical trial is assessing the safety and efficacy of the administration of perspinal etanercept plus dietary supplements (resveratrol, curcumin, omega-3, and quercetin) *versus* the administration of dietary supplements only. The University of Southampton recently finished a randomized, double-blind placebo-controlled Phase II clinical trial to evaluate the safety and tolerability of etanercept in patients with AD. As a secondary outcome, the study evaluated cognitive and behavioral functions, as well as systemic cytokine levels. In contrast to previous studies, etanarcept was administrated subcutaneously instead of perispinally. Etanercept was well tolerated and no new adverse events were recorded. There was no significant difference in cognition or behavior. However, the study found that the cognitive decline in patients treated with placebo was twice as bad as that expected and that patients treated with etanercept did not worsen or improve compared to baseline. This study suggests that a larger clinical trial should be performed (Butchart et al., 2015).

11.4.9.3 Thalidomide

Thalidomide was initially used to alleviate nausea and morning sickness during pregnancy. However, around 100,000 newborns suffered congenital malformations that in many cases resulted in death. For this reason, it was withdrawn in many countries, including the United States and United Kingdom. Despite its serious adverse events, the drug was investigated in other countries and was found to be a useful treatment for leprosy or multiple myeloma, finally approved by the FDA in 1998.

Thalidomide, as an immunomodulatory agent, inhibits the TNF-α cytokine, which plays a key role in producing an inflammatory response. When TNF-α is inhibited, the inflammatory response is reduced. Early preclinical studies on AD mouse models showed that thalidomide has neuroprotective effects and decreases microglial activation (Alkam et al., 2008; Tweedie et al., 2012). It also reduced tau phosphorylation, APP, and Aβ plaque load (Tweedie et al., 2012). In a later study on APP Swedish mutation transgenic APP23 mice, thalidomide decreased BACE1 activity (He et al., 2013). The evidence gathered in these studies supported the use of this drug in clinical trials. A Phase II study to evaluate the effects of thalidomide in patients with mild-to-moderate AD is ongoing in the United States. The primary endpoint of this study is to evaluate if the administration of the drug improves cognition and has an effect on CSF and plasma biomarkers.

11.4.10 Diabetes and Alzheimer's Disease

Type 2 diabetes mellitus (T2DM) and AD are common diseases in elderly people. This observation prompts the question of a possible relationship between these diseases. The Rotterdam study, that investigated the possible influence of TD2M on the risk of dementia and AD, found that TD2M almost doubled the risk of dementia (Ott et al., 1999), calling for more epidemiological studies of this relationship. Recently, a meta-analysis of these epidemiological studies confirmed that there is a correlation between T2DM and the risk of developing AD (Gudala et al., 2013). The possible biochemical basis of the relationship between T2DM and AD has been widely studied in *in vitro* models, *in vivo* models, animals, and humans. Insulin is well known for decreasing sugar levels in blood. However, it has been discovered that it has other important functions, in the brain in particular, including neuroprotective functions, and regulation of Aβ levels and phosphorylated tau (Carro and Torres-Aleman, 2004). It has also been demonstrated that low levels of insulin cause cognitive impairment and reduce LTP in the hippocampus (Trudeau et al., 2004).

11.4.10.1   Glucagon-Like Peptide-1

In patients with both AD and T2DM, insulin receptors in the brain seem desensitized, impairing insulin signaling (Carro and Torres-Aleman, 2004; Hoyer, 2004). Based on this premise, restoring the insulin receptors' sensitivity could be a good alternative target for AD. Glucagon-like peptide-1 (GLP-1) is an incretin hormone that regulates insulin secretion and levels of glucose in blood. As GLP-1 agonists, such as liraglutide, exenatide, and lixisenatide, are the standard treatments for T2DM, they as well might be an interesting option to treat AD. In addition, other treatments used to treat diabetes like insulin or metformin have been considered and tested in patients with AD.

11.4.10.2   Insulin

In preclinical trials, rats administrated with insulin enhanced memory in a passive-avoidance task (Park et al., 2000). In humans, intranasal administration of insulin was proposed as an administrative route as it could improve the delivery of insulin to the brain. Phase I clinical trials in healthy subjects demonstrated that this route of administration was safe. In this trial, insulin also improved cognition and memory in the treated subjects (Holscher, 2014). After these encouraging results, several Phase II and III trials have been carried out.

In a pilot study, insulin or placebo was randomly administered to a group of 40 patients diagnosed with AD and 64 with MCI for a 4-month period. This study measured the patients' cognition at different time points, collected CSF samples, and performed positron emission tomography with fludeoxyglucose. The results showed that cognition and memory was improved in patients treated with insulin. Though there was no significant difference in the biomarker levels between groups, improvements in memory and cognition were attributed to changes in $A\beta_{42}$ levels (Craft et al., 2012). The AD cooperative study had a series of clinical trials named the Study of Nasal Insulin to Fight Forgetfulness (SNIFF) whose purpose is to assess the effects of insulin in AD patients. In the SNIFF-LONG 21 trial, none of the groups meet the expected primary outcome measure, as they did not improve on the verbal composition test, though participants in a higher dose group did improve in visual retention and working memory. APOE carriers assigned to the higher dose group performed better in verbal memory while noncarriers declined.

11.4.10.3   Metformin

Metformin is a biguanidine antihyperglycemic drug, which is used as first-line treatment for TD2M. Its mechanism consists in decreasing the hepatic glucose production and intestinal absorption of glucose. It is believed that these effects are because metformin activates AMP-activated protein kinase (AMPK) that regulates lipid and glucose metabolism and increases insulin sensitivity (Yarchoan and Arnold, 2014).

An *in vitro* study on the N2a695 cell line showed that metformin upregulates BACE1, thus increasing the levels of Aβ. The authors hypothesized that this effect might be mediated by AMPK. They also found that when cells were treated with a combination of metformin and insulin Aβ decreased, suggesting that metformin alone might worsen the effects of AD (Chen et al., 2009). In a Phase II study, 80 overweight subjects aged 55–90 years diagnosed with amnestic MCI were treated with metformin and placebo. The purpose of this study was to evaluate if the administration of metformin improves memory and cognition. Though the trial is terminated, there are no published results yet. An ongoing Phase II pilot study carried on by the University of Pennsylvania is assessing the effect of different doses of metformin *versus* placebo for 4 weeks in patients with MCI and AD. This study will evaluate cognition and CSF biochemical biomarkers of AD.

11.5 CONCLUSION

Drug repositioning is a highly promising strategy for the discovery of new therapeutics and there is considerable interest in its application to AD, given the immense importance of AD and the continuing failure of conventional drug discovery programs. Hits can first be identified by screening against known targets or phenotypes related to AD, using drugs known to have useful modes of action, or more serendipitously from epidemiological or other data. Following up hits can be by confirming activity, if the target is known, or by target identification and determining the compound's mode of action, possibly revealing novel biology. The drug can be tested against numerous phenotypes in cellular or biophysical assays, before testing in an animal (nearly always mouse) model for AD. As these are known drugs, it may be possible to bypass many of the standard ADMET tests before testing in man.

Despite the considerable promise of the drug repositioning strategy, drug discovery for AD will remain a very challenging task, as shown by the recent failure of the monoamine oxidase B inhibitor sembragiline in a Phase II trial. Sembragiline was first developed to help people stop smoking, because MAO-B inhibitors mimic some of the antidepressant effects of nicotine (Weinberger et al., 2010). It was hypothesized that sembragiline might delay progression of AD by reducing oxidative stress in the brain and lessening dopamine-related symptoms, but no benefits were seen on cognition (http://www.evotec.com/uploads/cms_article/2739/PR_2015-06-30_Roche_e. pdf). Nevertheless, the above examples show that there is extensive current work on drug repositioning for AD and we await new developments with great interest.

REFERENCES

AbdAlla, S. et al., ACE inhibition with captopril retards the development of signs of neurodegeneration in an animal model of Alzheimer's disease. *International Journal of Molecular Sciences*, 2013. **14**(8): 16917–16942.

Aisen, P.S. et al., Tramiprosate in mild-to-moderate Alzheimer's disease—A randomized, double-blind, placebo-controlled, multi-centre study (the Alphase Study). *Archives of Medical Science*, 2011. **7**(1): 102–111.

Alkam, T. et al., Restraining tumor necrosis factor-alpha by thalidomide prevents the Amyloid beta-induced impairment of recognition memory in mice. *Behavioral Brain Research*, 2008. **189**(1): 100–106.

Allinson, T.M.J. et al., ADAMs family members as amyloid precursor protein α-secretases. *Journal of Neuroscience Research*, 2003. **74**(3): 342–352.

Araki, W., Potential repurposing of oncology drugs for the treatment of Alzheimer's disease. *BMC Medicine*, 2013. **11**: 82–84.

Bakker, A. et al., Reduction of hippocampal hyperactivity improves cognition in amnestic mild cognitive impairment. *Neuron*, 2012. **74**(3): 467–474.

Bansode, S.B. et al., Molecular investigations of protriptyline as a multi-target directed ligand in Alzheimer's disease. *PLoS One*, 2014. **9**(8): e105196.

Bennett, R.G., W.C. Duckworth, and F.G. Hamel, Degradation of amylin by insulin-degrading enzyme. *Journal of Biological Chemistry*, 2000. **275**(47): 36621–36625.

Bertram, L. and R.E. Tanzi, The genetic epidemiology of neurodegenerative disease. *Journal of Clinical Investigation*, 2005. **115**(6): 1449–1457.

Birks, J., Cholinesterase inhibitors for Alzheimer's disease. *Cochrane Database of Systematic Reviews*, 2006. 1: CD005593.

Borchelt, D.R. et al., Accelerated amyloid deposition in the brains of transgenic mice coexpressing mutant presenilin 1 and amyloid precursor proteins. *Neuron*, 1997. **19**(4): 939–945.

Burns, A., R. Howard, and W. Pettit, *Alzheimer's Disease: A Medical Companion*. 1995, Blackwell Sciences: London, U.K.

Butchart, J. et al., Etanercept in Alzheimer disease: A randomized, placebo-controlled, double-blind, phase 2 trial. *Neurology*, 2015. **84**(21): 2161–2168.

Carro, E. and I. Torres-Aleman, The role of insulin and insulin-like growth factor I in the molecular and cellular mechanisms underlying the pathology of Alzheimer's disease. *European Journal of Pharmacology*, 2004. **490**(1–3): 127–133.

Chen, Y.M. et al., Antidiabetic drug metformin (Glucophage(R)) increases biogenesis of Alzheimer's amyloid peptides via up-regulating BACE1 transcription. *Proceedings of the National Academy of Sciences of the United States of America*, 2009. **106**(10): 3907–3912.

Citron, M. et al., Mutation of the β-amyloid precursor protein familial Alzheimer's disease increases β-protein production. *Nature*, 1992. **360**: 672–674.

Corcoran, J.P.T., P.L. So, and M. Maden, Disruption of the retinoid signalling pathway causes a deposition of amyloid beta in the adult rat brain. *European Journal of Neuroscience*, 2004. **20**(4): 896–902.

Craft, S. et al., Intranasal insulin therapy for Alzheimer disease and amnestic mild cognitive impairment a pilot clinical trial. *Archives of Neurology*, 2012. **69**(1): 29–38.

Crespi, G.A.N. et al., Molecular basis for mid-region amyloid-β capture by leading Alzheimer's disease immunotherapies. *Scientific Reports*, 2015. **5**: 9649.

Cumbo, E. and L.D. Ligori, Levetiracetam, lamotrigine, and phenobarbital in patients with epileptic seizures and Alzheimer's disease. *Epilepsy & Behavior*, 2010. **17**(4): 461–466.

Cummings, J.L., T. Morstorf, and K. Zhong, Alzheimer's disease drug-development pipeline: Few candidates, frequent failures. *Alzheimers Research & Therapy*, 2014. **6**(4): 37–44.

Deane, R. et al., RAGE mediates amyloid-beta peptide transport across the blood-brain barrier and accumulation in brain. *Nature Medicine*, 2003. **9**(7): 907–913.

Dickson, D.W. et al., Correlations of synaptic and pathological markers with cognition of the elderly. *Neurobiology of Aging*, 1995. **16**(3): 285–298.

Dong, Y.F. et al., Perindopril, a centrally active angiotensin-converting enzyme inhibitor, prevents cognitive impairment in mouse models of Alzheimer's disease. *FASEB Journal*, 2011. **25**(9): 2911–2920.

Doody, R.S. et al., Phase 3 trials of solanezumab for mild-to-moderate Alzheimer's disease. *New England Journal of Medicine*, 2014. **370**(4): 311–321.

Elcioglu, H.K. et al., Effects of systemic thalidomide and intracerebroventricular etanercept and infliximab administration in a streptozotocin induced dementia model in rats. *Acta Histochemica*, 2015. **117**(2): 176–181.

Endres, K. et al., Increased CSF APPs-alpha levels in patients with Alzheimer disease treated with acitretin. *Neurology*, 2014. **83**(21): 1930–1935.

Fan, R. et al., Minocycline reduces microglial activation and improves behavioral deficits in a transgenic model of cerebral microvascular amyloid. *Journal of Neuroscience*, 2007. **27**(12): 3057–3063.

Farris, W. et al., Insulin-degrading enzyme regulates the levels of insulin, amyloid beta-protein, and the beta-amyloid precursor protein intracellular domain in vivo. *Proceedings of the National Academy of Sciences of the United States of America*, 2003. **100**(7): 4162–4167.

Feuerstein, G.Z., X.K. Wang, and F.C. Barone, The role of cytokines in the neuropathology of stroke and neurotrauma. *Neuroimmunomodulation*, 1998. **5**(3–4): 143–159.

Fillit, H. et al., Elevated circulating tumor-necrosis-factor levels in Alzheimer's disease. *Neuroscience Letters*, 1991. **129**(2): 318–320.

Forloni, G. et al., Anti-amyloidogenic activity of tetracyclines: Studies in vitro. *FEBS Letters*, 2001. **487**(3): 404–407.

Frost, J.L. et al., Pyroglutamate-3 amyloid-beta deposition in the brains of humans, non-human primates, canines, and Alzheimer disease-like transgenic mouse models. *American Journal of Pathology*, 2013. **183**(2): 369–381.

Gold, M. et al., Rosiglitazone monotherapy in mild-to-moderate Alzheimer's disease: Results from a randomized, double-blind, placebo-controlled phase III study. *Dementia and Geriatric Cognitive Disorders*, 2010. **30**(2): 131–146.

Goodman, A.B. and A.B. Pardee, Evidence for defective retinoid transport and function in late onset Alzheimer's disease. *Proceedings of the National Academy of Sciences of the United States of America*, 2003. **100**(5): 2901–2905.

Goodman, Y. and M.P. Mattson, Secreted form of β-amyloid precursor protein protect hippocampal neurons against amyloid-β peptide induced oxidative injury. *Experimental Neurology*, 1994. **128**(1): 1–12.

Gudala, K. et al., Diabetes mellitus and risk of dementia: A meta-analysis of prospective observational studies. *Journal of Diabetes Investigation*, 2013. **4**(6): 640–650.

Hardy, J. and D.J. Selkoe, The amyloid hypothesis of Alzheimer's disease: Progress and problems on the road to therapeutics. *Science*, 2002. **297**: 353–356.

Hardy, J.A. and G.A. Higgins, Alzheimer's disease: The amyloid cascade hypothesis. *Science*, 1992. **256**: 184–185.

Hayes, C.D. et al., Striking reduction of amyloid plaque burden in an Alzheimer's mouse model after chronic administration of carmustine. *BMC Medicine*, 2013. **11**: 81.

He, P. et al., Long-term treatment of thalidomide ameliorates amyloid-like pathology through inhibition of beta-secretase in a mouse model of Alzheimer's disease. *PLoS One*, 2013. **8**(2): e55091.

Hellweg, R. et al., Efficacy of memantine in delaying clinical worsening in Alzheimer's disease (AD): Responder analyses of nine clinical trials with patients with moderate to severe AD. *International Journal of Geriatric Psychiatry*, 2012. **27**(6): 651–656.

Hickman, S.E. and J. El Khoury, TREM2 and the neuroimmunology of Alzheimer's disease. *Biochemical Pharmacology*, 2014. **88**(4): 495–498.

Holscher, C., First clinical data of the neuroprotective effects of nasal insulin application in patients with Alzheimer's disease. *Alzheimer's & Dementia: The Journal of the Alzheimer's Association*, 2014. **10**(1 Suppl): S33–S37.

Hoyer, S., Glucose metabolism and insulin receptor signal transduction in Alzheimer disease. *European Journal of Pharmacology*, 2004. **490**(1–3): 115–125.

Kaneko, N. et al., Identification and quantification of amyloid beta-related peptides in human plasma using matrix-assisted laser desorption/ionization time-of-flight mass spectrometry. *Proceedings of the Japan Academy Series B—Physical and Biological Sciences*, 2014. **90**(3): 104–117.

Kaur, P., A. Muthuraman, and M. Kaur, The implications of angiotensin-converting enzymes and their modulators in neurodegenerative disorders: Current and future perspectives. *ACS Chemical Neuroscience*, 2015. **6**(4): 508–521.

Kehoe, P.G. and G.K. Wilcock, Is inhibition of the renin-angiotensin system a new treatment option for Alzheimer's disease? *Lancet Neurology*, 2007. **6**(4): 373–378.

Kennelly, S. et al., Apolipoprotein E genotype-specific short-term cognitive benefits of treatment with the antihypertensive nilvadipine in Alzheimer's patientsuan open-label trial. *International Journal of Geriatric Psychiatry*, 2012. **27**(4): 415–422.

Klafki, H.W. et al., Therapeutic approaches to Alzheimer's disease. *Brain*, 2006. **129**: 2840–2855.

Köpke, E. et al., Microtubule-associated protein tau. Abnormal phosphorylation of a nonpaired helical filament pool in Alzheimer disease. *Journal of Biological Chemistry*, 1993. **268**(32): 24374–24384.

Koronyo-Hamaoui, M. et al., ACE overexpression in myelomonocytic cells: Effect on a mouse model of Alzheimer's disease. *Current Hypertension Reports*, 2014. **16**(7): 444.

Lawlor, B. et al., NILVAD protocol: A European multicentre double-blind placebo-controlled trial of nilvadipine in mild-to-moderate Alzheimer's disease. *BMJ Open*, 2014. **4**(10): e006364.

Li, S.M. et al., Soluble A beta oligomers inhibit long-term potentiation through a mechanism involving excessive activation of extrasynaptic NR2B-containing NMDA receptors. *Journal of Neuroscience*, 2011. **31**(18): 6627–6638.

Liu, F. et al., Reduced O-GlcNAcylation links lower brain glucose metabolism and tau pathology in Alzheimer's disease. *Brain*, 2009. **132**(7): 1820–1832.

Madani, R. et al., Lack of neprilysin suffices to generate murine amyloid-like deposits in the brain and behavioral deficit in vivo. *Journal of Neuroscience Research*, 2006. **84**(8): 1871–1878.

Mandelkow, E.-M. and E. Mandelkow, Tau in Alzheimer's disease. *Trends in Cell Biology*, 1998. **8**(11): 425–427.

Mangialasche, F. et al., Alzheimer's disease: Clinical trials and drug development. *Lancet Neurology*, 2010. **9**(7): 702–716.

Massoud, F. and S. Gauthier, Update on the pharmacological treatment of Alzheimer's disease. *Current Neuropharmacology*, 2010. **8**(1): 69–80.

McLean, C.A. et al., Soluble pool of Abeta amyloid as a determinant of severity of neurodegeneration in Alzheimer's disease. *Annals of Neurology*, 1999. **46**: 860–866.

Medeiros, R., D. Baglietto-Vargas, and F.M. LaFerla, The role of Tau in Alzheimer's disease and related disorders. *CNS Neuroscience & Therapeutics*, 2011. **17**(5): 514–524.

Millington, C. et al., Chronic neuroinflammation in Alzheimer's disease: New perspectives on animal models and promising candidate drugs. *BioMed Research International*, 2014: 309129. doi:10.1155/2014/309129.

Miners, S. et al., Angiotensin-converting enzyme levels and activity in Alzheimer's disease: Differences in brain and CSF ACE and association with ACE1 genotypes. *American Journal of Translational Research*, 2009. **1**(2): 163–177.

Moreth, J., C. Mavoungou, and K. Schindowski, Passive anti-amyloid immunotherapy in Alzheimer's disease: What are the most promising targets? *Immunity & Ageing*, 2013. **10**: 18.

Ott, A. et al., Diabetes mellitus and the risk of dementia—The Rotterdam Study. *Neurology*, 1999. **53**(9): 1937–1942.

Parachikova, A. et al., Reductions in amyloid-beta-derived neuroinflammation, with minocycline, restore cognition but do not significantly affect tau hyperphosphorylation. *Journal of Alzheimers Disease*, 2010. **21**(2): 527–542.

Paris, D. et al., Nilvadipine antagonizes both A beta vasoactivity in isolated arteries, and the reduced cerebral blood flow in APPsw transgenic mice. *Brain Research*, 2004. **999**(1): 53–61.

Paris, D. et al., The spleen tyrosine kinase (Syk) regulates Alzheimer amyloid-beta production and tau hyperphosphorylation. *Journal of Biological Chemistry*, 2014. **289**(49): 33927–33944.

Park, C.R. et al., Intracerebroventricular insulin enhances memory in a passive-avoidance task. *Physiology & Behavior*, 2000. **68**(4): 509–514.

Phillips, M.I. and E.M. de Oliveira, Brain renin angiotensin in disease. *Journal of Molecular Medicine*, 2008. **86**(6): 715–722.

Prince, M. et al., The global prevalence of dementia: A systematic review and metaanalysis. *Alzheimers & Dementia*, 2013. **9**(1): 63–75.

Rafii, M.S. and P.S. Aisen, Recent developments in Alzheimer's disease therapeutics. *BMC Medicine*, 2009. **7**: 7–10.

Reisberg, B. et al., Memantine in moderate-to-severe Alzheimer's disease. *New England Journal of Medicine*, 2003. **348**(14): 1333–1341.

Rockwood, K., Epidemiological and clinical trials evidence about a preventive role for statins in Alzheimer's disease. *Acta Neurologica Scandinavica. Supplementum*, 2006. **185**: 71–77.

Roe, C.M. et al., Cancer linked to Alzheimer disease but not vascular dementia. *Neurology*, 2010. **74**(2): 106–112.

Ryu, J.K. et al., Minocycline inhibits neuronal death and glial activation induced by beta-amyloid peptide in rat hippocampus. *Glia*, 2004. **48**(1): 85–90.

Saavedra, J.M., Angiotensin II AT(1) receptor blockers as treatments for inflammatory brain disorders. *Clinical Science*, 2012. **123**(9–10): 567–590.

Sanchez, P.E. et al., Levetiracetam suppresses neuronal network dysfunction and reverses synaptic and cognitive deficits in an Alzheimer's disease model. *Proceedings of the National Academy of Sciences of the United States of America*, 2012. **109**(42): E2895–E2903.

Scheuner, D. et al., Secreted amyloid beta-protein similar to that in the senile plaques of Alzheimer's disease is increased in vivo by the presenilin 1 and 2 and APP mutations linked to familial Alzheimer's disease. *Nature Medicine*, 1996. **2**(8): 864–870.

Schuh, F.T., Molecular mechanism of action of galanthamine, an antagonist of nondepolarizing muscle-relaxants. *Anaesthesist*, 1976. **25**(9): 444–448.

Selkoe, D.J., Alzheimer disease: Mechanistic understanding predicts novel therapies. *Annals of Internal Medicine*, 2004. **140**(8): 627–638.

Shankar, G.M. et al., Amyloid-β protein dimers isolated directly from Alzheimer's brains impair synaptic plasticity and memory. *Nature Medicine*, 2008. **14**: 837–842.

Shi, J.Q. et al., Anti-TNF-alpha reduces amyloid plaques and tau phosphorylation and induces CD11c-positive dendritic-like cell in the APP/PS1 transgenic mouse brains. *Brain Research*, 2011. **1368**: 239–247.

Shi, J.Q. et al., Antiepileptics topiramate and levetiracetam alleviate behavioral deficits and reduce neuropathology in APPswe/PS1dE9 transgenic mice. *CNS Neuroscience & Therapeutics*, 2013. **19**(11): 871–881.

Shibata, M. et al., Clearance of Alzheimer's amyloid-beta(1-40) peptide from brain by LDL receptor-related protein-1 at the blood-brain barrier. *Journal of Clinical Investigation*, 2000. **106**(12): 1489–1499.

Siemers, E.R. et al., Effects of a gamma-secretase inhibitor in a randomized study of patients with Alzheimer disease. *Neurology*, 2006. **66**(4): 602–604.

Singh, B. et al., Attenuating effect of lisinopril and telmisartan in intracerebroventricular streptozotocin induced experimental dementia of Alzheimer's disease type: Possible involvement of PPAR-gamma agonistic property. *Journal of the Renin-Angiotensin-Aldosterone System*, 2013. **14**(2): 124–136.

Sramek, J.J., E.J. Frackiewicz, and N.R. Cutler, Review of the acetylcholinesterase inhibitor galantamine. *Expert Opinion on Investigational Drugs*, 2000. **9**(10): 2393–2402.

Suzuki, N. et al., An increased percentage of long Aβ protein secreted by familial APP precursor β-APP(717) mutants. *Science*, 1994. **264**: 1336–1340.

Tanzi, R.E. and L. Bertram, Twenty years of the Alzheimer's disease amyloid hypothesis: A genetic perspective. *Cell*, 2005. **120**(4): 545–555.

Tanzi, R.E., R.D. Moir, and S.L. Wagner, Clearance of Alzheimer's A beta peptide: The many roads to perdition. *Neuron*, 2004. **43**(5): 605–608.

Tariot, P.N. et al., Chronic divalproex sodium to attenuate agitation and clinical progression of Alzheimer disease. *Archives of General Psychiatry*, 2011. **68**(8): 853–861.

Thinakaran, G. and E.H. Koo, Amyloid precursor protein trafficking, processing, and function. *Journal of Biological Chemistry*, 2008. **283**(44): 29615–29619.

Tippmann, F. et al., Up-regulation of the alpha-secretase ADAM10 by retinoic acid receptors and acitretin. *FASEB Journal*, 2009. **23**(6): 1643–1654.

Tobinick, E. et al., TNF-alpha modulation for treatment of Alzheimer's disease: A 6-month pilot study. *Medscape General Medicine*, 2006. **8**(2): 25.

Tobinick, E.L. and H. Gross, Rapid cognitive improvement in Alzheimer's disease following perispinal etanercept administration. *Journal of Neuroinflammation*, 2008b. **5**: 2.

Tobinick, E.L. and H. Gross, Rapid improvement in verbal fluency and aphasia following perispinal etanercept in Alzheimer's disease. *BMC Neurology*, 2008a. **8**: 27.

Tristano, A.G., Neurological adverse events associated with anti-tumor necrosis factor alpha treatment. *Journal of Neurology*, 2010. **257**(9): 1421–1431.

Trudeau, F., S. Gagnon, and G. Massicotte, Hippocampal synaptic plasticity and glutamate receptor regulation: Influences of diabetes mellitus. *European Journal of Pharmacology*, 2004. **490**(1–3): 177–186.

Tweedie, D. et al., Tumor necrosis factor-alpha synthesis inhibitor 3,6'-dithiothalidomide attenuates markers of inflammation, Alzheimer pathology and behavioral deficits in animal models of neuroinflammation and Alzheimer's disease. *Journal of Neuroinflammation*, 2012. **9**: 106.

Vassar, R. et al., Beta-secretase cleavage of Alzheimer's amyloid precursor protein by the transmembrane aspartic protease BACE. *Science*, 1999. **286**(5440): 735–741.

Vellas, B. et al., EHT0202 in Alzheimer's disease: A 3-month, randomized, placebo-controlled, Double-Blind Study. *Current Alzheimer Research*, 2011. **8**(2): 203–212.

Villemagne, V.L. et al., Amyloid β deposition, neurodegeneration, and cognitive decline in sporadic Alzheimer's disease: A prospective cohort study. *Lancet Neurology*, 2013. **12**(4): 357–367.

Walsh, D.M. and D.J. Selkoe, Aβ Oligomers—A decade of discovery. *Journal of Neurochemistry*, 2007. **101**: 1172–1184.

Weinberg, J.M. et al., Biologic therapy for psoriasis: An update on the tumor necrosis factor inhibitors infliximab, etanercept, and adalimumab, and the T-cell-targeted therapies efalizumab and alefacept. *Journal of Drugs in Dermatology*, 2005. **4**(5): 544–555.

Weinberger, A.H. et al., A double-blind, placebo-controlled, randomized clinical trial of oral selegiline hydrochloride for smoking cessation in nicotine-dependent cigarette smokers. *Drug and Alcohol Dependence*, 2010. **107**(2–3): 188–195.

Welzel, A.T. et al., Secreted amyloid beta-proteins in a cell culture model include N-terminally extended peptides that impair synaptic plasticity. *Biochemistry*, 2014. **53**(24): 3908–3921.

Winterfield, L.S. et al., Psoriasis treatment: Current and emerging directed therapies. *Annals of the Rheumatic Diseases*, 2005. **64**: 87–90.

Wolfe, M.S., γ-Secretase inhibitors and modulators for Alzheimer's disease. *Journal of Neurochemistry*, 2012. **120**: 89–98.

World Health Organisation, The ICD-10 classification of mental and behavioral disorders: Clinical descriptions and diagnostic guidelines. 1992, World Health Organisation, Geneva.

Wright, J.W. and J.W. Harding, The brain RAS and Alzheimer's disease. *Experimental Neurology*, 2010. **223**(2): 326–333.

Yamada, K. et al., Effect of a centrally active angiotensin-converting enzyme inhibitor, perindopril, on cognitive performance in chronic cerebral hypo-perfusion rats. *Brain Research*, 2011. **1421**: 110–120.

Yamamoto, M. et al., Interferon-gamma and tumor necrosis factor-a regulate amyloid-alpha plaque deposition and beta-secretase expression in Swedish mutant APP transgenic mice. *American Journal of Pathology*, 2007. **170**(2): 680–692.

Yarchoan, M. and S.E. Arnold, Repurposing diabetes drugs for brain insulin resistance in Alzheimer disease. *Diabetes*, 2014. **63**(7): 2253–2261.

Younkin, S.G., The role of A beta 42 in Alzheimer's disease. *Journal of Physiology—Paris*, 1998. **92**(3–4): 289–292.

Yu, G. et al., Nicastrin modulates presenilin-mediated notch/glp-1 signal transduction and beta APP processing. *Nature*, 2000. **407**(6800): 48–54.

Zhang, C., Natural compounds that modulate BACE1-processing of amyloid-β precursor protein in Alzheimer's disease. *Discovery Medicine*, 2012. **76**: 189–197.

Zhao, Y. and B. Zhao, Oxidative stress and the pathogenesis of Alzheimer's disease. *Oxidative Medicine and Cellular Longevity*, 2013. **2013**: 316523.

12 Promising Candidates for Drug Repurposing in Huntington's Disease

Francesca Romana Fusco and Emanuela Paldino

CONTENTS

12.1 INTRODUCTION

Huntington's disease (HD) is a rare autosomal dominant neurodegenerative disorder, characterized by motor dysfunction, cognitive decline, and psychiatric disturbances. Motor symptoms are dominated by *chorea*, an involuntary muscle contraction that results from the impairment of the basal ganglia, which is the main target of HD. HD is caused by the mutation of the *IT15* gene that is located on the short arm of chromosome 4 and is characterized by a CAG expansion encoding a polyQ repeat at the N-terminus of *huntingtin* (HTT) protein (The Huntington's Disease Collaborative Research Group, 1993). The polyQ tract promotes the formation of toxic oligomers and aggregates. In physiological conditions, people have fewer than 36 glutamine repeats in the polyQ region resulting in the production of the cytoplasmatic protein HTT. A sequence of 36 or more CAG repeats result in the production of mutated huntingtin (mHTT) protein. Generally, the number of CAG repeats is related to the severity of the disease and accounts for about 60% of the variation of the age of the onset of symptoms. In fact, 36–39 repeats result in a reduced penetrance form of the disease with a later onset and slower progression of symptoms. Conversely, a large repeats' count determines a full penetrance of HD that might occur even before the age of 20, when it is then referred to as juvenile HD, and this accounts for about 7% of HD carriers (Albin and Tagle, 1995).

HTT interacts with over 100 other proteins and appears to have multiple biological functions. The behavior of this mutated protein is not completely understood, but it is toxic to certain cell types, particularly in the brain, because of the formation of neuronal intranuclear inclusions (NIIs) of mHTT (DiFiglia et al., 1997). An early neuronal damage is most evident in the striatal part of the basal ganglia in HD. In particular, medium spiny projection neurons, constituting about 95% of the striatum, degenerate massively (Auer et al., 1984; Smith et al., 1984; Kalimo et al., 1985). Interestingly, a similarly marked loss of the striatal projection neurons occurs in cerebral ischemia. Signs of neurodegeneration are observed also in the cortex, thalamus, and globus pallidus (in the later stages of the disease). Cortical pathology also occurs, contributing to the overall dramatic brain atrophy in the late stages of the disease (Hong et al., 2012; Unschuld et al., 2012; Gray et al., 2013; Samadi et al., 2013). Moreover, signs of cortical dysfunction are often observed before neuropathological signs are apparent.

One of the mechanisms underlying the vulnerability of striatum in HD is explained by the fact that these neurons do not synthetize sufficient amounts of brain-derived neurotrophic factor (BDNF). BDNF is very important for survival of mature neurons in the striatum (Zuccato and Cattaneo, 2007). Striatal BDNF depends on the cortex for its synthesis and release, as it is synthesized by cortical neurons and released in the striatum by corticostriatal anterograde transport. This microtubule-based transport depends on HTT and is altered in HD. Low levels of BDNF mRNA have been reported in the rat striatum (Baquet et al., 2004).

The cAMP response element-binding protein (CREB) is a transcription factor, and its function is impaired by mHTT (Altar et al., 1997; Sugars et al., 2004). This supports the hypothesis that the inhibition of cAMP response element (CRE)-mediated gene transcription contributes to HD. In fact, cAMP levels are decreased in the cerebrospinal fluid of HD patients and transcription of CREB-regulated genes is reduced in the R6/2 transgenic mouse model of HD (Nucifora et al., 2001; Wyttenbach et al., 2001).

HTT modulates the expression of neuron-restrictive silencer factor (NRSF)–controlled neuronal genes, including the *BDNF* gene (Zuccato et al., 2003). Therefore, wild-type HTT directly stimulates the production of BDNF, whereas mutant huntingtin inhibits it. In fact, BDNF is decreased in the brain of HD patients and in mice transgenic for mutant huntingtin (Ferrer et al., 2000; Duan et al., 2003; Zhang et al., 2003). Overexpression of BDNF showed to be neuroprotective in the R6/1 mouse model of HD (Gharami et al., 2008; Xie et al., 2010); however, mice overexpressing BDNF display higher susceptibility to seizure to kainic acid *in vivo* and hyper-excitability in the CA3 region of the hippocampus and entorhinal cortex *in vitro*, because of the effects of BDNF on epileptogenic regions, such as the entorhinal cortex and hippocampus (Papaleo et al., 2011). Moreover, the overexpression of BDNF in experimental animals leads to increased anxiety-like behavior and deficits in working memory (Bimonte et al., 2003). Thus, both excess and insufficient BDNF can be detrimental, and such issues have to be addressed before BDNF is used to treat HD patients. BDNF knockout mice have not only an earlier age of onset, but also more severe motor symptoms. Thus, a specific involvement of BDNF was demonstrated in the pathophysiology of the disease in several ways.

12.2 ANTIDEPRESSANTS AS AN APPROACH TO HD

Antidepressants are classified into five main types: selective serotonin reuptake inhibitors (SSRIs), serotonin and noradrenaline reuptake inhibitors, noradrenaline and specific serotoninergic antidepressants antagonizing alpha 2 receptors and selected serotonin receptors, tricyclics, and monoamine oxidase inhibitors. It was thought earlier that antidepressants acted by increasing levels of noradrenaline (NA) and serotonin in synaptic cleft. However, this process is not yet fully understood. Currently, it is suggested that cellular and molecular adaptations act at several levels of brain neurons in response to antidepressant treatment (Lauterbach, 2013). Antidepressants have neuroprotective effects by activating the mitogen-activated protein kinase (MAPK), extracellular signal–regulated kinase (ERK), phosphatidyl inositol 3-kinase (PI3K), and wingless-type MMTV integration site glycogen synthase kinase (GSK-3) signaling pathways. Moreover, antidepressants can upregulate the expression of neurotrophic/neuroprotective factors, such as BDNF, nerve growth factor, B-cell lymphoma-2 (Bcl-2)–associated athanogene 1, and inactivate proapoptotic molecules such as GSK-3 (Hunsberger et al., 2009; Lauterbach, 2013; Pla et al., 2014). In addition, they promote neurogenesis and are neuroprotective in animal models of neurodegenerative diseases (Kumar and Kumar, 2009a,b). Taken together, antidepressants have a potentially positive effect in neurodegenerative diseases like HD.

12.2.1 NEUROPROTECTIVE EFFECTS MEDIATED BY ANTIDEPRESSANTS

Administration of antidepressant fluoxetine to DBA/2J mice produces a 24% increase in *htt* expression and a higher neuronal proliferation rate in mouse hippocampus after 3 weeks of treatment (Miller et al., 2008; Lauterbach, 2013). It has been suggested that the strain-dependent effect of fluoxetine treatment on the rate of hippocampal cell proliferation is associated with the behavioral response to fluoxetine (Miller et al., 2008). There is currently no definite evidence that fluoxetine upregulates *HTT* gene expression even at very high doses. Intracellularly accumulated mHTT in neurons produces neurotoxicity (Hunsberger et al., 2009). HTT concentration can be modified by protein expression, aggregation, and clearance. Modulation of these processes by means of antidepressants would be beneficial in HD. However, there are currently no reports that antidepressants can directly influence HTT expression. The R6/1 and R6/2 transgenic mice were the first transgenic models developed to study HD. They both express exon 1 of the human HD gene with around 115 and 150 CAG repeats, respectively. The R6/2 mice have been the best characterized and the most widely used model to study the pathogenesis of HD and therapeutic interventions. Two different studies have reported that the clinical dose of the antidepressant sertraline on N171-82Q HD transgenic mice did not produce any effect on intranuclear aggregated HTT in striatum, hippocampus, or cortical neurons (Duan et al., 2008; Peng et al., 2008; Lauterbach, 2013). Moreover, there is increasing evidence for the role of mitochondrial disfunction in the pathophysiology and the neurodegenerative progression of HD. Several factors are involved in this mechanism: impaired bioenergetics leading to decreased ATP production, impaired calcium homeostasis, increased free radicals production, oxidative stress and initiation of an apoptotic process. Therefore, the protective properties of antidepressants on mitochondria could be of great therapeutic value in HD. It was demonstrated that fluoxetine interacts with a mitochondrial component and prevents mitochondria-mediated cell death. Fluoxetine has been reported to interact with the voltage-dependent anion channel (VDAC), modifying its conductance as well as inhibiting permeability transition pore (PTP) opening and release of cytochrome c. These data suggest that VDAC, located at the outer mitochondrial membrane, is a key player in apoptosis. VDAC is a component of the PTP, is involved in release of cytochrome c, and regulates apoptotic cell death (Nahon et al., 2005; Lauterbach, 2013). Antidepressants such as nortriptyline, desipramine, and maprotiline preserved mitochondrial integrity against the glutamate-induced mitochondrial PTP in the yeast artificial chromosome YAC128 mouse model of HD containing 128 CAG repeats (Lauterbach, 2013). YAC128 mice exhibit initial hyperactivity, followed by the onset of a motor deficit and finally hypokinesis (Tang et al., 2005; Lauterbach, 2013). Furthermore, nortriptyline has been reported to inhibit cytochrome c release in the same HD mouse model (Tang et al., 2005). Recently, it was reported that sertraline preserved striatal nitrite concentrations, lipid peroxidation, and mitochondrial enzyme dysfunction in the mitochondrial toxin 3-NP rat model of HD (Kumar et al., 2010).

The prevention/reversal of neuronal apoptotic death induced by mHTT protein might represent great therapeutic potential to delay the neurodegeneration in HD. The antidepressants nortriptyline, desipramine, and maprotiline have been reported to inhibit glutamate-induced apoptosis in YAC128 mice (Tang et al., 2005). YAC128 mice show a progressive decline on the rotarod test and are hyperkinetic on an open-field test beginning at 3 months, signs of hypokinesia at 6 months, cognitive dysfunction at 8.5 months, and inclusion bodies at 18 months (Tang et al., 2005). This mouse model is used for studying the HD pathophysiology because its life-span is longer than that of R6 mice. Therefore, the YAC mouse model is an attractive candidate for long-term therapeutic studies. There are many reports of antidepressants (desipramine, nortriptyline, and maprotiline) providing evidence for neuroprotective effects against apoptotic processes at clinically relevant doses/concentrations in HD striatal cell culture models. Therefore, the authors suggested that these findings might be translated to other HD models (Lauterbach, 2013).

12.2.2 EFFECTS OF ANTIDEPRESSANTS ON BRAIN-DERIVED NEUROTROPHIC FACTOR AND NEUROGENESIS

BDNF plays an important role in learning and memory, feeding, locomotion, stress responses, and affective behavior. BDNF regulates various aspects of developmental and adult neuroplasticity, including neurogenesis, neurite outgrowth and synaptogenesis, synaptic function, and cell survival. BDNF and 5-HT are known to regulate synaptic plasticity, neurogenesis, and neuronal survival in the adult brain. These two signal pathways co-regulate one another in such a manner that 5-HT stimulates the expression of BDNF, and BDNF enhances the growth and survival of 5-HT neurons. Impaired 5-HT and BDNF signaling is well implicated in the pathogenesis of depression, anxiety disorders, age-related disorders, including Alzheimer's disease and Huntington's disease. The function of wild-type HTT has been described to regulate production, transport, and release of BDNF (Zuccato and Cattaneo, 2009). Conversely, mHTT has been reported to disrupt CREB function, which is the activator of BDNF transcription (Zuccato and Cattaneo, 2009). Thus, mHTT indirectly inhibits the neuron-restrictive silencer element in the BDNF promoter region *via* sequestering its transcription factor RE1-silencing transcription factor (REST)/NRSF in the cytoplasm (Zuccato et al., 2003; Zuccato and Cattaneo, 2007). Nuclear accumulation of REST/NRSF, as seen in HD, is reported to impair BDNF transcription. Low levels of BDNF have been observed in the brain of HD patients, including the cortex, striatum, hippocampus, SNpc, and also cerebellum (Chen et al., 2013). In addition to this, low levels of BDNF protein and BDNF mRNA have been observed in various mouse models of HD. Wild-type HTT regulates BDNF vesicular trafficking, and disruption of this mechanism leads to a decreased release of BDNF from cortical and hippocampal neurons (Pla et al., 2014). The decreased release of BDNF leads, in turn, to the downregulation of Akt and ERK-1, both of which are downstream effectors of the BDNF receptor, namely, tyrosine kinase receptor type B (TrkB). Defects in transcription and trafficking of BDNF produce an alteration in density as well as the function of TrkB (Pla et al., 2014). Recently, a relation between the production of BDNF and depression-related behavior was confuted by some authors (Autry and Monteggia, 2012). A study on R6/1 mice has shown that the production of BDNF is mostly affected in females, because the number of BDNF isoforms (BDNF I, II, III, IV, and VI) is lower than that in males. In male mice, only BDNF I and VI transcripts are affected (Zajac et al., 2010). Antidepressants are widely used in the treatment of depression in HD patients (Sackley et al., 2011). Antidepressant treatment blocks the atrophy of CA3 pyramidal cells and increases neurogenesis of hippocampal granule cells (Malberg et al., 2000). Chronic antidepressant treatment upregulates the CRE-mediated gene expression in the rat cortex and hippocampus and the expression of CREB in both rodents and humans (Nibuya et al., 1996). Chronic treatment with SSRIs (fluoxetine or sertraline) in R6/1 mice results in increased hippocampal neurogenesis, ameliorated cognitive deficits and depression-like behavioral symptoms (Grote et al., 2005; Renoir et al., 2012), and increased BDNF levels and neurogenesis in R6/2 mice (Peng et al., 2008). In addition, chronic treatment with antidepressants results in the upregulation of CREB protein expression in depressed patients (Nibuya et al., 1996), CREB phosphorylation (Saarelainen et al., 2003), BDNF (Chen et al., 2005), and TrkB (Bayer et al., 2000) in the hippocampus. BDNF has been assigned to be a mediator of the effects of antidepressants by increasing the survival and differentiation of adult-born neurons in the dentate gyrus (Groves, 2007). Antidepressants such as SSRIs, tricyclics, or MAO-A inhibitors rapidly activate BDNF/TrkB signaling (Saarelainen et al., 2003; Rantamaki et al., 2007). Thus, the restoration of normal BDNF levels could be a successful therapy for HD patients. Fluoxetine and sertraline improved striatal neurogenesis, reversed volume loss in the hippocampal dentate gyrus, and increased BDNF brain levels in the R6/1 HD transgenic mouse and R6/2 transgenic mice, respectively (Grote et al., 2005; Peng et al., 2008). Similarly, sertraline at the same dose increased striatal neurogenesis and BDNF levels in the N171-82Q HD transgenic mouse model (Duan et al., 2008). In addition, chronic antidepressant treatment also increased the expression of CREB mRNA in the rat hippocampus (Nibuya et al., 1996), suggesting a potential regulatory mechanism for BDNF through CREB-mediated gene transcription. Interestingly, BDNF itself also possesses antidepressant-like effects in rodent models following direct infusion into midbrain (Siuciak et al., 1997) or hippocampus (Shirayama et al., 2002). It has been reported that fluoxetine prevents the neurotoxic effects of ecstasy (3,4-methylenedioxymethamphetamine) (Pan and Wang, 1991). Mechanistically, fluoxetine neuroprotective effects, in addition to restoring serotonin levels, may result from the activation of p38, MAPK, and BDNF (Mercier et al., 2004). However, the R6/1 transgenic mouse model of HD was found to display altered responses that reflect depression-related behavior, indicating that the HD mutation promotes a genetic susceptibility for developing depression in rodents. The depression-related behavioral phenotype of the R6/1 HD model was found to be associated with early downregulation in mRNA levels of the 5-HT 1A and 5-HT 1B receptors in the cortex and the hippocampus. The SSRI sertraline treatment decreases depressive-like behavior in female R6/1 mice (Renoir et al., 2012). Serotonergic signaling can also affect hippocampal neurogenesis. This effect depends on the serotonin receptor involved, although the overall effects are pro-neurogenic (Klempin et al., 2010). Therefore, the scarcity of various serotonin receptors partially explains the defects in neurogenesis.

Moreover, disruption of CREB-dependent transcription has been hypothesized to contribute to neuronal death and dysfunction in HD and other polyglutamine repeat disorders. Several standard antidepressant treatments (e.g., NA-reuptake inhibitors, selective 5-HT-reuptake inhibitors, and electroconvulsive seizures) upregulate CREB activity (Nibuya et al., 1996). Furthermore, these observations suggest a role of CREB-regulated expression in neural growth factors, for instance, the increased expression of BDNF in humans treated with antidepressants. Modulation of diverse protein kinases directly or indirectly converges to the activation of CREB, mainly through PKA, CaMKII, and MAPK, as well as PI3K and PKC activation. Moreover, these studies suggest that modulating CREB signaling is a preferable therapeutic approach to treat mood disorders. The ERK/MAPK signaling pathway plays an important role in cellular plasticity. In major depressive disorders, especially the prefrontal cortex and hippocampus are most likely affected in depressed patients, and recent work revealed hyperactivated ERK signaling in the rat prefrontal cortex after chronic stress (Di Benedetto et al., 2013). It has been reported that acute antidepressant treatment differently modulates ERK/MAPK activation in neurons and astrocytes of the adult mouse prefrontal cortex (Di Benedetto et al., 2013). Neurotrophic and neuroprotective effects of antidepressants and other endogenous molecules such as neurotrophins, neurotransmitters, and neuropeptides are exerted through the MAPK/ERK signaling pathway. These effects specifically act by promoting progenitor cell proliferation and differentiation, neuronal growth and regeneration, neuronal survival, and long-term synaptic remodeling and plasticity (Chen et al., 2005; Chen and Manji, 2006). The MAP/ERK signaling pathway acts through the GTP-bound RAS inducing rapidly accelerated fibrosarcoma protein, which phosphorylates and activates mitogen-activated protein kinase kinase (MEK), which in turn phosphorylates and activates MAPK/ERK. ERK regulates several downstream effector systems, including protein kinases such as RSK and MAPK, ion channels, neurotransmitter receptors, and transcription factors. RSK and MAPK are postulated to phosphorylate and activate CREB. CREB is a transcription factor and a common downstream target of both Pi3K/Akt and MEK/ERK pathways. Phosphorylated CREB regulates the expression of many different genes including Bcl-2 (Riccio et al., 1999; Creson et al., 2009) and BDNF (Tao et al., 1998) to enhance neuroprotection and neuronal survival mechanisms.

12.3   ANTIPSYCHOTIC DRUGS FOR HD

In 2007, FDA approved the use of the drug tetrabenazine (TZB) (dopamine-depleting agent that inhibits the vesicular dopamine transporter) specifically for the treatment of chorea in HD. TZB is a vesicular monoamine transporter 2 inhibitor that induces a strong depletion of monoamines, in particular dopamine in the brain. Antipsychotic drugs for HD have been used in the treatment of involuntary movements of several neurodisorders but are only palliative, leading to a temporarily limited improvement of clinical symptoms, and produce side effects like depression and sedation. Kegelmeyer and colleagues reported significant results obtained by a randomized clinical study consisting in the use of TZB in HD patients, which revealed improvements in their chorea scores (Kegelmeyer et al., 2014). However, while some patients showed beneficial effects after TZB administration, others reported deterioration in gait or increased falls for that the stop of therapy was necessary. Thus, it would be clinically useful to identify disease features that would predict patients who are more likely to respond well to TZB treatment.

12.4   CYCLIC NUCLEOTIDE PHOSPHODIESTERASES

Cyclic nucleotide phosphodiesterases (PDEs) are a group of enzymes that catalyze the hydrolysis of the $3'$ cyclic phosphate bonds in the second messenger molecules of adenosine and/or guanosine $3',5'$ cyclic monophosphate (cAMP and cGMP). PDEs can regulate the localization, duration, and amplitude of cyclic nucleotide signaling within subcellular domains. The second messengers cAMP and cGMP are responsible for the transduction of several extracellular signals, including hormones and neurotransmitters. The synthetized cAMP moves throughout the cell to sites where it can bind to and activate its target enzymes represented by cAMP- and cGMP-dependent protein kinases, such as protein kinase A (PKA) and protein kinase G (PKG). These kinases act by phosphorylating substrates such as ion channels, transcription factors, and contractile proteins that regulate key cellular functions. cAMP and cGMP signaling responses are distributed in different cellular regions, and this spatial compartmentalization suggests a specific regulation of the distinct pools of PKA and PKG. This idea was confirmed by observations of cAMP signaling in live cells by FRET that showed that the accumulation of this second messenger occurs in localized cAMP pools (Houslay, 1995). Such microdomains are created by physical interactions between different components of signaling cascades and structural elements of the cell. Critical processes are catalyzed by cAMP/cGMP hydrolyzing enzymes known as cyclic nucleotide PDEs. In particular, sequestration and anchoring of PDEs to distinct sites is the main mechanism responsible for cyclic nucleotide gradients allowing selective actions (Houslay and Milligan, 1997; Houslay and Adams, 2003).

The basis of this compartmentalization is that various PKA isoforms are bound with different specific intracellular sites by proteins called A-kinase anchoring proteins (AKAPs). It was postulated that AKAPs sequester PKA to distinct subcellular locations and allow specific enzymes to respond to changes in local cAMP concentrations (Rubin, 1994). High PDE activity reduces cellular cAMP levels and thus decreases the ability of anchored PKA to become active, whereas reduced PDE activity will favor PKA activation (Bauman and Scott, 2002). Inhibition of PDE activity in the brain can promote increased intracellular cAMP and/or cGMP levels, thereby modulating neuronal function. Twenty-one genes encode for the superfamily of PDEs, which is subdivided into 11 families according to structural and functional properties (Bender and Beavo, 2006). Each PDE family has several different isoforms and splice variants (Yan et al., 1994); they differ in their three-dimensional structure, mode of regulation, intracellular localization, cellular expression, pharmacological properties, and sensitivity to inhibitors. PDEs are restricted to specific intracellular sites such as cytosol, plasmatic membrane, and nuclear and cytoskeletal structures (Houslay, 1998, 2001). Individual isozymes modulate distinct regulatory pathways in the cell, and on the basis of substrate specificity they can be divided into three groups: cAMP-selective hydrolases (PDE 4, 7, and 8), cGMP-selective hydrolases (PDE 5, 6, and 9), and dual (cAMP and cGMP) hydrolases (PDE 1, 2, 3, 10, and 11).

12.4.1   PDEs in the Brain

Several PDEs are expressed in neurons and play different roles in cAMP and cGMP signaling. *In situ* hybridization and immunohistochemistry demonstrated that the PDE1A isoform is mostly expressed in cerebral cortex, striatum, and pyramidal cells of the hippocampus (Polli and Kincaid, 1994). PDE1B isoform is also expressed in several brain areas such as striatum, nucleus accumbens, dentate gyrus of hippocampus, medial thalamic nuclei, and brainstem (Yan et al., 1994; Menniti et al., 2006). Mice lacking PDE1B exhibit increased DARPP-32 phosphorylation at Thr34, thus indicating that PDE1B normally downregulates cAMP/PKA signaling in striatal neurons (Reed et al., 2002). PDE2A is typically localized in the cortex, hippocampus, and striatum (Repaske et al., 1993). PDE3A is relatively highly expressed in platelets, as well as in the vascular smooth muscle, cardiac myocytes, adipose tissue, liver, and in several cardiovascular tissues (Shakur et al., 2001). PDE4 family is the most clearly and best studied of PDEs. Four genes encoding different PDE4 enzymes (PDE4A, PDE4B, and PDE4D, but not PDE4C) are expressed in the CNS with high concentrations in the cortex, hippocampus, area postrema, and striatum (Cherry and Davis, 1999).

In the rodent brain, PDE5A mRNA was deeply studied in the Purkinje cells of the cerebellum, in the pyramidal cells of CA1, CA2, and CA3, as well as in the dentate gyrus of the hippocampus (Van Staveren et al., 2003). PDE6 was initially thought to be exclusively distributed to the retina; however, PDE6B mRNA expression was also described in mouse hippocampus (Jarnaess and Tasken, 2007). The PDE7 family is composed of two genes coding for the high-affinity, rolipram-insensitive cAMP-specific enzymes PDE7A and PDE7B. High mRNA concentrations of both PDE7A and PDE7B are expressed in rat brain and in numerous peripheral tissues, even if protein levels of these enzymes have not been reported. PDE7 mRNA are also found in the olfactory bulb and tubercle, the hippocampus, particularly in the granule cells of the dentate gyrus, and several brainstem nuclei as well as in cerebellum and several thalamic nuclei (Andreeva et al., 2001; Van Staveren et al., 2004).

The expression of mRNA of PDE9A in the rodent brain was described in the Purkinje cells and granule cells of the cerebellum, striatum olfactory bulb and tubercle, and CA1 and dentate gyrus of the hippocampus (Fujishige et al., 1999; Sasaki et al., 2002). In the human brain, PDE9 mRNA expression has been reported in the insula and the visual cortex; the CA1, CA2, and CA3 subfields; and the dentate gyrus of the hippocampus (Loughney et al., 1999). PDE10A is particularly expressed in the brain, with the highest levels in both the dorsal and ventral striatum (caudate nucleus, nucleus accumbens, and olfactory tubercle) and, to a lesser extent, in the cerebellum, thalamus, hippocampus, and spinal cord (Seeger et al., 2003; Hebb et al., 2004; Reyes-Irisarri et al., 2007). The presence of mRNA transcript PDE10A in the caudate region of the basal ganglia suggests a role in modulating striato nigral and striato pallidal pathways (Coskran et al., 2006).

12.4.2 Functions of PDEs in Relation to Their Distribution

PDE1B and PDE10A, as well as PDE2A, can catabolize both cAMP and cGMP, while PDE10A is membrane-bound in the vast majority of neurons and PDE1B is contained only in a soluble intracellular compartment. Moreover, membrane-bound PDE2A is specifically enriched in lipid rafts associated with high concentrations of adenylyl cyclase V/VI and PKA. Because of their distinct subcellular distribution in medium spiny neurons, they play a different role in regulating the excitability of medium spiny neurons (Siuciak et al., 2008; DiPilato et al., 2012). Moreover, PDEs, because of their ability to modulate cAMP/PKA signaling, can control the dopaminergic signaling in the striatum, where dopamine plays a key role in the regulation of motor and cognitive functions. Moreover, cAMP/PKA signaling cascade is essential for dopamine transmission (Zhu et al., 2004; Siuciak et al., 2006). Dopamine can have distinct effects in striatonigral or striatopallidal neurons. In fact, by acting on D1 receptors, dopamine stimulates cAMP/PKA signaling *via* the active G protein-mediated activation of adenylyl cyclase. Conversely, by acting on D2 receptors, dopamine inhibits cAMP/PKA signaling *via* the inactive G protein-mediated inactivation of adenylyl cyclase (Seino and Shibasaki, 2005). PDE10A and PDE4 are differently expressed in neuronal subtypes in the striatum, and such discrete cellular localization confers distinct roles in dopaminergic neurotransmission. Striatal PDE10A is localized proximally to the plasma membrane of postsynaptic sites in medium spiny neurons' dendritic spines (Stoof and Kebabian, 1981; Kotera et al., 2004). This particular localization allows PDE10A to regulate post-synaptic cyclic nucleotide signaling, which is involved in the integration of glutamatergic and dopaminergic neurotransmission. PDE10A is also highly expressed in medium spiny neurons' axons/terminals in the SNr and external globus pallidus.

In particular, PDE10A regulates cAMP/PKA signaling (Sano et al., 2008) as well as gene expression (Nishi et al., 2008) in both direct and indirect pathway neurons. In neurons of the direct pathway, PDE10A inhibition by papaverine upregulates cAMP/PKA signaling, thus leading to the potentiation of dopamine D1 receptor signaling by the phosphorylation of cAMP-dependent substrates, including the CREB and extracellular receptor kinase (ERK). PDE10A inhibition by papaverine is also able to upregulate cAMP/PKA signaling, in neurons of the indirect pathway, by potentiating adenosine A2A receptor signaling and inhibiting dopamine D2 receptor signaling simultaneously (Strick et al., 2010). Thus, PDE10A inhibition effectively counteracts dopamine D2 receptor signaling in striatopallidal neurons and potentiates D1 receptor signaling in striatonigral neurons, mainly *via* cAMP-mediated effects. Because the inhibition of conditioned avoidance response has been used as a measure of the antipsychotic activity of many drugs, PDE10A inhibitors have been suggested as therapeutic agents for schizophrenia. Indeed, the PDE10A inhibitor papaverine counteracts dopamine D2 receptor signaling and potentiates dopamine D1 receptor signaling, so that the pharmacological profile of papaverine resembles that of atypical antipsychotics (Siuciak et al., 2006). This observation supports the concept that PDE10A inhibition is beneficial for symptoms and cognitive deficits of psychosis.

On the other hand, PDE4B regulates cAMP/PKA signaling at striatal dopaminergic terminals, and inhibition of PDE4 by rolipram upregulates TH phosphorylation and dopamine synthesis, leading to an increase in dopaminergic tone (Menniti et al., 2007).

The expression level of PDE4B is higher in striatopallidal neurons than in striatonigral neurons where PDE4 inhibition selectively potentiates cAMP/PKA signaling. Rolipram treatment increases phosphorylation of Thr34 DARPP-32 in response to an adenosine A2A receptor agonist but has no effect on phosphorylation mediated by a dopamine D1 receptor agonist.

12.4.3   PDEs in Huntington's Disease

The intracellular cAMP and cGMP concentrations depend on the rate of their synthesis from ATP and GTP by adenylate and guanylate cyclase. PDEs hydrolyze cAMP and cGMP limiting both the duration and amplification of the cyclic nucleotide signal (Conti and Jin, 1999; Francis et al., 2000; Van Staveren et al., 2001). PDE1B levels were reduced in 12-week-old R6/2 HD mice where PDE4 distribution is mainly observed in the cortex (Giampà et al., 2009).

Regarding HD, the most interesting PDE is PDE10A, because it shows a very peculiar distribution in the striatum, which represents the main target of the disease. PDE10A is highly expressed in regions of the brain that are innervated by dopaminergic neurons such as the striatum, nucleus accumbens, and olfactory tubercle (Soderling et al., 1999). Moreover, it is strongly expressed in GABAergic spiny projection neurons with localization to the membrane of dendrites and dendritic spines. The impoverishment of PDE10A protein levels in the striatum has been associated with the impairment of motor functions in R6/1 and R6/2 mice. It has also been described that PDE10A expression levels are reduced in the postmortem brain of HD patients. Because cyclic nucleotides are important for intracellular signaling, these changes may contribute to the alteration of cell functions that cause motor, cognitive, or psychiatric disturbances observed in HD patients.

It has been shown that PDE inhibition has beneficial effects in HD animal models, leading to an apparent conflict between the decreased PDE levels associated with HD and the beneficial effect of PDE inhibitors in a genetic murine model of HD (Giampà et al., 2010). To address this issue, PDE10A protein expression levels in the R6/2 mice were investigated, placing particular attention to the different neuronal subpopulation of the striatum. The results demonstrated a dramatic increase in PDE10A in the medium spiny neurons of R6/2 transgenic HD mice compared to their wild-type littermates. Conversely, in striatal cholinergic interneurons, PDE10A levels were lower and were not significantly modified by disease progression. In the other subsets of striatal interneurons (parvalbuminergic, somatostatinergic, and calretininergic interneurons), PDE10A immunoreactivity was higher in the R6/2 compared to that in the wild-type mice. However, densitometric studies of the whole striatum showed that PDE10A immunoreactivity was lower in the R6/2 compared to that in the wild-type mice. Moreover, it was shown that PDE10A increases in the perikarya of projection neurons but is reduced in the whole striatum of the R6/2 mice. This suggests that, in HD, mutant huntingtin HTT protein disrupts PDE10A synthesis and trafficking, resulting in PDE10A accumulation in the perikarya of spiny projection neurons, which are vulnerable to the disease, thereby decreasing cAMP and cGMP locally. Therefore, even if levels of PDE10A are lower in the striatum *in toto*, the enzyme is too abundant in the somata of medium spiny neurons where it downregulates cyclic nucleotide signaling, which is dangerous for cell life. That study showed a particular resistance to HD neurodegeneration displayed by cholinergic interneurons, which contain a moderate amount of PDE10A in the early stages, both in the R6/2 and in the wild type (Fusco et al., 1999). Moreover, striatal cholinergic interneurons contain higher amounts of BDNF, compared to the more vulnerable medium spiny neurons, and are more enriched with phosphorylated CREB (Fusco et al., 2003). Therefore, it is possible that the low levels of PDE10A found in cholinergic interneurons are related to their selective resistance to HD neurodegeneration.

PDE10A immunoreactivity was observed in moderate amounts in the nuclei of all striatal interneurons except for cholinergic ones, and its levels were higher in the R6/2 than in the wild-type mice (Giampà et al., 2009). As mentioned earlier, the somewhat unexpected localization in interneurons can be explained by the observation that cyclic nucleotides and PKA have been described in the nuclei of brain cells (Van Staveren et al., 2002).

12.4.4   EFFECTS OF PDEs INHIBITION IN HUNTINGTON'S DISEASE

Phosphorylated CREB is differently expressed in several neuronal subpopulations of the striatum, both in control animals and in the murine model of HD. High levels of activated CREB are associated with the selective resistance of specific neuronal population to the neuronal damage (Lee et al., 2004; Giampà et al., 2006).

The design of drugs targeting the CREB loss of function could be considered as a powerful mean for the treatment of neurodegenerative disorders such as HD. The PDE4 inhibitor, rolipram, increases CREB phosphorylation, showing a neuroprotective effect in striatal spiny neurons, in the surgical model of HD (De March et al., 2006). The beneficial effects observed following rolipram treatment were also able to maintain BDNF protein expression levels. BDNF is, in fact, synthetized in the cortex and anterogradely transported to the striatum. Thus, the increased CREB phosphorylation exerted by rolipram in the quinolic acid (QA) model was likely responsible for the neuroprotection through an increase in cAMP levels.

In a later study, it was shown that rolipram is able to increase survival and could ameliorate clinical signs in the R6/2 mouse model of HD (De March et al., 2008). In that study, PDE4 inhibition through rolipram had neuroprotective effects by increasing both phosphorylated CREB and BDNF in the striatum. Rolipram prevented CREB binding protein sequestration into striatal NIIs, thus sparing parvalbuminergic interneurons of R6/2 mice and rescuing motor coordination and motor activity deficits (Giampà et al., 2009). Moreover, an increase in ERK phosphorylation was reported in the medium spiny neurons of the R6/2 mice after rolipram treatment. ERK phosphorylation has particular importance considering the altered activation of extracellular signal-regulated protein kinases in HD (Fusco et al., 2012).

Another possible target for neuroprotection in HD is PDE5, selective for cGMP. PDE5 is found in several brain regions, including the cortex, hippocampus, and basal ganglia (Marte et al., 2008; Puerta et al., 2009). In a recent study, Puerta and coworkers have shown that PDE5 inhibitors sildenafil and vardenafil were able to ameliorate neurological symptoms, reduce striatal projection neurons loss, and increase pCREB levels in the 3-nitroproprionic intoxication model of HD in rats (Puerta et al., 2010). Noteworthily, it was shown that mRNA and protein BDNF levels were significantly elevated in sildenafil-treated rat cortex, which accounted significantly for the neuroprotective effects.

These results provided a strong theoretical support for targeting cyclic nucleotides and CREB signaling through PDE inhibition. A PDE10 inhibitor (TP10, Pfizer) was administered in the QA rat surgical model of HD. The chronic administration of TP10 was able to reduce the QA lesion area by 52%, sparing medium spiny neurons, and to increase CREB levels in surviving striatal neurons (Giampà et al., 2010). Interestingly, TP10 treatment also had beneficial effects on cortical neurons. In fact, a decreased retrograde cortical neuron loss and increased levels of phosphorylated CREB and BDNF were observed, although the effect of TP10 on cortical levels of BDNF was moderate and only limited to the earlier time point. Because PDE10A is mostly expressed in striatal medium spiny neurons, it is possible that these effects on the survival of cortical neurons may be indirect. Following these results, PDE10A inhibition was further investigated by administering to the R6/2 mouse model of HD. Predictably, TP10 was able to rescue the neuronal loss, NIIs' formation, and microglial reaction and also TP-10 treatment was associated with a significant increase in phosphorylated CREB and BDNF in the cortex and striatum. The increase in the medium spiny neurons of cAMP signaling resulting from PDE10A inhibition may be trophic to these neurons *via* a number of downstream mechanisms. PDE10A inhibition in the wild-type mouse brain causes a robust increase in CREB phosphorylation downstream of cAMP, which is associated with a significant increase in BDNF levels in the striatum of R6/2 mice following TP-10 administration. Both CREB-mediated transcription and BDNF levels may contribute to the significant amelioration of striatal pathology resulting from treatment of the R6/2 mice with the PDE10A inhibitor TP-10. Moreover,

PDE10A inhibition also had a beneficial effect on cortical pathology in the R6/2 mice (Giampà et al., 2010). The beneficial effects observed after PDE10A inhibition on striatal pathology might contribute to maintain corticostriatal synaptic connections, which may reduce cortical neuron pathology by preventing retrograde degeneration. However, it is also conceivable that there is a direct effect of TP-10 treatment on cortical CREB phosphorylation and BDNF synthesis resulting from inhibition of the nuclear/perinuclear PDE10A present in the cortex.

Chronic inhibition of PDE10A promotes an up-regulation of mRNAs encoding genes such as PDE 1C prodynorphin, synaptotagmin 10, and diacylglycerol O-acyltransferase. Moreover, it produces a downregulation of mRNAs encoding choline acetyltransferase and Kv1.6, suggesting that the long-term suppression of PDE10A is associated with altered striatal excitability. These results support the hypothesis that PDE inhibitors could be considered a valid therapeutic approach to HD. However, more studies on the impact of PDEs inhibitors on patients' health are needed to promote a clinical trial for neurodegenerative diseases.

12.5 PARP-1 INHIBITION AS THERAPEUTIC APPROACH TO HD

Poly (ADP-ribose) polymerase 1 (PARP-1) is a nuclear enzyme involved in many physiological processes like DNA repair, genomic stability, and cellular programmed mechanism of death, apoptosis. In HD, PARP immunoreactivity was described in neurons and glial cells, demonstrating the role of apoptosis in this neurodegenerative disease and suggesting the use of a PARP-1 inhibitor as a possible treatment. In a very recent study, exciting data were described about the use, in the R6/2 mouse model of HD, of an anticancer drug, INO-1001, which acts as a PARP-1 inhibitor (Cardinale et al., 2015). This study reports beneficial effects of INO-1001 treatment in the R6/2 on survival, neurological impairment, and neuroprotection. Moreover, PARP-1 inhibitor INO-1001 showed an up-regulation of CREB activity and BDNF, and a beneficial effect on microglial activation, all contributing to the rescue of striatal neuronal cells.

We can conclude that a lot of progress has been made by therapeutic approaches to HD, compared to that based on dopamine depletion by antipsychotic drugs. Indeed, other mechanisms involved in the pathophysiology of HD neurodegeneration are being studied, and therefore new targets for drugs are being identified to promote neurorescue in HD.

REFERENCES

Albin RL, Tagle DA. Genetics and molecular biology of Huntington's disease. *Trends Neurosci* 1995; 18: 11–14.

Altar CA, Cai N, Bliven T et al. Anterograde transport of brain-derived neurotrophic factor and its role in the brain. *Nature* 1997; 389: 856–860.

Andreeva SG, Dikkes P, Epstein PM, Rosenberg PA. Expression of cGMP-specific phosphodiesterase 9A mRNA in the rat brain. *J Neurosci* 2001; 21: 9068–9076.

Auer RN, Olsson Y, Siesjö BK. Hypoglycemic brain injury in the rat. Correlation of density of brain damage with the EEG isoelectric time: A quantitative study. *Diabetes* 1984; 33(11): 1090–1098.

Autry AE, Monteggia LM. Brain-derived neurotrophic factor and neuropsychiatric disorders. *Pharmacol Rev* 2012; 64(2): 238–258.

Baquet ZC, Gorski JA, Jones KR. Early striatal dendrite deficits followed by neuron loss with advanced age in the absence of anterograde cortical brain-derived neurotrophic factor. *J Neurosci* 2004; 24: 4250–4258.

Bauman AL, Scott JD. Kinase- and phosphatase-anchoring proteins: Harnessing the dynamic duo. *Nat Cell Biol* 2002; 4: E203.

Bayer TE, Schramm M, Feldmann N, Knable MB, Falkai P. Antidepressant drug exposure is associated with mRNA levels of tyrosine receptor kinase B in major depressive disorder. *Prog Neuropsychopharmacol Biol Psychiatry* 2000; 24(6): 881–888.

Bender AT, Beavo JA. Cyclic nucleotide phosphodiesterases: Molecular regulation to clinical use. *Pharmacol Rev* 2006; 58: 488–520.

Bimonte HA, Nelson ME, Granholm AC. Age-related deficits as working memory load increases: Relationships with growth factors. *Neurobiol Aging* 2003; 24: 37–48.

Cardinale A, Paldino E, Giampà C, Bernardi G, Fusco FR. PARP-1 inhibition is neuroprotective in the R6/2 mouse model of Huntington's disease. *PLoS One* 2015; 10(8): e0134482.

Chen G, Creson T, Engel S, Hao Y, Wang G. Neurotrophic actions of moodstabilizers: A recent research discovery and its potential clinical applications. *Curr Psych Rev* 2005; 1(2): 173–185.

Chen G, Manji HK. The extracellular signal-regulated kinase pathway: An emerging promising target for mood stabilizers. *Curr Opin Psychiatry* 2006; 19(3): 313–323.

Chen JY, Wang EA, Cepeda C, Levine MS. Dopamine imbalance in Huntington's disease: A mechanism for the lack of behavioral flexibility. *Front Neurosci* 2013; 7: 114.

Cherry JA, Davis RL. Cyclic AMP phospho-diesterases are localized in regions of the mouse brain associated with reinforcement, movement, and affect. *J Comp Neurol* 1999; 407: 287–301.

Conti M, Jin SL. The molecular biology of cyclic nucleotide phosphodiesterases. *Prog Nucleic Acid Res Mol Biol* 1999; 63: 1–38.

Coskran TM, Morton D, Menniti FS et al. Immunohistochemical localization of phospho-diesterase 10A in multiple mammalian species. *J Histochem Cytochem* 2006; 54: 1205–1213.

Creson TK, Yuan P, Manji HK, Chen G. Evidence for involvement of ERK, PI3K, and RSK in induction of Bcl-2 by valproate. *J Mol Neurosci* 2009; 37(2): 123–134.

De March Z, Giampà C, Patassini S, Bernardi G, Fusco FR. Cellular localization of TRPC5 in the substantia nigra of rat. *Neurosci Lett* 2006; 402(1–2): 35–39.

De March Z, Giampà C, Patassini S, Bernardi G, Fusco FR. Beneficial effects of rolipram in the R6/2 mouse model of Huntington's disease. *Neurobiol Dis* 2008; 30(3): 375–387.

Di Benedetto B, Radecke J, Schmidt M, Rupprecht R. Acute antidepressant treatment differently modulates ERK/MAPK activation in neurons and astrocytes of the adult mouse prefrontal cortex. *Neuroscience* 2013; 232: 161–168.

DiFiglia M, Sapp E, Chase KO et al. Aggregation of huntingtin in neuronal intranuclear inclusions and dystrophic neurites in brain. *Science* 1997; 27: 1990–1993.

DiPilato LM, Yang JH, Ni Q, Saucerman JJ, Zhang J. Regulation of nuclear PKA revealed by spatiotemporal manipulation of cyclic AMP. *Nat Chem Biol* 2012; 8(4): 375–382.

Duan W, Guo Z, Jiang H et al. Dietary restriction normalizes glucose metabolism and BDNF levels, slows disease progression, and increases survival in huntingtin mutant mice. *Proc Natl Acad Sci USA* 2003; 100(5): 2911–2916.

Duan W, Peng Q, Masuda N, Ford E, Tryggestad E, Ladenheim B. Sertraline slows disease progression and increases neurogenesis in N171-82Q mouse model of Huntington's disease. *Neurobiol Dis* 2008; 30(3): 312–322.

Ferrer I, Goutan E, Marin C, Rey MJ, Ribalta T. Brain-derived neurotrophic factor in Huntington disease. *Brain Res* 2000; 866: 257–261.

Francis SH, Turko IV, Corbin JD. Cyclic nucleotide phosphodiesterases: Relating structure and function. *Prog Nucleic Acid Res Mol Biol* 2000; 65: 1–52.

Fujishige K, Kotera J, Michibata H, Yuasa K, Takebayashi S, Okumura K, Omori K. Cloning and characterization of a novel human phosphodiesterase that hydrolyzes both cAMP and cGMP (PDE10A). *J Biol Chem* 1999; 274: 18438–18445.

Fusco FR, Anzilotti S, Giampà C et al. Changes in the expression of extracellular regulated kinase (ERK 1/2) in the R6/2 mouse model of Huntington's disease after phosphodiesterase IV inhibition. *Neurobiol Dis* 2012; 46(1): 225–233.

Fusco FR, Chen Q, Lamoreaux WJ et al. Cellular localization of huntingtin in striatal and cortical neurons in rats: Lack of correlation with neuronal vulnerability in Huntington's disease. *J Neurosci* 1999; 19(4): 1189–1202.

Fusco FR, Zuccato C, Tartari M et al. Co-localization of brain-derived neurotrophic factor (BDNF) and wild-type huntingtin in normal and quinolinic acid-lesioned rat brain. *Eur J Neurosci* 2003; 18(5): 1093–1102.

Gharami K, Xie Y, An JJ, Tonegawa S, Xu B. Brain-derived neurotrophic factor over-expression in the forebrain ameliorates Huntington's disease phenotypes in mice. *J Neurochem* 2008; 105: 369–379.

Giampà C, DeMarch Z, D'Angelo V et al. Striatal modulation of cAMP-response-element-binding protein (CREB) after excitotoxic lesions: Implications with neuronal vulnerability in Huntington's disease. *Eur J Neurosci* 2006; 23(1): 11–20.

Giampà C, Laurenti D, Anzilotti S, Bernardi G, Menniti FS, Fusco FR. Inhibition of the striatal specific phosphodiesterase PDE10A ameliorates striatal and cortical pathology in R6/2 mouse model of Huntington's disease. *PLoS One* 2010; 5(10): e13417.

Giampà C, Middei S, Patassini S et al. Phosphodiesterase type IV inhibition prevents sequestration of CREB binding protein, protects striatal parvalbumin interneurons and rescues motor deficits in the R6/2 mouse model of Huntington's disease. *Eur J Neurosci* 2009; 29(5): 902–910.

Gray MA, Egan GF, Ando A, Churchyard A, Chua P, Stout JC, Georgiou-Karistianis N. Prefrontal activity in Huntington's disease reflects cognitive and neuropsychiatric disturbances: The IMAGE-HD study. *Exp Neurol* 2013; 239: 218–228.

Grote HE, Bull ND, Howard ML, Van Dellen A, Blakemore C, Bartlett PF. Cognitive disorders and neurogenesis deficits in Huntington's disease mice are rescued by fluoxetine. *Eur J Neurosci* 2005; 22(8): 2081–2088.

Groves JO. Is it time to reassess the BDNF hypothesis of depression? *Mol Psychiatry* 2007; 12(12): 1079–1088.

Hebb AL, Robertson HA, Denovan-Wright EM. Striatal phosphodiesterase mRNA and protein levels are reduced in Huntington's disease transgenic mice prior to the onset of motor symptoms. *Neuroscience* 2004; 123: 967–981.

Hong SL, Cossyleon D, Hussain WA, Walker LJ, Barton SJ, Rebec GV. Dysfunctional behavioral modulation of corticostriatal communication in the R6/2 mouse model of Huntington's disease. *PLoS One* 2012; 7(10): e47026.

Houslay MD. Compartmentalization of cyclic AMP phosphodiesterases, signaling 'crosstalk', desensitization and the phosphorylation of Gi-2 add cell specific personalization to the control of the levels of the second messenger cyclic AMP. *Adv Enzyme Regul* 1995; 35: 303–338.

Houslay MD. Adaptation in cyclic AMP signalling processes: A central role for cyclic AMP phosphodiesterases. *Semin Cell Dev Biol* 1998; 9: 161.

Houslay MD. PDE4 cAMP-specific phosphodiesterases. *Prog Nucleic Acid Res Mol Biol* 2001; 6: 249.

Houslay MD, Adams DR. PDE4 cAMP phosphodiesterases: Modular enzymes that orchestrate signalling cross-talk, desensitization and compartmentalization. *Biochem J* 2003; 370: 1–18.

Houslay MD, Milligan G. Tailoring cAMP-signalling responses through isoform multiplicity. *Trends Biochem Sci* 1997; 22: 217–224.

Hunsberger J, Austin DR, Henter ID, Chen G. The neurotrophic and neuroprotective effects of psychotropic agents. *Dialogues Clin Neurosci* 2009; 11(3): 333–348.

Jarnaess E, Tasken K. Spatiotemporal control of cAMP signalling processes by anchored signalling complexes. *Biochem Soc Trans* 2007; 35: 931–937.

Kalimo H, Auer RN, Siesjo BK. The temporal evolution of hypoglycemic brain damage. III. Light and electron microscopic findings in the rat caudato putamen. *Acta Neuropathol* 1985; 67: 37–50.

Kegelmeyer DA, Kloos AD, Fritz NE, Fiumedora MM, White SE, Kostyk SK. Impact of tetrabenazine on gait and functional mobility in individuals with Huntington's disease. *J Neurol Sci* 2014; 347(1–2): 219–223.

Klempin F, Babu H, Tonelli DDP, Alarcon E, Fabel K, Kempermann G. Oppositional effects of serotonin receptors 5-HT1a, 2, and 2c in the regulation of adult hippocampal neurogenesis. *Front Mol Neurosci* 2010; 3(14): 32.

Kotera J, Sasaki T, Kobayashi T, Fujishige K, Yamashita Y, Omori K. Subcellular localization of cyclic nucleotide phosphodiesterase type 10A variants, and alteration of the localization by cAMP-dependent protein kinase-dependent phosphorylation. *J Biol Chem* 2004; 279: 4366–4375.

Kumar P, Kalonia H, Kumar A. Nitric oxide mechanism in the protective effects of antidepressants against 3-Nitropropionic acid induced cognitive deficit, glutathione and mitochondrial alteration in animal model of Huntington's disease. *Behav Pharmacol* 2010; 21(3): 217–230.

Kumar P, Kumar A. Possible role of sertraline against 3-nitropropionic acid induced behavioral, oxidative stress and mitochondrial dysfunctions in rat brain. *Prog Neuropsychopharmacol Biol Psychiatry* 2009a; 33(1): 100–108.

Kumar P, Kumar A. Protective role of sertraline against 3-nitropropionic acid induced cognitive dysfunction and redox ratio in striatum, cortex and hippocampus of rat brain. *Indian J Exp Biol* 2009b; 47(9): 715–722.

Lauterbach EC. Neuroprotective effects of psychotropic drugs in Huntington's disease. *Int J Mol Sci* 2013; 14(11): 22558–22603.

Lee HT, Chang YC, Wang LY et al. cAMP response element-binding protein activation in ligation preconditioning in neonatal brain. *Ann Neurol* 2004; 56: 611–623.

Loughney K, Snyder PB, Uher L, Rosman GJ, Ferguson K, Florio VA. Isolation and characterization of PDE10A, a novel human 3′,5′-cyclic nucleotide phosphodiesterase. *Gene* 1999; 234: 109–117.

Malberg JE, Eisch AJ, Nestler EJ, Duman RS. Chronic antidepressant treatment increases neurogenesis in adult rat hippocampus. *J Neurosci* 2000; 20(24): 9104–9110.

Marte A, Pepicelli O, Cavallero A, Raiteri M, Fedele E. In vivo effects of phosphodiesterase inhibition on basal cyclic guanosine monophosphate levels in the prefrontal cortex, hippocampus and cerebellum of freely moving rats. *J Neurosci Res* 2008; 86: 3338–3347.

Menniti FS, Chappie TA, Humphrey JM, Schmidt CJ. Phosphodiesterase 10A inhibitors: A novel approach to the treatment of the symptoms of schizophrenia. *Curr Opin Investig Drugs* 2007; 8(1): 54–59. Review.

Menniti FS, Faraci WS, Schmidt CJ. Phosphodiesterases in the CNS: Targets for drug development. *Nat Rev Drug Discov* 2006; 8: 660–670.

Mercier G, Lennon AM, Renouf B, Dessouroux A, Ramaugé M, Courtin F, Pierre M. MAP kinase activation by fluoxetine and its relation to gene expression in cultured rat astrocytes. *J Mol Neurosci* 2004; 24(2): 207–216.

Miller BH, Schultz LE, Gulati A, Cameron MD, Pletcher MT. Genetic regulation of behavioral and neuronal responses to fluoxetine. *Neuropsychopharmacology* 2008; 33(6): 1312–1322.

Nahon E, Israelson A, Abu-Hamad S, Varda SB. Fluoxetine (Prozac) interaction with the mito-chondrial voltage-dependent anion channel and protection against apoptotic cell death. *FEBS Lett* 2005; 579(22): 5105–5110.

Nibuya M, Nestler EJ, Duman RS. Chronic antidepressant administration increases the expression of cAMP response element-binding protein (CREB) in rat hippocampus. *J Neurosci* 1996; 16(7): 2365–2372.

Nishi A, Kuroiwa M, Miller DB et al. Distinct roles of PDE4 and PDE10A in the regulation of cAMP/PKA signaling in the striatum. *J Neurosci* 2008; 28(42): 10460–10471.

Nucifora Jr FC, Sasaki M, Peters MF et al. Interference by huntingtin and atrophin-1 with cbp-mediated transcription leading to cellular toxicity. *Science* 2001; 291: 2423–2428.

Pan HS, Wang RY. MDMA: Further evidence that its action in the medial prefrontal cortex is mediated by the serotonergic system. *Brain Res* 1991; 539(2): 332–336.

Papaleo F, Silverman JL, Aney J, Tian Q, Barkan CL, Chadman KK, Crawley JN. Working memory deficits, increased anxiety-like traits, and seizure susceptibility in BDNF over-expressing mice. *Learn Mem* 2011; 18: 534–544.

Peng Q, Masuda N, Jiang M, Li Q, Zhao M, Ross CA, Duan W. The antidepressant sertraline improves the phenotype, promotes neurogenesis and increases BDNF levels in the R6/2 Huntington's disease mouse model. *Exp Neurol* 2008; 210(1): 154–163.

Pla P, Orvoen S, Saudou F, David DJ, Humbert S. Mood disorders in Huntington disease: From behaviour to cellular and molecular mechanism. *Front Behav Neurosci* 2014; 8: 135.

Polli JW, Kincaid RL. Expression of a calmodulin-dependent phosphodiesterase isoform (PDE1B1) correlates with brain regions having extensive dopaminergic innervation. *J Neurosci* 1994; 14: 1251–1256.

Puerta E, Hervias I, Barros-Miñones L et al. Sildenafil protects against 3-nitropropionic acid neurotoxicity through the modulation of calpain, CREB, and BDNF. *Neurobiol Dis* 2010; 38(2): 237–245.

Puerta E, Hervias I, Goñi-Allo B, Lasheras B, Jordan J, Aguirre N. Phosphodiesterase 5 inhibitors prevent 3,4-methylenedioxymethamphetamine-induced 5-HT deficits in the rat. *J Neurochem* 2009; 108(3): 755–766.

Rantamaki T, Hendolin P, Kankaanpaa A et al. Pharmacologically diverse antidepressants rap-idly activate brain derived neurotrophic factor receptor TrkB and induce phospholipase-C gamma signalling pathways in mouse brain. *Neuropsychopharmacology* 2007; 32(10): 2152–2162.

Reed TM, Repaske DR, Snyder GL, Greengard P, Vorhees CV. Phosphodiesterase 1B knock-out mice exhibit exaggerated locomotor hyperactivity and DARPP-32 phosphorylation in response to dopamine agonists and display impaired spatial learning. *J Neurosci* 2002; 22: 5188–5197.

Renoir T, Pang TY, Zajac MS et al. Treatment of depressive like behaviour in Huntington's disease mice by chronic sertraline and exercise. *Br J Pharmacol* 2012; 165(5): 1375–1389.

Repaske DR, Corbin JG, Conti M, Goy MF. A cyclic GMP-stimulated cyclic nucleotide phos-phodiesterase gene is highly expressed in the limbic system of the rat brain. *Neuroscience* 1993; 56: 673–686.

Reyes-Irisarri E, Markerink-Van Ittersum M, Mengod G, de Vente J. Expression of the cGMP-specific phosphodiesterases 2 and 9 in normal and Alzheimer's disease human brains. *Eur J Neurosci* 2007; 25: 3332–3338.

Riccio A, Ahn S, Davenport CM, Blendy JA, Ginty DD. Mediation by a CREB family tran-scription factor of NGF-dependent survival of sympathetic neurons. *Science* 1999; 286(5448): 2358–2361.

Rubin CS. A kinase anchor proteins and the intracellular targeting of signals carried by cAMP. *Biochim Biophys Acta* 1994; 224: 467–479.

Saarelainen T, Hendolin P, Lucas G et al. Activation of the TrkB neurotrophin receptor is induced by antidepressant drugs and is required for antidepressant induced behavioral effects. *J Neurosci* 2003; 23(1): 349–357.

Sackley C, Hoppitt TJ, Calvert M, Gill P, Eaton B, Yao G, Pall H. Huntington's disease: Current epidemiology and pharmacological management in UK primary care. *Neuroepidemiology* 2011; 37(3–4): 216–221.

Samadi P, Boutet A, Rymar VV et al. Relationship between BDNF expression in major striatal afferents, striatum morphology and motor behavior in the R6/2 mouse model of Huntington's disease. *Genes Brain Behav* 2013; 12(1): 108–124.

Sano H, Nagai Y, Miyakawa T, Shigemoto R, Yokoi M. Increased social interaction in mice deficient of the striatal medium spiny neuron-specific phos-phodiesterase 10A2. *J Neurochem* 2008; 105: 546–556.

Sasaki T, Kotera J, Omori K. Novel alternative splice variants of rat phosphodiesterase 7B showing unique tissue-specific expression and phosphorylation. *Biochem J* 2002; 361: 211–220.

Seeger TF, Bartlett B, Coskran TM et al. Immunohistochemical localization of PDE10A in the rat brain. *Brain Res* 2003; 985: 113–126.

Seino S, Shibasaki T. PKA-dependent and PKA-independent pathways for cAMP-regulated exocytosis. *Physiol Rev* 2005; 85: 1303–1342.

Shakur Y, Holst LS, Landstrom TR, Movsesian M, Degerman E, Manganiello V. Regulation and function of the cyclic nucleotide phosphodiesterase (PDE3) gene family. *Prog Nucleic Acid Res Mol Biol* 2001; 66: 241–277.

Shirayama Y, Chen AC-H, Nakagawa S, Russell DS, Duman RS. Brain derived neurotrophic factor produces antidepressant effects in behavioral models of depression. *J Neurosci* 2002; 22(8): 3251–3261.

Siuciak JA, Chapin DS, Harms JF, Lebel LA, McCarthy SA, Chambers L, Shrikhande A, Wong S, Menniti FS, Schmidt CJ. Inhibition of the striatum-enriched phosphodiesterase PDE10A: A novel approach to the treatment of psychosis. *Neuropharmacology* 2006; 51(2): 386–396.

Siuciak JA, Lewis DR, Wiegand SJ, Lindsay RM. Antidepressant-like effects of brain-derived neurotrophic factor (BDNF). *Pharmacol Biochem Behav* 1997; 56(1): 131–137.

Siuciak JA, McCarthy SA, Chapin DS, Martin AN. Behavioral and neurochemical characterization of mice deficient in the phosphodiesterase-4B (PDE4B) enzyme. *Psychopharmacology* 2008; 197: 115–126.

Smith ML, Auer RN, Siesjö BK. The density and distribution of ischemic brain injury in the rat following 2–10 min of forebrain ischemia. *Acta Neuropathol* 1984; 64(4): 319–332.

Soderling SH, Bayuga SJ, Beavo JA. Isolation and characterization of a dual-substrate phosphodiesterase gene family: PDE10A. *Proc Natl Acad Sci USA* 1999; 96: 7071–7076.

Stoof JC, Kebabian JW. Opposing roles for D-1 and D-2 dopamine receptors in efflux of cyclic AMP from rat neostriatum. *Nature* 1981; 294: 366–368.

Strick CA, James LC, Fox CB, Seeger TF, Menniti FS, Schmidt CJ. Alterations in gene regulation following inhibition of the striatum-enriched phosphodiesterase, PDE10A. *Neuropharmacology* 2010; 58(2): 444–445.

Sugars KL, Brown R, Cook LJ, Swartz J, Rubinsztein DC. Decreased cAMP response element-mediated transcription, an early event in exon 1 and full-length cell models of Huntington's disease that contributes to polyglutamine pathogenesis. *J Biol Chem* 2004; 279: 4988–4999.

Tang T-S, Slow E, Lupu V et al. Disturbed Ca^{2+} signalling and apoptosis of medium spiny neurons in Huntington's disease. *Proc Natl Acad Sci USA* 2005; 102(7): 2602–2607.

Tao X, Finkbeiner S, Arnold DB, Shaywitz AJ, Greenberg ME. Ca^{2+} influx regulates BDNF transcription by a CREB family transcription factor-dependent mechanism. *Neuron* 1998; 20(4): 709–726.

The Huntington's Disease Collaborative Research Group. A novel gene containing a trinucleotide repeat that is expanded and unstable on Huntington's disease chromosomes. *Cell* 1993; 72(6): 971–983.

Unschuld PG, Joel SE, Pekar JJ, Reading SA, Oishi K, McEntee J. Depressive symptoms in prodromal Huntington's disease correlate with Stroop-interference related functional connectivity in the ventromedial prefrontal cortex. *Psychiatry Res* 2012; 203(2–3): 166–174.

Van Staveren WC, Glick J, Markerink-van Ittersum M et al. Cloning and localization of the cGMP-specific phosphodiesterase type 9 in the rat brain. *J Neurocytol* 2002; 31(8–9): 729–741.

Van Staveren WCG, Markerink-van Ittersum M, Steinbusch HWM, de Vente J. The effects of phosphodiesterase inhibition on cyclic GMP and cyclic AMP accumulation in the hippocampus of the rat. *Brain Res* 2001; 888: 275–286.

Van Staveren WC, Steinbusch HW, Markerink-van Ittersum M, Behrends S, de Vente J. Species differences in the localization of cGMP-producing and NO-responsive elements in the mouse and rat hippocampus using cGMP immunocyto-chemistry. *Eur J Neurosci* 2004; 19: 2155–2168.

Van Staveren WC, Steinbusch HW, Markerink-Van Ittersum M et al. mRNA expression patterns of the cGMP-hydrolyzing phosphodiesterases types 2, 5, and 9 during development of the rat brain. *J Comp Neurol* 2003; 467: 566–580.

Wyttenbach A, Swartz J, Kita H et al. Polyglutamine expansions cause decreased CRE-mediated transcription and early gene expression changes prior to cell death in an inducible cell model of Huntington's disease. *Hum Mol Genet* 2001; 10: 1829–1845.

Xie Y, Hayden MR, Xu B. BDNF overexpression in the forebrain rescues Huntington's disease phenotypes in YAC128 mice. *J Neurosci* 2010; 30(44): 14708–14718.

Yan C, Bentley JK, Sonnenburg WK, Beavo JA. Differential expression of the 61 kDa and 63 kDa calmodulin-dependent phosphodiesterases in the mouse brain. *J Neurosci* 1994; 14: 973–984.

Zajac M, Pang T, Wong N, Weinrich B, Leang L, Craig J, Saffery R, Hannan A. Wheel running and environmental enrichment differentially modify exon specific BDNF expression in the hippocampus of wild-type and pre-motor symptomatic male and female Huntington's disease mice. *Hippocampus* 2010; 20(5): 621–636.

Zhang Y, Li M, Drozda M et al. Depletion of wild-type huntingtin in mouse models of neurologic diseases. *J Neurochem* 2003; 87(1): 101–106.

Zhu G, Okada M, Yoshida S, Hirose S, Kaneko S. Pharmacological discrimination of protein kinase associated exocytosis mechanisms between dopamine and 3,4-dihydroxyphenylalanine in rat striatum using in vivo microdialysis. *Neurosci Lett* 2004; 363: 120–124.

Zuccato C, Cattaneo E. Role of brain-derived neurotrophic factor in Huntington's disease. *Prog Neurobiol* 2007; 81(5–6): 294–330.

Zuccato C, Cattaneo E. Brain derived neurotrophic factor in neurodegenerative diseases. *Nat Rev Neurol* 2009; 5(6): 311–322.

Zuccato C, Tartari M, Crotti A, Goffredo D, Valenza M, Conti L, Cataudella T. Huntingtin interacts with REST/NRSF to modulate the transcription of NRSE-controlled neuronal genes. *Nat Genet* 2003; 35(1): 76–83.

13 Drug Repositioning Opportunities in Psychiatry

Alexander W. Charney, Joseph R. Scarpa,
Douglas M. Ruderfer, and Dennis S. Charney

CONTENTS

13.1 INTRODUCTION

Since the mid-twentieth century, "single neurotransmitter, single disease" hypotheses have shaped the understanding of the neurobiology of mental illness. Beginning with the catecholamine hypothesis of bipolar disorder (Schildkraut, 1965) and the dopamine hypothesis of schizophrenia (Van Rossum, 1966), causality has been proposed for most neurotransmitter/disorder combinations, with abnormalities in dopaminergic, adrenergic, noradrenergic, serotonergic, glutamatergic, GABAergic, and cholinergic neurotransmission having all at one time or another been posited as the underlying cause of depressive, anxiety, psychotic, manic, and autistic disorders (Schildkraut, 1965; Van Rossum, 1966; Emrich et al., 1980; Charney and Redmond, 1983; Kahn and Van Praag, 1988; Dilsaver and Coffman, 1989; Leiva, 1990; Hussman, 2001; Mahmood and Silverstone, 2001; Battaglia, 2002; Baumeister and Hawkins, 2004; Bergink et al., 2004; Previc, 2007; Yoo et al., 2007; Luscher et al., 2011; Choudhury et al., 2012; Egerton and Stone, 2012; Möhler, 2012; Sanacora et al., 2012; Harrington et al., 2013; Pålsson et al., 2015). As others have noted (Insel and Scolnick, 2006), these hypotheses have resulted in few new treatments over the past 5 decades.

The biopsychosocial model of human disease posits that illness results from ceaseless feedback between social and molecular forces (Engel, 1977).

Consistent with this idea, social (e.g., economic status), biological (e.g., neuroimaging, genetics), and socio-biological (e.g., gender, race) variables have all been shown to impact the risk of mental illness. Critically, no single data dimension can fully explain how a mental illness develops. Consider, for instance, recent advances in psychiatric genetics. A large-scale genome-wide association study (GWAS) of schizophrenia recently reported 108 independent genomic loci contributing to disease risk, spanning over 300 genes on 20 chromosomes (Schizophrenia Working Group of the Psychiatric Genomics Consortium, 2014). This and earlier such international collaborations provide direct evidence for the biological basis of schizophrenia. However, genetics alone is unable to paint a nuanced picture of the neurobiology underlying psychosis. To achieve a complete, definitive understanding requires integrating all dimensions of biopsychosocial data. Predictive modeling using a suite of statistical and graph theoretical tools makes this possible in principle (Zhu et al., 2012). By inferring causal relationships between the variables in the system (e.g., genetics → gene expression rather than gene expression → genetics), predictions can be made regarding how higher-order phenotypes, such as psychiatric symptoms, will change when these variables are perturbed (Schadt, 2009).

A natural application of predictive modeling in psychiatry is computational drug repositioning. In simple terms, pipelines for computational repositioning take as input high-dimensional data about diseases and drugs and predict novel therapeutic interventions. While yet to result in new treatments for psychiatric conditions, this method is poised to feature prominently in clinical research in the years ahead. As such, in this chapter we aim to survey the current state of repositioning opportunities in psychiatry. Specifically, our aims are three-fold. First, we will attempt to illustrate the types of large-scale biological data sets that are of value for repositioning efforts in psychiatry. Second, we aim to emphasize the translational potential of this method by profiling select efforts that have already been made to integrate psychiatric data types. Finally, we will anticipate where such efforts might fall short and propose expanding the notion of repositioning to include components of the therapeutic discovery process beyond the active molecule.

13.2   DRUG INDICATIONS, SIDE EFFECTS, AND MECHANISMS

The most influential biological theories of mental illness have been extrapolated from pharmacological observations of drugs that cause and treat psychiatric phenomena. The dopamine and glutamate hypotheses of schizophrenia, for instance, arose from observations of anti- and pro-psychotic agents, respectively (Moghaddam, 2004; Kendler and Schaffner, 2011). While predictive models of psychiatric illness, in contrast, do not rely on drug indications/side effects as the sole means of inferring pathogenesis, these data types still play a critical role in the prediction process. The resources currently available for finding known drug–phenotype links in psychiatry are therefore worth considering in some detail.

Contrary to intuition, grouping drugs by psychiatric indication is a formidable task. The most obvious method is to utilize a standardized drug classification scheme that groups drugs by their indications. The Anatomical Therapeutic Chemical Classification (ATCC) (http://www.whocc.no/atc_ddd_index/) is a hierarchical, indication-based categorization of pharmacological compounds. There are five levels in the hierarchy, level one is the most broad (e.g., "nervous system"), level three contains indication information (e.g., "antipsychotics"), and level five is the drug itself (e.g., "clozapine"). The ATCC categorizes psychiatric medications into indications using terms such as "antipsychotics" (64 drugs), "anxiolytics" (35 drugs), and "antidepressants" (65 drugs). Unfortunately, the ATCC and similar systems are not comprehensive. Drugs are included only at the request of manufacturers, regulatory agencies, researchers, and other users. In some cases, drug classifications are not evidence-based. Lithium, for instance, is classified as an antipsychotic in the ATCC despite no evidence for its effectiveness in treating psychosis (Leucht et al., 2007). Additionally, drugs are not always classified according to all of their indications. Selective serotonin reuptake inhibitors (SSRIs) are the first-line agents for the treatment of generalized anxiety disorder, yet none of the SSRIs are classified under the level three group "anxiolytics" in the ATCC (all are classified as "antidepressants"). Finally, in some instances, entire classes of medications do not have a corresponding category in the classification scheme. For example, there is no "mood stabilizer" category in level three of the ATCC.

Some of these limitations can be addressed using databases maintained by drug regulatory agencies. FDALabel (http://www.fda.gov/) is a web-based application that allows users to search a database of 70,000 manufacturer drug labels. Each label contains information such as drug name, approved indications, dosing, and side effects. A search for labels with obsessive-compulsive disorder (OCD) listed as an indication returned 323 results for eight unique drugs. The ATCC, in contrast, does not contain a category for drugs used to treat OCD. A major limitation of this resource, however, is the lack of a standardized medical lexicon across drug labels. Consider the task of identifying drugs in FDALabel that treat bipolar disorder. The number of drug labels returned by the search differs depending on if the text used to search is "bipolar disorder" (394 labels), "bipolar I disorder" (614 labels), "bipolar II disorder" (102 labels), "manic-depression" (571 labels), or "manic episode" (163 labels), despite the fact that all of these terms describe the same illness.

Neither standardized classification schemes nor regulatory databases are sufficient for identifying instances where preliminary evidence suggests for a drug's effectiveness, but formal approval or general acceptance by practitioners has yet to be established. Finding these putative indications requires clinical trial registries, such as clinicaltrials.gov. Started in 1997 by the FDA, this registry currently contains data on 200,000 registered trials. Each entry in the register contains basic trial information such as disease, intervention, and trial phase. By definition, drugs that are the subject of phase III trials must have shown some degree of efficacy in phase II. Using this fact, this database can be harnessed to detect putative drug indications despite the fact that the actual study results are rarely reported (Anderson et al., 2015). For example, at the time of writing, the register contains 425 phase III trials of drugs to treat schizophrenia. In nearly half, the drug under study is not a recognized antipsychotic.

Notable among these are omega-3 fatty acids, which have been found to be protective against developing psychosis in at-risk individuals (Amminger et al., 2010). As the pharmacological features of these less established indications may be entirely different than those of known treatments, they provide unique biological information to incorporate into predictive models.

Even more challenging than linking drugs to psychiatric indications is linking them to psychiatric side effects. A clinical trial is the first step in evaluating the safety of a new drug but cannot be expected to identify all side effects because only a small number of patients are studied and only for a brief length of time (Karimi et al., 2015). Passive and active post-marketing surveillance methods are therefore relied upon for drug side effect profiling. "Passive" primarily refers to unsolicited side effect reports made by medical professionals, patients, and drug companies. These reports are made through regulatory agency programs such as the FDA Adverse Event Reporting System (FAERS) (http://www.fda.gov). Moore et al. (2010) mined FAERS reports and found an association between antidepressants and violent behavior, a link that had been suggested in case reports but never tested systematically. Generally speaking, however, drawing conclusions from passive surveillance is highly problematic due to reporter bias, with estimates that up to 94% of side effects go unreported (Hazell and Shakir, 2006). "Active" surveillance refers to the systematic detection of side effects through mining of administrative health databases, electronic medical records, online forums, social media, web-search logs, medical literature, and other resources. Electronic health records of thousands of psychiatric patients, for instance, were analyzed to find a previously unreported association between the antipsychotic levomepromazine and the development of nightmares (Eriksson et al., 2014). Active surveillance methods also have limitations, mainly in that the resources being mined consist of large amounts of free text (e.g., physicians notes, twitter posts) that natural language processing algorithms have difficulty parsing because of the unique spelling and grammatical patterns characteristic of the language used in these contexts (Karimi et al., 2015; Raja and Jonnalagadda, 2015).

When combined with information from large drug chemistry databases, the relationships identified through the resources outlined above make it possible to identify pharmacological features (e.g., compound structures) characteristic of modulating a disease or symptom. By overlaying these features with molecular data on the illness itself in predictive models, repositioning opportunities with a greater likelihood of succeeding may be detected.

13.3   GENOMICS

Genetic variation plays a central role in predictive modeling of human disease because it is a natural source of perturbation in the biopsychosocial system. Due to this property, causal links between gene expression, protein function, metabolic activity, and higher-order phenotypic traits such as psychiatric symptoms can be inferred statistically (Schadt, 2009). Both common and rare genetic variants have been utilized to predict repositioning opportunities for nonpsychiatric disorders (St. Hilaire et al., 2011; Sanseau et al., 2012).

In the last decade, the contribution of common genetic variation to several psychiatric diseases has come to light through GWAS. The largest identified a total of 108 unique common variant associations with schizophrenia (Schizophrenia Working Group of the Psychiatric Genomics Consortium, 2014). Large studies of bipolar disorder have also implicated several common variants (Charney et al., 2017), as have smaller studies of autism (Glessner et al., 2014), and work ongoing for disorders such as depression, OCD, and anorexia nervosa promise similar advances. While a majority of the common variant genetic studies have focused on a single disorder, a shared genetic overlap has repeatedly been shown across disorders (Cross-Disorder Group of the Psychiatric Genomics Consortium, 2013a,b).

The contribution of rare variation to psychiatric disease has been studied through analyses of copy number and single-nucleotide variants (CNVs and SNVs, respectively). The effects of large CNVs on cognition were first described in the 1950s in cytogenetic studies of severe developmental disorders (Jacobs et al., 1959). More recently, techniques have been developed to detect CNVs with better resolution, leading to a more nuanced understanding of their association with these severe cognitive phenotypes. Specifically, present estimates are that 15% of intellectual disability cases result from large chromosomal abnormalities visible by cytogenetics (Leonard and Wen, 2002; Rauch et al., 2006) and an additional 15%–20% from smaller CNV events only observable using these more refined methods (Hochstenbach et al., 2009; Cooper et al., 2011). CNVs have also been found to contribute to the risk of autism and schizophrenia, which are both characterized in part by cognitive dysfunction (Stone et al., 2008; Levy et al., 2011; Malhotra and Sebat, 2012; Poultney et al., 2013). CNVs that arise *de novo* in the germ line (not inherited from either parent) make a particularly strong contribution in this regard (Sebat et al., 2007; Kirov et al., 2012). In contrast to CNVs, the contribution of rare SNVs to psychiatric disease has only recently come to light through next-generation sequencing. Searching for *de novo* point mutations in exome sequences of patients with intellectual disability and autism has resulted in a robust set of candidate genes that for some patients may be causal (Iossifov et al., 2012; Neale et al., 2012; O'Roak et al., 2012; Sanders et al., 2012). Sequencing studies of schizophrenia have generally failed to point to single genes that contribute strongly to risk but have identified sets of functionally related genes with higher rates of damaging SNVs in cases compared to controls (Fromer et al., 2014; Purcell et al., 2014).

13.4 GENE EXPRESSION

As with genetic variation, differential gene expression is a powerful component of predictive modeling of biological systems. Indeed, genetic variation often exerts its effects through the regulation of expression (Schadt, 2009). While the genome is generally considered to be identical from cell to cell, the transcriptome varies. A recent study comparing expression across organs found that variation was greater between tissue types than between individuals (Melé et al., 2015). Therefore, it is critical to choose the correct tissue for expression assays. Due to the inability to sample brain tissue from living patients, gene expression in blood is often used to study psychiatric conditions. This approach has been called into question since intertissue expression differences appear to be greatest between blood and solid tissues, and, amongst solid tissues, the brain appears to be the most distinct (Melé et al., 2015). However, there are multiple mechanisms by which neural-derived genetic material could, in principle, enter the peripheral circulation (Huang et al., 2013; Goetzl et al., 2015), suggesting that blood expression studies may not be entirely devoid of predictive value. One of the largest such studies compared patients diagnosed with depression to unaffected controls. The genes most differentially expressed between acutely depressed cases and controls were enriched for genes with known function in the immune system (Jansen et al., 2015).

Assays of postmortem human brain (PMHB) from affected (aPMHB) and unaffected (uPMHB) cohorts provide a more direct measure of neurobiology. Multiple small studies comparing matched aPMHB and uPMHB have reported disease-related transcriptional changes in sets of functionally related genes. In schizophrenia, for instance, neurodevelopmental, immune system and stress response gene sets have all been implicated (Sequeira et al., 2012). Immune system and neurodevelopmental sets have also been implicated in autism (Voineagu et al., 2011; Chow et al., 2012), while changes in stress response have been implicated in depression (Malki et al., 2015). Studies of bipolar disorder have pointed toward a role for changes in neuropeptide transcription (Seifuddin et al., 2013). Due to the small sample sizes characteristic of these studies, large-scale efforts are currently being conducted by collaborative consortia that promise to significantly further understanding of the contribution of brain gene expression to the pathogenesis of schizophrenia and bipolar disorder (http://commonmind.org/WP/). The results of these efforts should guide brain gene expression studies of other neuropsychiatric disorders for which insufficient numbers of PMHB specimens exist.

Knowledge of brain gene expression profiles gained from the characterization of uPMHB alone also can factor into predictive modeling of neuropsychiatric disorders. Built from microarray profiles of over 100 brain regions from 6 uPMHB specimens, the Allen Brain Atlas is currently the most comprehensive atlas of gene expression in the human brain (Hawrylycz et al., 2012). BrainSpan (http://www.brainspan.org) and BrainCloud (http://braincloud.jhmi.edu), while containing data on fewer regions, offer the opportunity to evaluate gene expression profiles in the human brain at specific points in the lifespan. BrainSpan characterizes multiple regions from 42 uPMHB samples ranging in age from 4 gestational weeks to late adulthood. BrainCloud characterizes the prefrontal cortex of 269 uPMHB samples ranging from 14 gestational weeks to late adulthood.

13.5   ADDITIONAL MOLECULAR DATA TYPES

In large part through studies of animal models of addiction, an extraordinarily intricate landscape of epigenetic regulatory mechanisms influencing psychiatric phenotypes has come to light (Nestler, 2014) and predictive models would not be complete without taking these phenomena into account. The epigenetic landscape of the healthy and diseased human brain remains largely uncharted (Akbarian and Nestler, 2013).

Large, consortia-led epigenetic studies are underway for autism, bipolar disorder, and schizophrenia (http://psychencode.org), promising comprehensive maps of the active regulatory regions (e.g., promoters, enhancers) of specific cell types (e.g., neurons, glia) in the near future.

Proteome- and metabolome-wide studies of mental illness have also yet to be carried out on a large scale. Initial human proteome atlases were only recently assembled (Kim et al., 2014; Uhlén et al., 2015). Like the small-scale studies of gene expression, small-scale proteome and metabolome studies have pointed to broad concepts such as "energy metabolism" and "immune system function" as playing a role in various psychiatric conditions (Turck and Filiou, 2015). For instance, abnormal levels of cytokine IL-6 were found in two separate cohorts of patients with treatment-refractory depression (Hodes et al., 2014). A growing body of metabolomics research is concerned with the role of the gut microbiome in behavior. These experiments, performed using germ-free mice, have shown that, by exerting metabolomic changes in the blood, the gut microbiome alters anxiety, depressive, learning, and memory phenotypes (Hsiao et al., 2013; Sampson and Mazmanian, 2015). Preliminary work suggests these findings may extend to human populations (Messaoudi et al., 2011; Tillisch et al., 2013).

13.6 MOLECULAR DATA INTEGRATION

As the earlier discussion demonstrates, large-scale molecular data sets needed for robust predictive models of psychiatric illness have yet to be assembled for most conditions. However, steps have already been taken toward the eventual construction of these complex computational models. In this section, we will highlight select examples of these efforts to integrate different data dimensions.

Roussos et al. (2014) integrated the schizophrenia GWAS results with uPMHB-derived genetic, gene expression, and epigenetic data to begin describing the mechanisms through which genomic regions implicated by GWAS increase disease risk. First, uPMHB genetic and gene expression data were assayed to identify variants regulating brain gene expression. Active cis-regulatory elements in uPMHB were then determined using epigenetic assays. It was shown that variants increasing risk for schizophrenia are enriched for variants regulating gene expression and that this enrichment is largely driven by variants falling within active cis-regulatory elements. Since most GWAS loci fall within noncoding regions, this work begins to unravel the mechanisms by which these loci cause disease. Previous work in mammalian models has shown that incorporating cis-regulatory annotations in this manner improves the reliability of predictive models (Zhu et al., 2007).

The spatiotemporal components of psychiatric disease processes have been inferred through integrating disease genetics with uPMHB gene expression. Parikshak et al. (2013) combined data from patients with neurodevelopmental disorders with uPMHB data in a five-step process. First, lists of genes implicated in neurodevelopmental disorders were generated using online databases and the primary literature. Second, uPMHB from the BrainSpan data set was analyzed to generate networks of gene co-expression across different stages of human neo-cortical development. Third, additional uPMHB (and nonhuman primate brain) specimens were assayed to generate networks of gene co-expression for different cortical layers. Fourth, the groups of co-expressed genes (or "modules") in the resulting networks were annotated for the enrichment of sets of functionally related genes (e.g., "immune system"). Fifth, modules were evaluated for the enrichment of the neurodevelopmental gene lists generated in step 1. It was found that modules enriched for the functional category "synaptic development" and modules derived from "superficial cortical layers" were both enriched for the neurodevelopmental genes. Another group taking a similar approach found that modules derived from the fetal frontal cortex were enriched for both the neurodevelopmental genes and the genes highly expressed in cortical layers 5 and 6 (Willsey et al., 2013). A third study used this approach to study schizophrenia, reporting that modules derived from the fetal frontal cortex were enriched for schizophrenia risk genes and genes involved in brain development (Gulsuner et al., 2013). Collectively, these three studies demonstrate how through the integration of disease genetics with region- and age-specific brain gene expression data it is possible to infer brain regions, developmental windows, and biological pathways involved in disease pathogenesis. This information may prove of considerable value in predicting not only the biological networks therapeutic interventions should target, but also the age at which these interventions will be most effective and the brain regions to which the drugs should be delivered.

Progress has also recently been made in integrating large-scale psychiatric genetic data with drug mechanisms. Ruderfer et al. (2014) collated drug target data from two large databases, one that contains information on known drug targets (Law et al., 2014) and the other that contains information on computationally predicted drug targets (Keiser et al., 2009). Of the 167 ATCC level-three drug classes whose target sets were evaluated for overlap with schizophrenia risk genes (Schizophrenia Working Group of the Psychiatric Genomics Consortium, 2014; Purcell et al., 2014), the only class that showed a significant overlap was "antipsychotics." This enrichment was observed even when the known drug targets, which included dopamine and serotonin receptors, were removed from the drug target sets. One interpretation of these findings is that there is a complex biological overlap between disease pathogenesis and drug mechanisms and that it extends beyond classical hypotheses. This approach demonstrates how disease and drug molecular data can be integrated to develop a more nuanced understanding of the biology of a given phenotype, and future work aiming to predict repositioning opportunities may benefit from incorporating these insights into computational models.

Disease biology and drug mechanisms have also been integrated recently in order to arrive at a novel treatment hypothesis for bipolar disorder. The use of calcium channel blockers (CCBs) to treat bipolar disorder was first suggested over 30 years ago, based on the finding that increased calcium levels in the blood of acutely ill patients corrected with lithium-induced remission (Dubovsky and Buzan, 1995). Subsequent clinical trials of various CCBs yielded inconclusive results (Levy and Janicak, 2000). Interest in the therapeutic potential of calcium receptors, however, has been renewed since GWAS variants linked to the L-type calcium channel receptor gene *CACNA1C* have been found to increase risk for bipolar disorder (Sklar et al., 2011). It has been hypothesized that previous CCB trials were inconclusive because the agents utilized were not selective for L-type calcium channels (Casamassima et al., 2010). A pilot study of a more selective agent, isradipine, in the treatment of bipolar depression was recently published showing promising results (Ostacher et al., 2014). To our knowledge, this is the first (and perhaps only) clinical trial where a drug was repositioned for a psychiatric condition as a direct result of findings from large-scale genetic studies. It is worth considering, however, whether the availability of the appropriate gene expression data could have resulted in a different therapeutic hypothesis. The *CACNA1C* risk allele identified in GWAS of bipolar disorder has been found to be associated with a decreased level of *CACNA1C* expression in uPMHB specimens (Gershon et al., 2014; Roussos et al., 2014). L-type calcium channel agonism, rather than antagonism, might therefore be a more effective treatment strategy, and drugs with this pharmacologic profile, such as ibutilide (Doggrell and Hancox, 2005), are a more promising treatment strategy. As appropriate molecular data sets for building predictive models of bipolar disorder become available, repositioning opportunities will be arrived at more systematically.

While it did not directly lead to a drug being repositioned, recent work on Fragile X syndrome (FXS) highlights the translational potential of integrative approaches. FXS is the most common single-gene cause of autism and intellectual disability, resulting from inheritance of an abnormal version of the *FMR1* gene and the subsequent lack of the fragile X mental retardation protein (FMRP). Darnell et al. (2011) demonstrated in a rodent model that under normal conditions FMRP regulates the translation of hundreds of proteins in the brain. Subsequently, Kwan et al. (2012) found that in the unaffected developing human cortex, FMRP regulates translation of the protein nitric oxide synthase 1 (NOS1). Interestingly, this protein was not observed to be under FMRP regulation in rodents. These researchers then obtained PMHB specimens from pre- and post-natal FXS cases and showed that NOS1 translation is nearly absent in the disease state. These findings make NOS1 the only protein validated as being both targeted by FMRP in the developing human brain and altered in the brains of FXS patients. NOS1 functions as an enzyme in the synthesis of nitric oxide, which is involved in a variety of neural processes (e.g., neuroplasticity) through multiple biochemical pathways, including the inhibition of matrix metalloproteinase-9, an FMRP target (Berry-Kravis, 2014). Coincidentally, several drugs believed to be acting through these pathways had already been repurposed for the treatment of FXS and other forms of autism to modest effect (Colvin and Kwan, 2014), including the NO donor and the matrix metalloproteinase-9 inhibitor minocycline (Paribello et al., 2010; Leigh et al., 2013), originally an antibiotic. While these repurposing opportunities were not pursued as a result of molecular insights, it is not unreasonable to suggest that minocycline may have been predicted had models incorporating drug mechanisms and the genetic, gene expression, and proteomic FXS data been developed. That this drug shows efficacy, albeit minimal, is an encouraging sign that such approaches in the future may yield truly novel, clinically actionable predictions.

To date, arguably the most significant contribution of predictive modeling to the neurobiology of mental illness was reported by Zhang et al. (2013), who utilized an integrated systems biology approach to identify key driver genes in the pathogenesis of Alzheimer's disease (AD). For the study, PMHBs from about 400 AD cases and 200 unaffected controls were analyzed. Extensive neuropathological characterization of each brain was performed, documenting traits such as Braak staging of neurofibrillary tangles (Braak and Braak, 1995). Genetic variation and gene expression data were then generated from each case specimen, and expression single nucleotide polymorphisms (eSNPs) were calculated. From the AD-case brain expression data, co-expression network analysis was performed. The resulting modules were then ranked in terms of their relevance to disease pathology, which was calculated using a novel method that accounted for factors such as the number of associated neuropathological traits and the number of eSNPs in the module. Bayesian inference analysis, a method of predictive modeling, was then employed on this specific module to compute a list of genes predicted to be driving the activity of the module at large. Finally, these genes were ranked according to a measure of their regulatory strength, and the gene *TYROBP* was identified as a key driver of AD pathogenesis. Though never before associated with AD, this gene had previously been found to cause a rare form of early dementia in humans (Paloneva et al., 2000) and represents a novel target for drug repositioning efforts.

In this section, we have elected to highlight in detail a few select integrative analyses rather than provide a cursory review of all such efforts as they relate to psychiatry. Additional work on phenotypes such as posttraumatic stress disorder (PTSD) (Breen et al., 2015), sleep/stress (Jiang et al., 2015), and bipolar disorder (Hunsberger et al., 2015) has also been performed.

13.7   AN EXPANDED VIEW OF REPOSITIONING

Given the relative ease with which computational repositioning approaches identify new treatment hypotheses, we anticipate that many initially promising repositioning opportunities will ultimately disappoint. A recent retrospective analysis of administrative health records, for instance, found that the anticonvulsant topiramate did not appear to reduce inflammatory bowel disease flares (Crockett et al., 2014), as had been expected based on findings from both computational and rodent models (Dudley et al., 2011). When clinical trials fail, variables other than the drug itself may be responsible. As such, expanding the concept of repositioning to include the components of the clinical trial beyond the active molecule may lead to more predictive models.

The importance of choosing the correct population to study is underscored by recent experiences targeting metabotropic glutamate receptor (mGluR) in autism and schizophrenia. The scope, cost, and generally negative findings have led many to declare these trials as enormous disappointments (Li et al., 2015; Mullard, 2015). These perceived failures, however, may be due to the incorrect choice of study populations. For instance, mGluR5 antagonists, while not effective in treating autism at large, did show efficacy when the study population was limited only to those FXS patients with the epigenetic modification of the *FMR1* gene (Jacquemont et al., 2011). Similarly, trials of mGluR2/3 agonists in schizophrenia that were halted due to the lack of efficacy (Adams et al., 2014) may have been perceived more favorably had the initial study population been limited to those patients early in the illness, as in these patients the drugs were significantly more effective than placebo (Kinon et al., 2015). Just as repositioning opportunities for the active molecule can be predicted using computational models integrating multiple data dimensions, so can the appropriate patient population too be predicted. For instance, by integrating knowledge of disease-related biology and drug mechanisms, researchers studying long-chain fatty acids in psychosis hypothesized the intervention would be most effective in the prevention, rather than the treatment, of psychotic illness, and by limiting their study population to at-risk individuals they found this to be the case (Amminger et al., 2010). Predictive modeling has the potential to make the process of identifying such connections more systematic.

The method by which a drug is delivered can also determine efficacy. The incorporation of oil-based preparations initially studied for use in anesthesia (Kelly, 1947) and hormone replacement therapy (Junkmann, 1956) into the design of antipsychotics (Kline and Simpson, 1964) led to long-lasting injections that are often the mainstay of current treatment regimens, and by some measures superior to oral formulations (Kirson et al., 2013). In the coming years, the mapping of brain connectivity will demand further innovation in drug delivery methods. These multibillion dollar endeavors arose from the premise that the complete knowledge of neural connectivity will translate into better treatments for poorly understood diseases of the brain. Current conceptions of the brain structure and function will likely in the near future be completely reshaped by the findings from these efforts. For example, a recent report suggested that structural abnormalities in individual axons may underlie human diseases such as autism (Chung et al., 2013), a level of detail entirely hidden from the neuroimaging modalities such as functional magnetic resonance imaging that currently shape the understanding of brain activity. As the picture of the human brain comes into greater focus, the pharmacological challenge will not only be to identify the appropriate therapeutic molecule but also the appropriate method to deliver it to the appropriate cells in the brain. Studies in a rodent model of amyotrophic lateral sclerosis have already shown that such tools are within reach. Disease progress was halted by delivering the active molecule to specific cell types in the brain *via* intramuscular injection of a vector that harnessed the retrograde axonal transport capabilities of the rabies virus (Azzouz et al., 2004). In this case, the method of delivery was critical to efficacy since there was a need to circumvent the alteration of the half-life of the drug that occurs in plasma (Genç and Özdinler, 2014).

One of the advantages of predictive modeling is that data types that today are unavailable can be incorporated into existing models as they develop. Thus, as comprehensive data sets of the structural and functional connectivity of the brain in different psychiatric disease states become available, integrating this information into predictive models will enable the systematic identification of repositioning opportunities that extend beyond the active molecule.

13.8 CONCLUSIONS

From the first report of the antipsychotic properties of the anesthetic agent chlorpromazine (Hamon et al., 1952) to the first report of the antidepressant properties of the anesthetic agent ketamine 50 years later (Berman et al., 2000), the history of psychopharmacology has been defined by drug repositioning. The next generation of discovery will more systematically harness the potential of existing drugs using data-driven predictive models of disease. In this chapter, we have attempted to survey the rapidly growing landscape of the large-scale, multidimensional data sets that currently exist for building these models. While we note that these methods have yet to bring forth new treatments for mental illness, we are hopeful that these tools will allow the field to build on the single-neurotransmitter models that have led to our current state of knowledge in order to achieve a more complex understanding of the human brain and diseases by which it is afflicted.

REFERENCES

Adams DH, Zhang L, Millen BA, Kinon BJ, and Gomez JC. Pomaglumetad methionil (LY2140023 monohydrate) and aripiprazole in patients with schizophrenia: A phase 3, multicenter, double-blind comparison. *Schizophr Res Treat* 2014 (2014): 758212.

Akbarian S and Nestler EJ. Epigenetic mechanisms in psychiatry. *Neuropsychopharmacology* 38 (2013): 1–2.

Amminger GP, Schäfer MR, Papageorgiou K, Klier CM, Cotton SM, Harrigan SM, Mackinnon A, McGorry PD, and Berger GE. Long-chain ω-3 fatty acids for indicated prevention of psychotic disorders: A randomized, placebo-controlled trial. *Arch Gen Psychiatry* 67 (2010): 146–154.

Anderson L, Chiswell K, Peterson ED, Tasneem A, Topping J, and Califf RM. Compliance with results reporting at ClinicalTrials.gov. *N Engl J Med* 372 (2015): 1031–1039.

Azzouz M, Ralph GS, Storkebaum E, Walmsley LE, Mitrophanous KA, Kingsman SM, Carmeliet P, and Mazarakis ND. VEGF delivery with retrogradely transported lentivector prolongs survival in a mouse ALS model. *Nature* 429 (2004): 413–417.

Battaglia M. Beyond the usual suspects: A cholinergic route for panic attacks. *Mol Psychiatry* 7 (2002): 239–246.

Baumeister AA and Hawkins MF. The serotonin hypothesis of schizophrenia: A historical case study on the heuristic value of theory in clinical neuroscience. *J Hist Neurosci* 13 (2004): 277–291.

Bergink V, van Megen HJ, and Westenberg HG. Glutamate and anxiety. *Eur Neuropsychopharmacol* 14 (2004): 175–183.

Berman RM, Cappiello A, Anand A, Oren DA, Heninger GR, Charney DS, and Krystal JH. Antidepressant effects of ketamine in depressed patients. *Biol Psychiatry* 47 (2000): 351–354.

Berry-Kravis E. Mechanism-based treatments in neurodevelopmental disorders: Fragile X syndrome. *Pediatr Neurol* 50 (2014): 297–302.

Boraska V, Franklin CS, Floyd JA, Thornton LM, Huckins LM, Southam L, Rayner NW et al. A genome-wide association study of anorexia nervosa. *Mol Psychiatry* 19 (2014): 1085–1094.

Braak H and Braak E. Staging of Alzheimer's disease-related neurofibrillary changes. *Neurobiol Aging* 16 (1995): 271–278.

Breen MS, Maihofer AX, Glatt SJ, Tylee DS, Chandler SD, Tsuang MT, Risbrough VB et al. Gene networks specific for innate immunity define post-traumatic stress disorder. *Mol Psychiatry* 20 (2015): 1538–1545.

Casamassima F, Hay AC, Benedetti A, Lattanzi L, Cassano GB, and Perlis RH. L-type calcium channels and psychiatric disorders: A brief review. *Am J Med Genet B: Neuropsychiatr Genet* 153 (2010): 1373–1390.

Charney AW, Ruderfer DM, Stahl EA, Moran JL, Chambert K, Belliveau RA, Forty L et al. Evidence for genetic heterogeneity between clinical subtypes of bipolar disorder. *Transl Psychiat* 7(1) (2017): e993.

Charney DS and Redmond DE Jr. Neurobiological mechanisms in human anxiety evidence supporting central noradrenergic hyperactivity. *Neuropharmacology* 22 (1983): 1531–1536.

Choudhury PR, Lahiri S, and Rajamma U. Glutamate mediated signaling in the pathophysiology of autism spectrum disorders. *Pharmacol Biochem Behav* 100 (2012): 841–849.

Chow ML, Pramparo T, Winn ME, Barnes CC, Li HR, Weiss L, Fan JB et al. Age-dependent brain gene expression and copy number anomalies in autism suggest distinct pathological processes at young versus mature ages. *PLoS Genet* 8 (2012): e1002592.

Chung K, Wallace J, Kim SY, Kalyanasundaram S, Andalman AS, Davidson TJ, Mirzabekov JJ et al. Structural and molecular interrogation of intact biological systems. *Nature* 497 (2013): 332–337.

Colvin SM and Kwan KY. Dysregulated nitric oxide signaling as a candidate mechanism of fragile X syndrome and other neuropsychiatric disorders. *Front Genet* 5 (2014): 239.

CONVERGE Consortium. Sparse whole-genome sequencing identifies two loci for major depressive disorder. *Nature* 523 (2015): 588–591.

Cooper GM, Coe BP, Girirajan S, Rosenfeld JA, Vu TH, Baker C, Williams C et al. A copy number variation morbidity map of developmental delay. *Nat Genet* 43 (2011): 838–846.

Costello EJ, Compton SN, Keeler G, and Angold A. Relationships between poverty and psychopathology: A natural experiment. *JAMA* 290 (2003): 2023–2029.

Crockett, SD, Schectman R, Stürmer T, and Kappelman MD. Topiramate use does not reduce flares of inflammatory bowel disease. *Dig Dis Sci* 59(7) (2014): 1535–1543.

Cross-Disorder Group of the Psychiatric Genomics Consortium. Identification of risk loci with shared effects on five major psychiatric disorders: A genome-wide analysis. *Lancet* 381(9875) (2013a): 1371–1379.

Cross-Disorder Group of the Psychiatric Genomics Consortium. Genetic relationship between five psychiatric disorders estimated from genome-wide SNPs. *Nat Genet* 45(9) (2013b): 984–994.

Darnell JC, Van Driesche SJ, Zhang C, Hung KYS, Mele A, Fraser CE, Stone EF et al. FMRP stalls ribosomal translocation on mRNAs linked to synaptic function and autism. *Cell* 146(2) (2011): 247–261.

Dilsaver SC and Coffman JA. Cholinergic hypothesis of depression: A reappraisal. *J Clin Psychopharmacol* 9(3) (1989): 173–179.

Doggrell SA and Hancox JC. Ibutilide-recent molecular insights and accumulating evidence for use in atrial flutter and fibrillation. *Expert Opin Investig Drugs* 14(5) (2005): 655–669.

Dubovsky SL and Buzan R. The role of calcium channel blockers in the treatment of psychiatric disorders. *CNS Drugs* 4(1) (1995): 47–57.

Dudley JT, Sirota M, Shenoy M, Pai RK, Roedder S, Chiang AP, Morgan AA, Sarwal MM, Pasricha PJ, and Butte AJ. Computational repositioning of the anticonvulsant topiramate for inflammatory bowel disease. *Sci Transl Med* 3(96) (2011): 96ra76.

Egerton A and Stone JM. The glutamate hypothesis of schizophrenia: Neuroimaging and drug development. *Curr Pharm Biotechnol* 13(8) (2012): 1500–1512.

Emrich HM, Zerssen DV, Kissling W, Möller H-J, and Windorfer A. Effect of sodium valproate on mania. *Archiv für Psychiatrie und Nervenkrankheiten* 229(1) (1980): 1–16.

Engel GL. The need for a new medical model: A challenge for biomedicine. *Science* 196(4286) (1977): 129–136.

Eriksson R, Werge T, Jensen LJ, and Brunak S. Dose-specific adverse drug reaction identification in electronic patient records: Temporal data mining in an inpatient psychiatric population. *Drug Safety* 37(4) (2014): 237–247.

Fromer M, Pocklington AJ, Kavanagh DH, Williams HJ, Dwyer S, Gormley P, Georgieva L et al. De novo mutations in schizophrenia implicate synaptic networks. *Nature* 506(7487) (2014): 179–184.

Genç B and Özdinler PH. Moving forward in clinical trials for ALS: Motor neurons lead the way please. *Drug Discov Today* 19(4) (2014): 441–449.

Gershon ES, Grennan K, Busnello J, Badner JA, Ovsiew F, Memon S, Alliey-Rodriguez N, Cooper J, Romanos B, and Liu C. A rare mutation of CACNA1C in a patient with bipolar disorder, and decreased gene expression associated with a bipolar-associated common SNP of CACNA1C in brain. *Mol Psychiatry* 19(8) (2014): 890–894.

Glessner JT, Connolly JJ, and Hakonarson H. Genome-wide association studies of Autism. *Curr Behav Neurosci Rep* 1(4) (2014): 234–241.

Goetzl EJ, Boxer A, Schwartz JB, Abner EL, Petersen RC, Miller BL, and Kapogiannis D. Altered lysosomal proteins in neural-derived plasma exosomes in preclinical Alzheimer disease. *Neurology* 85(1) (2015): 40–47.

Gulsuner S, Walsh T, Watts AC, Lee MK, Thornton AM, Casadei S, Rippey C et al. Spatial and temporal mapping of de novo mutations in schizophrenia to a fetal prefrontal cortical network. *Cell* 154(3) (2013): 518–529.

Hallak JEC, Maia-de-Oliveira JP, Abrao J, Evora PR, Zuardi AW, Crippa JAS, Belmonte-de-Abreu P, Baker GB, and Dursun SM. Rapid improvement of acute schizophrenia symptoms after intravenous sodium nitroprusside: A randomized, double-blind, placebo-controlled trial. *JAMA Psychiatry* 70(7) (2013): 668–676.

Hamon J, Paraire J, and Velluz J. Remarques sur l'action du 4560 RP sur l'agitation maniaque. *Ann Méd Psychol (Paris)* 110 (1952): 331–335.

Harrington RA, Lee L-C, Crum RM, Zimmerman AW, and Hertz-Picciotto I. Serotonin hypothesis of autism: Implications for selective serotonin reuptake inhibitor use during pregnancy. *Autism Res* 6(3) (2013): 149–168.

Hawrylycz MJ, Lein ES, Guillozet-Bongaarts AL, Shen EH, Ng L, Miller JA, van de Lagemaat LN et al. An anatomically comprehensive atlas of the adult human brain transcriptome. *Nature* 489(7416) (2012): 391–399.

Hazell L and Shakir SAW. Under-reporting of adverse drug reactions. *Drug Safety* 29(5) (2006): 385–396.

Hochstenbach R, van Binsbergen E, Engelen J, Nieuwint A, Polstra A, Poddighe P, Ruivenkamp C, Sikkema-Raddatz B, Smeets D, and Poot M. Array analysis and karyotyping: Workflow consequences based on a retrospective study of 36,325 patients with idiopathic developmental delay in the Netherlands. *Eur J Med Genet* 52(4) (2009): 161–169.

Hodes GE, Pfau ML, Leboeuf M, Golden SA, Christoffel DJ, Bregman D, Rebusi N et al. Individual differences in the peripheral immune system promote resilience versus susceptibility to social stress. *Proc Natl Acad Sci USA* 111(45) (2014): 16136–16141.

Hsiao EY, McBride SW, Hsien S, Sharon G, Hyde ER, McCue T, Codelli JA et al. Microbiota modulate behavioral and physiological abnormalities associated with neurodevelopmental disorders. *Cell* 155(7) (2013): 1451–1463.

Huang X, Yuan T, Tschannen M, Sun Z, Jacob H, Du M, Liang M et al. Characterization of human plasma-derived exosomal RNAs by deep sequencing. *BMC Genomics* 14(1) (2013): 319.

Hunsberger JG, Chibane FL, Elkahloun AG, Henderson R, Singh R, Lawson J, Cruceanu C et al. Novel integrative genomic tool for interrogating lithium response in bipolar disorder. *Transl Psychiatry* 5(2) (2015): e504.

Hussman JP. Letters to the editor: Suppressed GABAergic inhibition as a common factor in suspected etiologies of autism. *J Autism Dev Disord* 31(2) (2001): 247–248.

Insel TR and Scolnick EM. Cure therapeutics and strategic prevention: Raising the bar for mental health research. *Mol Psychiatry* 11(1) (2006): 11–17.

Iossifov I, Ronemus M, Levy D, Wang Z, Hakker I, Rosenbaum J, Yamrom B et al. De novo gene disruptions in children on the autistic spectrum. *Neuron* 74(2) (2012): 285–299.

Jacobs PA, Court Brown WM, Baikie AG, and Strong JA. The somatic chromosomes in mongolism. *Lancet* 273(7075) (1959): 710.

Jacquemont S, Curie A, Des Portes V, Torrioli MG, Berry-Kravis E, Hagerman RJ, Ramos FJ et al. Epigenetic modification of the FMR1 gene in fragile X syndrome is associated with differential response to the mGluR5 antagonist AFQ056. *Sci Transl Med* 3(64) (2011): 64ra1.

Jansen R., Penninx BWJH, Madar V, Xia K, Milaneschi Y, Hottenga JJ, Hammerschlag AR et al. Gene expression in major depressive disorder. *Mol Psychiatry* 21(3) (2016): 339–347.

Jiang P, Scarpa JR, Fitzpatrick K, Losic B, Gao VD, Hao K, Summa KC et al. A systems approach identifies networks and genes linking sleep and stress: Implications for neuropsychiatric disorders. *Cell Rep* 11(5) (2015): 835–848.

Junkmann K. Long-acting steroids in reproduction. *Recent Prog Horm Res* 13 (1956): 389–419.

Kahn RS and Van Praag HM. A serotonin hypothesis of panic disorder. *Hum Psychopharmacol Clin Exp* 3(4) (1988): 285–288.

Karimi S, Wang C, Metke-Jimenez A, Gaire R, and Paris C. Text and data mining techniques in adverse drug reaction detection. *ACM Comput Surveys (CSUR)* 47(4) (2015): 56.

Keiser MJ, Setola V, Irwin JJ, Laggner C, Abbas AI, Hufeisen SJ, Jensen NH et al. Predicting new molecular targets for known drugs. *Nature* 462(7270) (2009): 175–181.

Kelly M. Failure of oil-soluble anesthetics to give prolonged analgesia. *Lancet* 249(6456) (1947): 710–711.

Kendler KS and Schaffner KF. The dopamine hypothesis of schizophrenia: An historical and philosophical analysis. *Philos Psychiatry Psychol* 18(1) (2011): 41–63.

Kim M-S, Pinto SM, Getnet D, Nirujogi RS, Manda SS, Chaerkady R, Madugundu AK et al. A draft map of the human proteome. *Nature* 509(7502) (2014): 575–581.

Kinon BJ, Millen BA, Zhang L, and McKinzie DL. Exploratory analysis for a targeted patient population responsive to the metabotropic glutamate 2/3 receptor agonist pomaglumetad methionil in schizophrenia. *Biol Psychiatry* 78(11) (2015): 754–762.

Kirov G, Pocklington AJ, Holmans P, Ivanov D, Ikeda M, Ruderfer D, Moran J et al. De novo CNV analysis implicates specific abnormalities of postsynaptic signalling complexes in the pathogenesis of schizophrenia. *Mol Psychiatry* 17(2) (2012): 142–153.

Kirson NY, Weiden PJ, Yermakov S, Huang W, Samuelson T, Offord SJ, Greenberg PE, and Wong BJ. Efficacy and effectiveness of depot versus oral antipsychotics in schizophrenia: Synthesizing results across different research designs. *J Clin Psychiatry* 74(6) (2013): 568–575.

Kline NS and Simpson GM. A long-acting phenothiazine in office practice. *Am J Psychiatry* 120(10) (1964): 1012–1014.

Kwan KY, Lam MMS, Johnson MB, Dube U, Shim S, Rašin M-R, Sousa AMM et al. Species-dependent posttranscriptional regulation of NOS1 by FMRP in the developing cerebral cortex. *Cell* 149(4) (2012): 899–911.

Law V, Knox C, Djoumbou Y, Jewison T, Guo AC, Liu Y, Maciejewski A et al. DrugBank 4.0: Shedding new light on drug metabolism. *Nucleic Acids Res* 42(D1) (2014): D1091–D1097.

Leigh MJS, Nguyen DV, Mu Y, Winarni TI, Schneider A, Chechi T, Polussa J et al. A randomized double-blind, placebo-controlled trial of minocycline in children and adolescents with fragile x syndrome. *J Dev Behav Pediatr* 34(3) (2013): 147.

Leiva DB. The neurochemistry of mania: A hypothesis of etiology and a rationale for treatment. *Progr Neuro-Psychopharmacol Biol Psychiatry* 14(3) (1990): 423–429.

Leonard H and Wen X. The epidemiology of mental retardation: Challenges and opportunities in the new millennium. *Mental Retard Dev Disabil Res Rev* 8(3) (2002): 117–134.

Leucht S, Helfer B, Dold M, Kissling W, and McGrath JJ. Lithium for schizophrenia. *Cochrane Database Syst Rev* 2015(10) (2015): CD003834. doi: 10.1002/14651858. CD003834.pub3.

Levy D, Ronemus M, Yamrom B, Lee Y-H, Leotta A, Kendall J, Marks S et al. Rare de novo and transmitted copy-number variation in autistic spectrum disorders. *Neuron* 70(5) (2011): 886–897.

Levy NA and Janicak PG. Calcium channel antagonists for the treatment of bipolar disorder. *Bipolar Disord* 2(2) (2000): 108–119.

Li M-L, Hu X-Q, Li F, and Gao W-J. Perspectives on the mGluR2/3 agonists as a therapeutic target for schizophrenia: Still promising or a dead end? *Progr Neuro-Psychopharmacol Biol Psychiatry* 60 (2015): 66–76.

Luscher B, Shen Q, and Sahir N. The GABAergic deficit hypothesis of major depressive disorder. *Mol Psychiatry* 16(4) (2011): 383–406.

Mahmood T and Silverstone T. Serotonin and bipolar disorder. *J Affect Disord* 66(1) (2001): 1–11.

Malhotra D and Sebat J. CNVs: Harbingers of a rare variant revolution in psychiatric genetics. *Cell* 148(6) (2012): 1223–1241.

Malki K, Pain O, Tosto MG, Du Rietz E, Carboni L, and Schalkwyk LC. Identification of genes and gene pathways associated with major depressive disorder by integrative brain analysis of rat and human prefrontal cortex transcriptomes. *Transl Psychiatry* 5(3) (2015): e519.

Melé M, Ferreira PG, Reverter F, DeLuca DS, Monlong J, Sammeth M, Young TR et al. The human transcriptome across tissues and individuals. *Science* 348(6235) (2015): 660–665.

Messaoudi M, Lalonde R, Violle N, Javelot H, Desor D, Nejdi A, Bisson J-F et al. Assessment of psychotropic-like properties of a probiotic formulation (Lactobacillus helveticus R0052 and Bifidobacterium longum R0175) in rats and human subjects. *Br J Nutr* 105(05) (2011): 755–764.

Moghaddam B. Targeting metabotropic glutamate receptors for treatment of the cognitive symptoms of schizophrenia. *Psychopharmacology* 174(1) (2004): 39–44.

Möhler H. The GABA system in anxiety and depression and its therapeutic potential. *Neuropharmacology* 62(1) (2012): 42–53.

Moore TJ, Glenmullen J, and Furberg CD. Prescription drugs associated with reports of violence towards others. *PLoS One* 5(12) (2010): e15337.

Mullard A. Fragile X disappointments upset autism ambitions. *Nat Rev Drug Discov* 14(3) (2015): 151–153.

Neale BM, Kou Y, Liu L, Ma'Ayan A, Samocha KE, Sabo A, Lin C-F et al. Patterns and rates of exonic de novo mutations in autism spectrum disorders. *Nature* 485(7397) (2012): 242–245.

Neale BM, Medland SE, Ripke S, Asherson P, Franke B, Lesch K-P, Faraone SV et al. Meta-analysis of genome-wide association studies of attention-deficit/hyperactivity disorder. *J Am Acad Child Adolesc Psychiatry* 49(9) (2010): 884–897.

Nestler EJ Epigenetic mechanisms of drug addiction. *Neuropharmacology* 76 (2014): 259–268.

O'Roak BJ, Vives L, Girirajan S, Karakoc E, Krumm N, Coe BP, Levy R et al. Sporadic autism exomes reveal a highly interconnected protein network of de novo mutations. *Nature* 485(7397) (2012): 246–250.

Ostacher MJ, Iosifescu DV, Hay A, Blumenthal SR, Sklar P, and Perlis RH. Pilot investigation of isradipine in the treatment of bipolar depression motivated by genome-wide association. *Bipolar Disord* 16(2) (2014): 199–203.

Paloneva J, Kestilä M, Wu J, Salminen A, Böhling T, Ruotsalainen V, Hakola P et al. Loss-of-function mutations in TYROBP (DAP12) result in a presenile dementia with bone cysts. *Nat Genet* 25(3) (2000): 357–361.

Pålsson E, Jakobsson J, Södersten K, Fujita Y, Sellgren C, Ekman C-J, Ågren H, Hashimoto K, and Landén M. Markers of glutamate signaling in cerebrospinal fluid and serum from patients with bipolar disorder and healthy controls. *Eur Neuropsychopharmacol* 25(1) (2015): 133–140.

Paribello C, Tao L, Folino A, Berry-Kravis E, Tranfaglia M, Ethell IM, and Ethell DW. Open-label add-on treatment trial of minocycline in fragile X syndrome. *BMC Neurol* 10(1) (2010): 91.

Parikshak NN, Luo R, Zhang A, Won H, Lowe JK, Chandran V, Horvath S, and Geschwind DH. Integrative functional genomic analyses implicate specific molecular pathways and circuits in autism. *Cell* 155(5) (2013): 1008–1021.

Poultney CS, Goldberg AP, Drapeau E, Kou Y, Harony-Nicolas H, Kajiwara Y, De Rubeis S et al. Identification of small exonic CNV from whole-exome sequence data and application to autism spectrum disorder. *Am J Hum Genet* 93(4) (2013): 607–619.

Previc FH. Prenatal influences on brain dopamine and their relevance to the rising incidence of autism. *Med Hypotheses* 68(1) (2007): 46–60.

Purcell SM, Moran JL, Fromer M, Ruderfer D, Solovieff N, Roussos P, O'Dushlaine C et al. A polygenic burden of rare disruptive mutations in schizophrenia. *Nature* 506(7487) (2014): 185–190.

Raja K and Jonnalagadda SR. Natural language processing and data mining for clinical text. *Healthcare Data Anal* 36 (2015): 219.

Rauch A, Hoyer J, Guth S, Zweier C, Kraus C, Becker C, Zenker M et al. Diagnostic yield of various genetic approaches in patients with unexplained developmental delay or mental retardation. *Am J Med Genet A* 140(19) (2006): 2063–2074.

Ripke S, Wray NR, Lewis CM, Hamilton SP, Weissman MM, Breen G, Byrne EM et al. A mega-analysis of genome-wide association studies for major depressive disorder. *Mol Psychiatry* 18(4) (2013): 497–511.

Roussos P, Mitchell AC, Voloudakis G, Fullard JF, Pothula VM, Tsang J, Stahl EA et al. A role for noncoding variation in schizophrenia. *Cell Rep* 9(4) (2014): 1417–1429.

Ruderfer DM, Charney AW, Readhead B, Kidd BA, Kähler AK, Kenny PJ, and Sullivan PF. Polygenic overlap between schizophrenia risk and antipsychotic response: A genomic medicine approach. *Lancet Psychiat* 3(4) (2016): 350–357.

Sampson TR and Mazmanian SK. Control of brain development, function, and behavior by the microbiome. *Cell Host Microbe* 17(5) (2015): 565–576.

Sanacora G, Treccani G, and Popoli M. Towards a glutamate hypothesis of depression: An emerging frontier of neuropsychopharmacology for mood disorders. *Neuropharmacology* 62(1) (2012): 63–77.

Sanders SJ, Murtha MT, Gupta AR, Murdoch JD, Raubeson MJ, Jeremy Willsey A, Gulhan Ercan-Sencicek A et al. De novo mutations revealed by whole-exome sequencing are strongly associated with autism. *Nature* 485(7397) (2012): 237–241.

Sanseau P, Agarwal P, Barnes MR, Pastinen T, Brent Richards J, Cardon LR, and Mooser V. Use of genome-wide association studies for drug repositioning. *Nat Biotechnol* 30(4) (2012): 317–320.

Schadt EE. Molecular networks as sensors and drivers of common human diseases. *Nature* 461(7261) (2009): 218–223.

Schildkraut JJ. The catecholamine hypothesis of affective disorders: A review of supporting evidence. *Am J Psychiatry* 122(5) (1965): 509–522.

Schizophrenia Working Group of the Psychiatric Genomics Consortium. Biological insights from 108 schizophrenia-associated genetic loci. *Nature* 511(7510) (2014): 421–427.

Sebat J, Lakshmi B, Malhotra D, Troge J, Lese-Martin C, Walsh T, Yamrom B et al. Strong association of de novo copy number mutations with autism. *Science* 316(5823) (2007): 445–449.

Seifuddin F, Pirooznia M, Judy JT, Goes FS, Potash JB, and Zandi PP. Systematic review of genome-wide gene expression studies of bipolar disorder. *BMC Psychiatry* 13(1) (2013): 213.

Sequeira PA, Martin MV, and Vawter MP. The first decade and beyond of transcriptional profiling in schizophrenia. *Neurobiol Dis* 45(1) (2012): 23–36.

Sklar P, Ripke S, Scott LJ, Andreassen OA, Cichon S, Craddock N, Edenberg HJ et al. Large-scale genome-wide association analysis of bipolar disorder identifies a new susceptibility locus near ODZ4. *Nat Genet* 43(10) (2011): 977.

St. Hilaire C, Ziegler SG, Markello TC, Brusco A, Groden C, Gill F, Carlson-Donohoe H et al. NT5E mutations and arterial calcifications. *N Engl J Med* 364(5) (2011): 432–442.

Stone JL, O'Donovan MC, Gurling H, Kirov GK, Blackwood DHR, Corvin A, Craddock NJ et al. Rare chromosomal deletions and duplications increase risk of schizophrenia. *Nature* 455(7210) (2008): 237–241.

Tillisch K, Labus J, Kilpatrick L, Jiang Z, Stains J, Ebrat B, Guyonnet D et al. Consumption of fermented milk product with probiotic modulates brain activity. *Gastroenterology* 144(7) (2013): 1394–1401.

Turck CW and Filiou MD. What have mass spectrometry-based proteomics and metabolomics (not) taught us about psychiatric disorders. *Mol Neuropsychiatry* 1(2) (2015): 69–75.

Uhlén M, Fagerberg L, Hallström BM, Lindskog C, Oksvold P, Mardinoglu A, Sivertsson Å et al. Tissue-based map of the human proteome. *Science* 347(6220) (2015): 1260419.

Van Rossum JM. The significance of dopamine-receptor blockade for the mechanism of action of neuroleptic drugs. *Archives internationales de pharmacodynamie et de thérapie* 160(2) (1966): 492.

Voineagu I, Wang X, Johnston P, Lowe JK, Tian Y, Horvath S, Mill J, Cantor RM, Blencowe BJ, and Geschwind DH. Transcriptomic analysis of autistic brain reveals convergent molecular pathology. *Nature* 474(7351) (2011): 380–384.

Willsey AJ, Sanders SJ, Li M, Dong S, Tebbenkamp AT, Muhle RA, Reilly SK et al. Coexpression networks implicate human midfetal deep cortical projection neurons in the pathogenesis of autism. *Cell* 155(5) (2013): 997–1007.

Yoo JH, Valdovinos MG, and Williams DC. Relevance of donepezil in enhancing learning and memory in special populations: A review of the literature. *J Autism Dev Disord* 37(10) (2007): 1883–1901.

Zhang B, Gaiteri C, Bodea L-G, Wang Z, McElwee J, Podtelezhnikov AA, Zhang C et al. Integrated systems approach identifies genetic nodes and networks in late-onset alzheimer's disease. *Cell* 153(3) (2013): 707–720.

Zhu J, Sova P, Xu Q, Dombek KM, Xu EY, Vu H, Tu Z, Brem RB, Bumgarner RE, and Schadt EE. Stitching together multiple data dimensions reveals interacting metabolomic and transcriptomic networks that modulate cell regulation. *PLoS Biol* 10(4) (2012): e1001301.

Zhu J, Wiener MC, Zhang C, Fridman A, Minch E, Lum PY, Sachs JR, and Schadt EE. Increasing the power to detect causal associations by combining genotypic and expression data in segregating populations. *PLoS Comput Biol* 3(4) (2007): e69.

14 Pharmacology of Amyotrophic Lateral Sclerosis

Old Strategies and New Perspectives

Tiziana Petrozziello, Valentina Tedeschi,
Alba Esposito, and Agnese Secondo

CONTENTS

14.1   INTRODUCTION: REPURPOSING DRUGS IN AMYOTROPHIC LATERAL SCLEROSIS

Repurposing drugs already approved for the treatment of other diseases represent an advance strategy to reduce time frame for developing new drugs, decrease costs, and improve success rates. Drug repositioning studies, by involving the integration of translational bioinformatics resources, statistical methods, and experimental techniques, have demonstrated success in several diseases lacking appropriate therapy. For these drugs, a rapid integration into health care normally occurs, since detailed information on their formulation and potential toxicity is already available. Building upon previous research and development, drug repositioning is now used to search new treatments in amyotrophic lateral sclerosis (ALS). Several trials are currently in progress with tamoxifen, approved for breast cancer treatment, and ceftriaxone, a semisynthetic third-generation cephalosporin antibiotic, both used as add-on therapy to riluzole. Furthermore, rasagiline, a potent irreversible inhibitor of monoamine oxidase B and an antiapoptotic drug already approved for the treatment of Parkinson's disease, fingolimod, approved by the U.S. Food and Drug Administration (FDA) for multiple sclerosis, and tocilizumab, an anti-inflammatory drug currently approved to treat rheumatoid arthritis, are undergoing Phase II placebo-controlled clinical trials. However, further studies are warranted to determine their efficacy in ALS. On the other hand, one of the main reasons for the failure in pharmacotherapy is the lack of a proper knowledge of the causes underlying the pathology. Therefore, the identification of new molecular targets toward direct candidate molecules is necessary and should be emphasized. In this review, we summarize some of the most relevant mechanisms involved in ALS pathogenesis that could be pharmacologically modulated in the future.

14.2   PHYSIOPATHOLOGY OF ALS

ALS is a motor neuron disease, first described in 1869 by the French neurologist Jean-Martin Charcot. This complex neurodegenerative disease involves perturbation of some cellular pathways in several cell types (Boillee et al., 2006) and significantly reduces patients' quality of life. In fact, ALS rapidly progresses from mild motor symptoms to paralysis and premature death, often by respiratory failure within 2–5 years of the first symptoms. Spasticity, one of the major clinical features of ALS, seems to be related to the loss of the pyramidal neurons of layer V of the motor cortex, which are the origin of the descending corticospinal tracts. However, motor neurons of the ventral horn of the spinal cord are particularly compromised in ALS (Mitchell and Borasio, 2007). The sporadic form (sALS) comprises 90%–95% of all cases, the familial form (fALS) accounts for 5%–10% of all cases, whereas the Guamanian form has the highest incidence in Guam and the Trust Territories of the Pacific. The incidence of sporadic ALS is between 1.5 and 2 per 100,000 population per year, giving a prevalence of around 6 per 100,000. The cause of the motor neuron loss in ALS remains unknown, but theories include autoimmunity, abnormal protein aggregation, glutamate-induced excitotoxicity, astrocyte dysfunction, free radicals, and viral infection (Rowland, 1996; Al-Chalabi et al., 2012). In this regard, several studies focused on excitotoxicity and oxidative stress. Concerning the former hypothesis, excitotoxins including glutamate, alpha-amino-3-hydroxy-5-methyl-4-isoxasolepropionic acid (AMPA) receptor, and kainate are thought to induce neuronal death by triggering excessive Na$^+$ and Ca^{2+} influx and free-radical production into motor neurons that, in turn, stimulate transductional cascades resulting in neuronal death. The *oxidative stress hypothesis* was formulated after the publication of Rosen's paper (1993) indicating the presence of the mutation G93A on the chromosome 21 superoxide dismutase 1 (*SOD1*) gene in some families with a history of familial ALS. Specifically, this SOD1 mutation accounts respectively for 1% or 2% of ALS cases and for 20% of familial cases. In physiological conditions, SOD1 plays an important role in the metabolism and inactivation of potentially neurotoxic free radicals. Evidence supports Rosen's hypothesis, since transgenic animals for the mutant SOD1 do develop progressive motor neuron disease. In these animals, neuronal degeneration seems to involve mitochondrial dysfunction and oxidative stress related to iNOS dysfunction (Martin, 2007). Based on the SOD1 mutations observed in ALS, trials on the efficacy of free-radical scavengers and *SOD1* replacement gene therapy are in progress. However, despite the intense preclinical research on this issue, the putative role played by mutated SOD1 remains elusive.

Moreover, in the past decade, other pathogenetic hypotheses have been formulated. Among these, the most qualified have taken into consideration some molecular features of ALS such as dysregulation of intracellular ionic homeostasis, axonal transport defects, protein aggregation, reduction of vascular endothelial growth factor circulating levels (Oosthuyse et al., 2001), and growth hormone (GH) secretion (Morselli et al., 2006).

14.3 CURRENT PHARMACOLOGY OF ALS

14.3.1 RILUZOLE

There is currently no cure for this devastating disease. The pathogenetic theories on ALS led to various clinical trials most of which have failed. Surprisingly, the "excitotoxic hypothesis" has led to the identification of riluzole, a drug probably inhibiting glutamate release that prevents neuronal damage occurring during the neurodegenerative process. At the moment, riluzole is the only drug approved by the FDA in 1995 for the treatment of ALS (Miller, 1999). Two adequate and well-controlled trials on riluzole efficacy were conducted on patients with either familial or sporadic ALS. The first study was performed in France and Belgium on 155 ALS patients (Miller, 1999) and the second one in both Europe and North America on 959 ALS patients followed up to 18 months (Miller, 1999). Riluzole extends survival of 2/6 months and/or time to tracheostomy in ALS patients with no effects on muscle strength and neurological function. However, although the drug is well tolerated at high doses with low side effects, it does not slow or halt the progression of ALS.

The mechanism underlying riluzole efficacy is actually unknown but seems that the drug interacts with glutamate receptors (Owen, 2012) blocking the activity of mGluR1a (Malgouris et al., 1994). Furthermore, although it appears to antagonize electrophysiological currents flowing through N-methyl-D-aspartate and kainate, it does not bind to these ionotropic receptors (Doble, 1996).

14.4 SYMPTOMATIC DRUGS

14.4.1 BACLOFEN

Many drugs are available to help control symptoms of the pathology, thus alleviating patients suffering. Among these drugs, baclofen relieves rigidity in the limbs and throat. This molecule is a partial agonist at the level of pre- and postsynaptic gamma aminobutyric acid (GABA) B receptor (GABA$_B$) in the spinal cord. GABA$_B$ is a metabotropic G$_i$-coupled receptor whose activation leads to membrane hyperpolarization, reduction of Ca^{2+} influx, and inhibition of endogenous excitatory neurotransmitter release. This mechanism underlines the inhibition of mono- and polysynaptic spinal reflexes (Simon and Yelnik, 2010). Baclofen has several adverse effects including potential hepatotoxicity, systemic muscle relaxation, and sedation. It has also described a deleterious effect on brain plasticity (McDonnell et al., 2007). To avoid the vast systemic side effects induced by the oral administration of baclofen, administration of the drug has been developed *via* an intrathecal catheter. This method allows a higher CNS concentration at the lower spinal cord level. However, besides its efficacy for the therapy of lower limb spasticity, this implantable device has been associated with overdose and withdrawal. Therefore, baclofen must be gradually tapered off to avoid adverse effects such as hyperthermia, seizures, altered mental status, hyperthermia, rhabdomyolysis, and disseminated intravascular coagulation related to withdrawing administration.

14.4.2 PHENYTOIN AND ANTIDEPRESSANT

Another example of drug repositioning in ALS therapy was the use of phenytoin and tricyclic antidepressants as symptomatic drugs. In fact, the antiepileptic drug phenytoin may alleviate cramps, while tricyclic antidepressants can help control saliva overproduction that often accompanies the severe form of the pathology.

14.5 OTHER TREATMENTS

Sialorrhea represents a frequent problem in ALS patients. Several anticholinergic drugs such as atropine, glycopyrrolate, amitriptyline, hyoscyamine, and scopolamine are often used, but their effectiveness in patients with ALS is questioned. Botulinum toxin injections and/or salivary glands' radiation therapy can be used when anticholinergic drugs are not effective. Finally, although no specific studies have been conducted, surgical therapies represent a valid option for the treatment of this problem. On the other hand, muscle decline and weight loss in ALS patients are treated with branched-chain amino acids.

14.6   FAILED CLINICAL TRIALS

Several clinical trials on drugs and treatments already approved for other uses have been performed in ALS. These include topiramate, lamotrigine, IGF-I, methionyl growth hormone, vitamin E, pentoxifylline, and creatinine. Furthermore, double-blind trials of some drugs used as add-on therapy to riluzole have been also performed with the aim to determine the efficacy in reducing neuronal loss in the motor cortex and, possibly, mortality and motor dysfunction of ALS patients. However, none of these trials reached statistically significant results on survival, disease progression, and muscular strength. It has been proposed that some of these studies are incomplete or too small in order to draw clear conclusions. Therefore, an update of some studies has been done in an attempt to better direct future research on the therapy of this pathology. For instance, Dal Bello-Haas et al. (2008) conducted a systematic review on three trials involving 386 participants treated with creatinine, an organic acid involved in adenosine triphosphate (ATP) production able to increase survival in preclinical models. They concluded that in patients already diagnosed with ALS, creatinine did not have a statistically significant effect on survival and on other secondary outcomes taken into consideration. Considering nonpharmacological therapy, the same group (Dal Bello-Haas and Florence, 2013) identified two randomized controlled trials examining the effects of exercise in ALS. The first trial examined in 25 people a twice-daily exercise program of moderate exercise *versus* "usual activities" with ALS, and the second study examined in 27 people with ALS the effects of thrice weekly moderate exercises in comparison with stretching exercises. Although no statistically significant differences in quality of life were found, the authors concluded that these studies were too small to draw conclusions on the lack of beneficial effects. In conclusion, more clinical trials should be done to clarify the effect of pharmacological and nonpharmacological treatments in ALS patients. Some ongoing trials are reported next.

14.7   REPURPOSING DRUGS IN ALS

14.7.1   TAMOXIFEN

Tamoxifen, together with raloxifene and toremifene, is a selective estrogen receptor modulator (SERM). Like other SERMs, tamoxifen is a competitive inhibitor of estrogen binding to receptors in breast tissue but displays an activity of partial agonist in endometrium. Therefore, its active metabolite, 4-Hydroxytamoxifen, binds to estrogen receptors competitively in breast tissue, producing a nuclear complex that inhibits estrogen effects on DNA synthesis. Tamoxifen is currently used for the treatment of both early and advanced estrogen-receptor-positive breast cancer and for the prevention of breast cancer in women at a high risk of developing the disease. Interestingly, tamoxifen enhances the proteasome and autophagy pathway, both involved in the degradation of proteins. These two clearance systems are severely impaired in ALS. One of the proteins involved in ALS, TDP-43, is normally degraded either by the proteasome or autophagy pathway system. Therefore, TDP-43 accumulation and inclusion body formation occurs in ALS and it has been considered one of the pathogenetic mechanisms of the disease (Lagier-Tourenne et al., 2010). It has been shown that tamoxifen reduces TDP-43 accumulation and rescues motor function in animal models of ALS (Wang et al., 2012). Therefore, clinical trials aimed to establish the effect of tamoxifen in ALS patients with or without the regular usage of riluzole are now in progress (ClinicalTrials.gov, U.S. National Institutes of Health).

14.7.2 CEFTRIAXONE

Ceftriaxone is approved for treating bacterial infections. ALS patients have reduced glial glutamate transporter 1 (GLT1) levels, which is involved in the elimination of glutamate from neuromuscular synapses (Guo et al., 2003). Interestingly, it has been indicated that many beta-lactam antibiotics and cephalosporins upregulated the levels of GLT1 reducing glutamate excitotoxicity *via* the stimulation of the promoter sequence for GLT1 (Rothstein et al., 2005). Therefore, cephalosporin and beta-lactam antibiotics are now considered potential drugs for ALS therapy. Phase I–II clinical trials were conducted in order to determine pharmacokinetics, safety, tolerability, and efficacy of ceftriaxone in this neurodegenerative disease (Berry et al., 2013).

14.7.3 RASAGILINE

Acting as a potent irreversible inhibitor of monoamine oxidase B, rasagiline (*N*-propargyl-1-R-aminoindan) is an antiapoptotic drug. It is already approved by the U.S. FDA for the treatment of Parkinson's disease. Rasagiline increases mitochondrial survival, thus helping slow motor neuron death (Weinreb et al., 2010). Currently, this drug is undergoing a Phase II placebo-controlled clinical trial (ClinicalTrials.gov identifier: NCT0123273).

14.7.4 FINGOLIMOD AND TOCILIZUMAB

Neuroinflammation seems to have a relevant role in the etiopathogenesis of ALS (Evans et al., 2013). Therefore, one of the possible targets for ALS therapy could be the activation of microglia, astrocytes, and T lymphocytes that infiltrate the nervous system in the pathology. Fingolimod, approved by the U.S. FDA for multiple sclerosis, is an agonist of sphingosine 1 phosphate receptor blocking T cells in the secondary lymphoid tissue (Chun and Brinkmann, 2011). It is being tested in ALS patients in an ongoing Phase IIa double-blind, placebo-controlled study (ClinicalTrials.gov identifier: NCT01786174). Furthermore, a Phase II clinical trial is ongoing with tocilizumab, an anti-inflammatory drug currently approved to treat rheumatoid arthritis. Specifically, it is an anti-interleukin 6 receptor antibody that reduces the activation of macrophages, monocytes, and T cells. The blockade of interleukin 6 receptor signaling by tocilizumab may protect motor neurons from damage by decreasing the production of proinflammatory cytokines, thereby reducing ALS progression (Mizwicki et al., 2012).

14.8 PERSPECTIVES IN PHARMACOLOGY: NEW MECHANISMS AND POTENTIAL DRUGS

The identification of novel molecular targets involved in motor neuron degeneration is a necessary step to draw selective drugs useful to the development of more effective therapeutic strategies for ALS. The principal transduction cascades involved in ALS pathogenesis and those molecular targets potentially useful for designing new therapies against ALS have been listed in the following sections.

14.8.1 MODULATION OF HEAT SHOCK PROTEINS IN ALS

ALS is characterized by the aggregation of misfolded proteins that induce neurotoxicity. In some familial forms of ALS linked to *SOD1* gene mutations, SOD1 forms aggregates and impairs proteasome activity. Interestingly, the clearance of mutant SOD1 is increased by heat shock proteins (Hsps) *via* autophagy, particularly in muscle (Crippa et al., 2013). However, some results suggest that upregulation of a single specific Hsp is not sufficient to improve the clearance of mutant SOD1 rather than the interaction among proteins of a more complex chaperone-co-chaperone network (Patel et al., 2005). For instance, although the small Hsp27 has been shown to reduce aggregation of mutant SOD1 *in vitro* (Yerbury et al., 2013), genetic overexpression of Hsp70 in transgenic SOD1 mice did not affect disease survival (Liu et al., 2005). In fact, the overexpression of Hsp27 in SOD1^{G93A} transgenic mice induced transient protective effects on the neuromuscular system, lasting up to only 70 days of age (Sharp et al., 2008). On the other hand, single overexpression of the small heat shock protein B8 (HspB8) promotes autophagic removal of misfolded proteins in *in vivo* models of ALS (Crippa et al., 2010). In fact, HspB8 decreases aggregation and increases clearance of mutant SOD1, but not of wild-type SOD1. This small Hsp acts in the same way also on other proteins involved in both fALS and sALS. Mutant SOD1 seems to interact with the specific complex of HspB8 activating removal of its misfolded form. Thus, HspB8 increases mutant SOD1 clearance *via* autophagy. Therefore, pharmacological modulation of Hsp expression is an attractive therapeutic strategy for ALS therapy. Accordingly, it has been shown that co-inducers of Hsp expression play neuroprotection in a number of preclinical models of ALS. Among these compounds, the hydroxylamine derivative arimoclomol has been found to rescue motor neurons, improves neuromuscular function, and extends the life span in SOD1^{G93A} transgenic mice (Kalmar et al., 2014). This drug is currently under investigation in a Phase II trial on fALS patients. Molecularly, arimoclomol increases the expression of a number of Hsps, including Hsp60, Hsp70, Hsp90, and Grp94 (Vígh et al., 1997), with a transductional cascade initiated by the cell membrane at the level of lipid rafts (Batulan et al., 2003). This mechanism seems to sensitize cells to external stress *stimuli*. Compared to 17-AAG and celastrol, two drugs able to modulate Hsp expression, arimoclomol is able to act in both unstressed and stressed conditions (Kalmar et al., 2014). Finally, arimoclomol, like BGP-15, exerts antioxidant activity (Kieran et al., 2004; Kalmar et al., 2008), a mechanism potentially useful in ALS therapy. However, further studies are warranted to determine its efficacy.

14.8.2 Modulation of Purinergic System in ALS

Several research groups showed that the complex purinergic signaling involving ATP release, receptor activation, and ectonucleotide enzyme activity play a crucial role in neurodegeneration, neuroprotection, and regeneration (Franke and Illes, 2006; Franke et al., 2006; Abbracchio et al., 2009, Burnstock and Verkhratsky, 2012; Illes et al., 2012; Ulrich et al., 2012; Volonté and Burnstock, 2012; Burnstock, 2015). Furthermore, neuroprotective mechanisms involving purinergic signaling play important roles in some neurodegenerative diseases including ALS. In particular, it has been proposed that purinergic receptor activation may constitute an important mechanism involved in ALS progression (Amadio et al., 2011). At this time, there are four subtypes of the adenosine P1 receptor (A1, A2A, A2B, and A3), seven subtypes of the P2X ion channel receptor (P2X1-7), and eight subtypes of G protein-coupled P2Y receptor (P2Y1, P2Y2, P2Y4, P2Y6, P2Y11, P2Y12, P2Y13, and P2Y14) (Burnstock, 2007). Among these receptors, P2X4 and P2X7 subtypes are particularly involved in the physiopathology of ALS. Accordingly, the density of P2X7 in microglial cells was significantly increased in human ALS specimens (Yiangou et al., 2006). It has been reported that the upregulation of P2X4, P2X7, and P2Y6 receptors or the downregulation of ATP-hydrolyzing activities occur in microglia from transgenic mice overexpressing human SOD1 (D'Ambrosi et al., 2009). Furthermore, in astrocytes obtained from SOD1 transgenic mice, repetitive stimulation of P2X7 receptor with ATP or its derivative induces motor neuron death (Gandelman et al., 2010), whereas the high level of P2X4 receptor expression is selectively localized on degenerating motor neurons in the ventral horn of the spinal cord (Casanovas et al., 2008). Interestingly, in neurons, but not in glial cells, SOD1^{G93A} forms conformers with the P2X4 receptor that probably plays a role in neuroinflammation (Hernández et al., 2010). Many other evidence-based studies suggest that the purinergic system represents a useful route to direct therapy of ALS. In fact, the loss of motor neurons in the pathology can be reduced by adenosine A2A receptor antagonists (Mojsilovic-Petrovic et al., 2006). Furthermore, the P2X7 receptor is also involved in ALS pathogenesis. In fact, P2X7 activation increases the pro-inflammatory actions of microglia in SOD1^{G93A} transgenic mice (Apolloni et al., 2013). Accordingly, spinal cord injury is reduced by P2X7 receptor antagonists in some preclinical models of ALS (Apolloni et al., 2014).

14.8.3 Modulation of Tumor Necrosis Factor Alpha in ALS

Tumor necrosis factor α (TNFα), expressed by glia and neurons, acts through the membrane receptors TNFR1 and TNFR2 that have opposite effects in neurodegeneration. Interestingly, TNFα dysfunction occurs in experimental models of ALS and, more importantly, in ALS patients. However, the contribution of TNFα to ALS development is still debated (Poloni et al., 2000; Gowing et al., 2006). In particular, TNFR2, but not TNFR1, could be considered a new target for multi-intervention therapies of ALS (Tortarolo et al., 2015) since it is implicated in motor neuron loss in SOD1^{G93A} mice. Accordingly, TNFR2 knocking down in SOD1^{G93A} mice partially protects spinal motor neurons. However, TNFR2 knocking down does not improve motor function and survival in transgenic mice. In fact, TNFR2 deletion partially protects motor neurons and sciatic nerves in SOD1^{G93A} mice, but does not improve ALS symptoms and survival. More research must be performed to clarify the role of TNFα in ALS pathogenesis.

14.8.4 Modulation of Endoplasmic Reticulum–Resident Sigma-1 Receptor in ALS

Sigma-1 receptor (S1R), an endoplasmic reticulum–resident protein with chaperone-like activity, is enriched in cholinergic postsynaptic densities in spinal cord motor neurons (Mavlyutov et al., 2013). S1R is involved in several processes leading to acute and chronic neurodegeneration, including ALS (Peviani et al., 2014). Mutations in S1R result in earlier onset of the autosomal recessive form of ALS. In fact, a mutation in S1R causes juvenile ALS (Al-Saif et al., 2011). SOD1^{G93A} transgenic mice without S1R exhibit earlier loss of body weight, earlier signs of motor decline, and reduced longevity compared to control mice expressing S1R. Furthermore, the treatment with PRE-084, a specific S1R agonist, improves locomotor function and motor neuron survival in either presymptomatic or early symptomatic mutant SOD1^{G93A} transgenic mice (Peviani et al., 2014). However, Peviani et al. (2014) showed that, during ALS progression, increased staining for S1R is detectable in morphologically spared cervical spinal cord motor neurons and reactive microglial cells obtained from the Wobbler mouse, a model of spontaneous motor neuron degeneration. Therefore, S1R may be considered a key therapeutic target also for ALS not linked to SOD1 mutation. Interestingly, chronic treatment with PRE-084 significantly improved motor neuron survival by increasing the levels of the BDNF (brain-derived neurotrophic factor) in the gray matter. This S1R agonist significantly reduced the number of reactive astrocytes. Thus, pharmacological manipulation of S1R may represent another promising strategy to cure ALS by increasing release of growth factors and modulating the molecular function of astrocytes and microglia.

14.8.5 Modulation of DNA Methylation in ALS

It has been reported that ALS could be linked to the interplay among several factors, including genes, environment, and metabolism. In this respect, it has been hypothesized that epigenetic mechanisms may contribute to the pathogenesis of the disease (Martin and Wong, 2013). DNA methylation represents a dynamic gene regulatory mechanism that may occur within minutes to hours. It has been reported that aberrant DNA methylation driving apoptosis occurred in both *in vitro* and *in vivo* models of ALS (Chestnut et al., 2011). Interestingly, several studies identified methylated genes in ALS different from control cases (Morahan et al., 2009). In particular, CACNA1B and CACNA1C were hypermethylated, whereas NRXN1, the gene for neurexin-1, and GFRA1-2, encoding for glial cell–derived neurotrophic factor receptors, were hypomethylated. This suggests that ion channels and other plasma membrane receptors resulted dysregulated in ALS by aberrant DNA methylation of the relative genes. In another study, it has been reported that many genes associated with immune and inflammatory responses were differentially methylated in the spinal cord of sporadic

ALS patients (Figueroa-Romero et al., 2012). This suggests that epigenetic mechanisms involving DNA methylation might be an exciting potential field useful to identify a new therapeutic strategy for ALS.

14.8.6 MODULATION OF AMPA RECEPTORS IN ALS

There is evidence that an excess of extracellular glutamate, especially *via* AMPA receptors, is a key factor in ALS neuropathogenesis (Tortarolo et al., 2006). Talampanel is an orally active noncompetitive antagonist of AMPA receptors (Pascuzzi et al., 2010) showing a good efficacy in ALS experimental models including SOD1^{G93A} mice (Paizs et al., 2011). Likewise, a Phase II study showed that talampanel has beneficial effects on the rate of functional decline and the progression of symptoms in ALS patients.

14.8.7 MODULATION OF TYROSINE KINASES AND RHO KINASE IN ALS

The inhibition of kinases involved in apoptosis and inflammation has been proposed as a novel target in ALS therapy. Masitinib (AB1010) is a new tyrosine kinase inhibitor, already evaluated to treat multiple sclerosis and other inflammatory diseases showing a lower toxicity profile compared to the other tyrosine kinase inhibitors (Dubreuil et al., 2009). A Phase III clinical trial is currently ongoing for the treatment of ALS (EudraCT Number: 2010-024423-24). Furthermore, other preclinical studies considered fasudil hydrochloride, a Rho-kinase inhibitor, as a therapeutic approach in ALS (Tönges et al., 2014). In fact, combining both neuroprotection and immunomodulation (Tönges et al., 2014), fasudil hydrochloride exerts a strong survival effect on damaged motor neurons *in vitro* reducing the release of TNFα and interleukin 6. Furthermore, fasudil hydrochloride displays the prolonged survival of SOD1^{G93A} mice and improved motor functions (Takata et al., 2013). A current Phase II clinical trial with fasudil is in progress in ALS patients (ClinicalTrials.gov identifier: NCT01935518).

REFERENCES

Abbracchio, M.P., Burnstock, G., Verkhratsky, A., and Zimmermann, H. Purinergic signalling in the nervous system: An overview. *Trends Neurosci* 32 (2009): 19–29.

Al-Chalabi, A., Jones, A., Troakes, C. et al. The genetics and neuropathology of amyotrophic lateral sclerosis. *Acta Neuropathol* 124 (2012): 339–352.

Al-Saif, A., Al-Mohanna, F., and Bohlega, S.A. Mutation in sigma-1 receptor causes juvenile amyotrophic lateral sclerosis. *Ann Neurol* 70(6) (2011): 913–919.

Amadio, S., Apolloni, S., D'Ambrosi, N., and Volonté, C. Purinergic signalling at the plasma membrane: A multipurpose and multidirectional mode to deal with amyotrophic lateral sclerosis and multiple sclerosis. *J Neurochem* 116 (2011): 796–805.

Apolloni, S., Amadio, S., Montilli, C., Volonté, C., and D'Ambrosi, N. Ablation of P2X7 receptor exacerbates gliosis and motoneuron death in the SOD1-G93A mouse model of amyotrophic lateral sclerosis. *Hum Mol Genet* 22 (2013): 4102–4116.

Apolloni, S., Amadio, C., Parisi, A. et al. Spinal cord pathology is ameliorated by P2X7 antagonism in a SOD1-mutant mouse model of amyotrophic lateral sclerosis. *Dis Model Mech* 7 (2014): 1101–1109.

Batulan, Z., Shinder, G.A., Minotti, S. et al. High threshold for induction of the stress response in motor neurons is associated with failure to activate HSF1. *J Neurosci* 23(13) (2003): 5789–5798.

Berry, J.D., Shefner, J.M., Conwit, R. et al. Design and initial results of a multi-phase randomized trial of ceftriaxone in amyotrophic lateral sclerosis. *PLoS One* 8(4) (2013): e61177.

Boillee, S., Vande Velde, C., and Cleveland, D.W. ALS: A disease of motor neurons and their nonneuronal neighbours. *Neuron* 52 (2006): 39–59.

Burnstock, G. Purine and pyrimidine receptors. *Cell Mol Life Sci* 64 (2007): 1471–1483.

Burnstock, G. Physiopathological roles of P2X receptors in the central nervous system. *Curr Med Chem* 22 (2015): 819–844.

Burnstock, G. and Verkhratsky, A. *Purinergic Signalling and the Nervous System*. Springer, Heidelberg, Germany, 2012.

Casanovas, A., Hernández, S., Tarabal, O., Rosselló, J., and Esquerda, J.E. Strong P2X4 purinergic receptor-like immunoreactivity is selectively associated with degenerating neurons in transgenic rodent models of amyotrophic lateral sclerosis. *J Comp Neurol* 506 (2008): 75–92.

Chestnut, B.A., Chang, Q., Price, A., Lesuisse, C., Wong, M., and Martin, L.J. Epigenetic regulation of motor neuron cell death through DNA methylation. *J Neurosci* 31 (2011): 16619–16636.

Chun, J. and Brinkmann, V. A mechanistically novel, first oral therapy for multiple sclerosis: The development of fingolimod (FTY720, Gilenya). *Discov Med* 12 (2011): 213–228.

Crippa, V., Boncoraglio, A., Galbiati, M. et al. Differential autophagy power in the spinal cord and muscle of transgenic ALS mice. *Front Cell Neurosci* 7 (2013): 234–246.

Crippa, V., Sau, D., Rusmini, P. et al. The small heat shock protein B8 (HspB8) promotes autophagic removal of misfolded proteins involved in amyotrophic lateral sclerosis (ALS). *Hum Mol Genet* 19(17) (2010): 3440–3456.

D'Ambrosi, N., Finocchi, P., and Apolloni, S. The proinflammatory action of microglial P2 receptors is enhanced in SOD1 models for amyotrophic lateral sclerosis. *J Immunol* 183 (2009): 4648–4656.

Dal Bello-Haas, V. and Florence, J.M. Therapeutic exercise for people with amyotrophic lateral sclerosis or motor neuron disease. *Cochrane Database Syst Rev* 31(5) (2013): CD005229.

Dal Bello-Haas, V., Florence, J.M., and Krivickas, L.S. Therapeutic exercise for people with amyotrophic lateral sclerosis or motor neuron disease. *Cochrane Database Syst Rev* 16(2) (2008): CD005229.

Doble, A. The pharmacology and mechanism of action of riluzole. *Neurology* 47 (1996): S233–S241.

Dubreuil, P., Letard, S., Ciufolini, M. et al. Masitinib (AB1010), a potent and selective tyrosine kinase inhibitor targeting KIT. *PLoS One* 30(4) (2009): e7258.

Evans, M.C., Couch, Y., Sibson, N., and Turner, M.R. Inflammation and neurovascular changes in amyotrophic lateral sclerosis. *Mol Cell Neurosci* 53 (2013): 34–41.

Figueroa-Romero, C., Hur, J., Bender, D.E. et al. Identification of epigenetically altered genes in sporadic amyotrophic lateral sclerosis. *PLoS One* 7 (2012): e52672.

Franke, H. and Illes, P. Involvement of P2 receptors in the growth and survival of neurons in the CNS. *Pharmacol Ther* 109 (2006): 297–324.

Franke, H., Krügel, U., and Illes, P. P2 receptors and neuronal injury. *Pflugers Arch* 452 (2006): 622–644.

Gandelman, M., Peluffo, H., Beckman, J.S., Cassina, P., and Barbeito, L. Extracellular ATP and the P2X7 receptor in astrocyte-mediated motor neuron death: Implications for amyotrophic lateral sclerosis. *J Neuroinflammation* 7 (2010): 33–42.

Gowing, G., Dequen, F., Soucy, G., and Julien, J.P. Absence of tumor necrosis factor-alpha does not affect motor neuron disease caused by superoxide dismutase 1 mutations. *J Neurosci* 26 (2006): 11397–11402.

Guo, H., Lai, L., Butchbach, M.E. et al. Increased expression of the glial glutamate transporter EAAT2 modulates excitotoxicity and delays the onset but not the outcome of ALS in mice. *Hum Mol Genet* 12(19) (2003): 2519–2532.

Hernández, S., Casanovas, A., Piedrafita, L., Tarabal, O., and Esquerda, J.E. Neurotoxic species of misfolded SOD1G93A recognized by antibodies against the P2X4 subunit of the ATP receptor accumulate in damaged neurons of transgenic animal models of amyotrophic lateral sclerosis. *J Neuropathol Exp Neurol* 69 (2010): 176–187.

Illes, P., Verkhratsky, A., Burnstock, G., and Franke, H. P2X receptors and their role in astroglia in the central and peripheral nervous system. *Neuroscientist* 18 (2012): 422–438.

Kalmar, B., Lu, C.H., and Greensmith, L. The role of heat shock proteins in Amyotrophic Lateral Sclerosis: The therapeutic potential of Arimoclomol. *Pharmacol Ther* 141(1) (2014): 40–54.

Kalmar, B., Novoselov, S., Gray, A., Cheetham, M.E., Margulis, B., and Greensmith, L. Late stage treatment with arimoclomol delays disease progression and prevents protein aggregation in the SOD1 mouse model of ALS. *J Neurochem* 107(2) (2008): 339–350.

Kieran, D., Kalmar, B., Dick, J.R., Riddoch-Contreras, J., Burnstock, G., and Greensmith, L. Treatment with arimoclomol, a coinducer of heat shock proteins, delays disease progression in ALS mice. *Nat Med* 10(4) (2004): 402–405.

Lagier-Tourenne, C., Polymenidou, M., and Cleveland, D.W. TDP-43 and FUS/TLS: Emerging roles in RNA processing and neurodegeneration. *Hum Mol Genet* 15(19) (2010): R46–R64.

Liu, J., Shinobu, L.A., Ward, C.M., Young, D., and Cleveland, D.W. Elevation of the Hsp70 chaperone does not effect toxicity in mouse models of familial amyotrophic lateral sclerosis. *J Neurochem* 93(4) (2005): 875–882.

Malgouris, C., Daniel, M., and Doble, A. Neuroprotective effects of riluzole on *N*-methyl-D-aspartate- or veratridine-induced neurotoxicity in rat hippocampal slices. *Neurosci Lett* 177 (1994): 95–99.

Martin, L.J. Transgenic mice with human mutant genes causing Parkinson's disease and amyotrophic lateral sclerosis provide common insight into mechanisms of motor neuron selective vulnerability to degeneration. *Rev Neurosci* 18(2) (2007): 115–136.

Martin, L.J. and Wong, M. Aberrant regulation of DNA methylation in amyotrophic lateral sclerosis: A new target of disease mechanisms. *Neurotherapeutics* 10(4) (2013): 722–733.

Mavlyutov, T., Epstein, M.L., Verbny, Y.I. et al. Lack of sigma-1 receptor exacerbates ALS progression in mice. *Neuroscience* 240 (2013): 129–134.

McDonnell, M.N., Orekhov, Y., and Ziemann, U. Suppression of LTP-like plasticity in human motor cortex by the GABAB receptor agonist baclofen. *Exp Brain Res* 180(1) (2007): 181–186.

Miller, R.G. Carrell-Krusen symposium invited lecture. Clinical trials in motor neuron diseases. *J Child Neurol* 14(3) (1999): 173–179.

Mitchell, J.D. and Borasio, G.D. Amyotrophic lateral sclerosis. *Lancet* 369(9578) (2007): 2031–2041.

Mizwicki, M.T., Fiala, M., Magpantay, L. et al. Tocilizumab attenuates inflammation in ALS patients through inhibition of IL6 receptor signaling. *Am J Neurodegener Dis* 1(3) (2012): 305–315.

Mojsilovic-Petrovic, J., Jeong, G.B., Crocker, A., and Arneja, A. Protecting motor neurons from toxic insult by antagonism of adenosine A2a and Trk receptors. *J Neurosci* 26 (2006): 9250–9263.

Morahan, J.M., Yu, B., Trent, R.J., and Pamphlett, R. A genome-wide analysis of brain DNA methylation identifies new candidate genes for sporadic amyotrophic lateral sclerosis. *Amyotroph Lateral Scler* 10 (2009): 418–429.

Morselli, L.L., Bongioanni, P., Genovesi, M.R. et al. Hormone secretion is impaired in amyotrophic lateral sclerosis. *Clin Endocrinol* 65 (2006): 385–388.

Oosthuyse, B., Moons, L., Storkebaum, E. et al. Deletion of the hypoxia-response element in the vascular endothelial growth factor promoter causes motor neuron degeneration. *Nat Genet* 28(2) (2001): 131–138.

Owen, R.T. Glutamatergic approaches in major depressive disorder: Focus on ketamine, memantine and riluzole. *Drugs Today* 48 (2012): 469–478.

Paizs, M., Tortarolo, M., Bendotti, C., Engelhardt, J.I., and Siklos, L. Talampanel reduces the level of motoneuronal calcium in transgenic mutant SOD1 mice only if applied presymptomatically. *Amyotroph Lateral Scler* 12(5) (2011): 340–344.

Pascuzzi, R.M., Shefner, J., Chappell, A.S. et al. A phase II trial of talampanel in subjects with amyotrophic lateral sclerosis. *Amyotroph Lateral Scler* 11(3) (2010): 266–271.

Patel, Y.J., Payne Smith, M.D., de Belleroche, J., and Latchman, D.S. Hsp27 and Hsp70 administered in combination have a potent protective effect against FALS-associated SOD1-mutant-induced cell death in mammalian neuronal cells. *Brain Res Mol Brain Res* 134(2) (2005): 256–274.

Peviani, M., Salvaneschi, E., Bontempi, L. et al. Neuroprotective effects of the Sigma-1 receptor (S1R) agonist PRE-084, in a mouse model of motor neuron disease not linked to SOD1 mutation. *Neurobiol Dis* 62 (2014): 218–232.

Poloni, M., Facchetti, D., Mai, R. et al. Circulating levels of tumour necrosis factor-alpha and its soluble receptors are increased in the blood of patients with amyotrophic lateral sclerosis. *Neurosci Lett* 287 (2000): 211–214.

Rosen, D.R., Siddique, T., Patterson, D. et al. Mutations in Cu/Zn superoxide dismutase gene are associated with familial amyotrophic lateral sclerosis. *Nature* 362(6415) (1993): 59–62.

Rothstein, J.D., Patel, S., Regan, M.R. et al. Beta-lactam antibiotics offer neuroprotection by increasing glutamate transporter expression. *Nature* 433(7021) (2005): 73–77.

Rowland, L.P. Controversies about amyotrophic lateral sclerosis. *Neurologia* 11(Suppl. 5) (1996): 72–74.

Sharp, P.S., Akbar, M.T., Bouri, S. et al. Protective effects of heat shock protein 27 in a model of ALS occur in the early stages of disease progression. *Neurobiol Dis* 30(1) (2008): 42–55.

Simon, O. and Yelnik, A.P. Managing spasticity with drugs. *Eur J Phys Rehabil Med* 46(3) (2010): 401–410.

Takata, M., Tanaka, H., Kimura, M. et al. Fasudil, a rho kinase inhibitor, limits motor neuron loss in experimental models of amyotrophic lateral sclerosis. *Br J Pharmacol* 170 (2013): 341–351.

Tönges, L., Günther, R., Suhr, M. et al. Rho kinase inhibition modulates microglia activation and improves survival in a model of amyotrophic lateral sclerosis. *Glia* 62 (2014): 217–232.

Tortarolo, M., Grignaschi, G., Calvaresi, N. et al. Glutamate AMPA receptors change in motor neurons of SOD1G93A transgenic mice and their inhibition by a noncompetitive antagonist ameliorates the progression of amytrophic lateral sclerosis-like disease. *J Neurosci Res* 83(1) (2006): 134–146.

Tortarolo, M., Vallarola, A., Lidonnici, D. et al. Lack of TNF-alpha receptor type 2 protects motor neurons in a cellular model of amyotrophic lateral sclerosis and in mutant SOD1 mice but does not affect disease progression. *J Neurochem* 135(1) (2015): 109–124.

Ulrich, H., Abbracchio, M.P., and Burnstock, G. Extrinsic purinergic regulation of neural stem/progenitor cells: Implications for CNS development and repair. *Stem Cell Revs Rep* 8 (2012): 755–767.

Vígh, L., Literáti, P.N., Horváth, I. et al. Bimoclomol: A nontoxic, hydroxylamine derivative with stress protein-inducing activity and cytoprotective effects. *Nat Med* 3(10) (1997): 1150–1154.

Volonté, C. and Burnstock, G. Editorial: Pharmacology and therapeutic activity of purinergic drugs for disorders of the nervous system. *CNS Neurol Disord Drug Targets* 11 (2012): 649–651.

Wang, I.F., Guo, B.S., Liu, Y.C. et al. Autophagy activators rescue and alleviate pathogenesis of a mouse model with proteinopathies of the TAR DNA-binding protein 43. *Proc Natl Acad Sci USA* 109(37) (2012): 15024–15029.

Weinreb, O., Amit, T., Bar-Am, O., and Youdim, M.B. Rasagiline: A novel anti-Parkinsonian monoamine oxidase-B inhibitor with neuroprotective activity *Prog Neurobiol* 92 (2010): 330–344.

Yerbury, J.J., Gower, D., Vanags, L., Roberts, K., Lee, J.A., and Ecroyd, H. The small heat shock proteins αB-crystallin and Hsp27 suppress SOD1 aggregation in vitro. *Cell Stress Chaperones* 18(2) (2013): 251–257.

Yiangou, Y., Facer, P., and Durrenberger, P. COX-2, CB2 and P2X7-immunoreactivities are increased in activated microglial cells/macrophages of multiple sclerosis and amyotrophic lateral sclerosis spinal cord. *BMC Neurol* 6 (2006): 12–26.

15 Repositioning Clinic-Ready Compounds for the Treatment of Spinal Muscular Atrophy

Faraz Farooq

CONTENTS

15.1 INTRODUCTION

Proximal spinal muscular atrophy (SMA) is an autosomal, recessive neuromuscular disease characterized by the degeneration of motor neurons from the parts of the central nervous system (Mercuri et al. 2007). The loss of motor neurons results in generalized muscle weakness and progressive muscle atrophy (Mercuri et al. 2007). With the incidence rate of about 1 in 10,000 and a carrier frequency of 1 in 50 (Feldkotter et al. 2002; Ogino et al. 2002, 2004; Pearn 1978; Sugarman et al. 2012), SMA is one of the common orphan genetic diseases and a leading cause of infant death globally (Roberts et al. 1970). The deletion or mutation of the disease causing the *SMN1* gene results in a low amount of survival motor neuron (SMN) protein (Lefebvre et al. 1995), which leads to motor neuron degeneration and progressive skeletal muscle loss.

15.2   CLINICAL CLASSIFICATION

Based on the age of onset and clinical severity, SMA is divided into five major clinical types (Dubowitz 1995; Iannaccone et al. 2000; Wang et al. 2007; Zerres and Davies 1999).

1. SMA type 0 or congenital type SMA is the most severe form where the disease manifests prenatally. Affected patients die within the first 6 months after birth due to respiratory failure, as they are born with congenital hypotonia and have very weak respiratory muscles (Zerres and Davies 1999).
2. SMA type I (also known as Werdnig–Hoffmann Disease) is the most prevalent disease type. The age of onset of the disease is between 0 and 6 months postnatally. In this severe form of the disease, the patients have hypotonia, generalized muscle weakness, and profound skeletal muscle loss. The patients have poor head control and difficulty in swallowing and suckling with impaired bulbar function. They never sit without support and succumb to death within the first 2–5 years (O'Hagen et al. 2007; Thomas and Dubowitz 1994; Zerres and Davies 1999).
3. SMA type II (also known as Dubowitz type) is the intermediate form of the disease. The age of onset of the disease is between 6 and 18 months after birth, with approximately 70% survival to adulthood. In this form of the disease, patients have proximal limb weakness in infancy and progressive generalized muscle weakness in childhood. They sit without support but never stand or walk unsupported (Zerres and Davies 1999).
4. SMA type III (also known as Kugelberg–Welander disease) is the mild form of the disease. The age of onset of the disease is within 3 years after birth, with a normal life span. In this form of the disease, patients have proximal muscle weakness in childhood with tremors and joint contractures. Patients have a high risk of fractures and scoliosis. They mostly stand and walk unsupported, but in some cases they require wheelchair assistance (Zerres et al. 1997a,b; Zerres and Davies 1999).
5. SMA type IV (also known as Adult type) is a very mild form of the disease, which sometimes can go undiagnosed. In this form of the disease, patients have very mild muscle weakness and cramps in adulthood. They have normal mobility and a life span (Zerres and Davies 1999).

15.3   GENETICS OF THE DISEASE

The reduced amount of functional SMN protein due to the *SMN1* gene deletion or mutation is the cause of SMA (Lefebvre et al. 1995). Complete loss of functional full-length SMN protein is embryonically lethal (Kariya et al. 2009; Schrank et al. 1997; Simic 2008). Due to an evolutionary duplication event at the same chromosome (5q13), humans possess an inverted duplicate gene, *SMN2*, and thus uniquely among all species can survive the loss of *SMN1*.

Although translationally, there are very few nucleotide differences between *SMN1 and SMN2*; exon 7 of *SMN2* is alternatively spliced due to C to T transition at

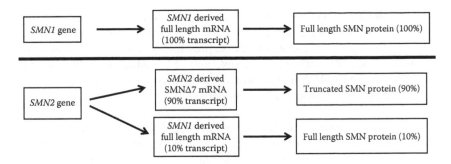

FIGURE 15.1   Schematic diagram of *SMN1* and *SMN2* gene products in humans.

position 6 of the exon. This results in the production of only ~10% full-length functional SMN protein; the alternatively spliced mRNA is translated into a truncated, unstable protein called SMNΔ7, which is quickly degraded by the cell (Figure 15.1).

All SMA patients have at least one copy of the disease modifier *SMN2*, which produces low levels of functional SMN protein. There is an inverse correlation between clinical severity and *SMN2* copy number (Campbell et al. 1997; Harada et al. 2002). SMA type I patients usually have 1–2 copies of *SMN2*, whereas type II and type III patients typically have 2–4 copies. Individuals with more than 5 copies of the *SMN2* are completely asymptomatic.

The primary function of SMN protein is the assembly of small nuclear ribonucleoproteins (snRNPs) in splicing machinery (Buhler et al. 1999; Fischer et al. 1997; Friesen and Dreyfuss 2000; Jones et al. 2001; Meister et al. 2000; Mourelatos et al. 2001; Narayanan et al. 2004; Pellizzoni et al. 1999, 2001a,b). Although the precise pathogenic molecular mechanism of SMA is still under investigation, it is believed that the lack of SMN protein within motor neurons might lead to synaptopathy and apoptotic death of these specialized neurons (Kariya et al. 2008; Simic 2008).

15.4   THERAPEUTIC STRATEGIES FOR SMA

Presently, only supportive care is available for most children with SMA (Wang et al. 2007). Although these interventions have improved both quality and longevity of life, an effective cure for SMA is eagerly awaited by the whole SMA community. Some approaches that are currently being pursued are briefly discussed.

Gene replacement therapy: Recent reports of gene therapy as an SMA therapeutic have been very promising (Dominguez et al. 2010; Foust et al. 2010; Passini et al. 2010; Valori et al. 2010). A phase I clinical trial is currently underway to evaluating safety and efficacy of *SMN1* gene replacement in SMA type I patients. Despite these advances, there are some major challenges (clinical safety, the cost of virus, and the possibility of an immune response neutralizing the adeno-associated virus (AAV) that need to be addressed before bringing this treatment into clinics (MacKenzie 2010).

SMN2-dependent therapies: Since *SMN2* is a disease-modifying gene for SMA, several strategies are used to develop drugs to target the gene. These strategies include (1) inducing the expression of *SMN2*, (2) modulating splicing of *SMN2*-derived transcript, and (3) stabilizing the full-length *SMN2*-derived mRNA and/or protein (Farooq 2012; Farooq et al. 2013b; Farooq and MacKenzie 2015; Lorson et al. 2010).

There are some new small drug compounds that are developed and tested for SMA treatment using one or more of the earlier mentioned strategies. Currently, antisense oligonucleotides (ASOs) are the most promising therapeutic compound for SMA (Aoki et al. 2013; Hua et al. 2011) and are currently in Phase 3 clinical trial for patients with infantile onset and later onset of SMA.

15.5   NEED FOR REPURPOSING CLINIC-READY COMPOUNDS

There are a significant number of orphan diseases that are still awaiting effective therapy. With the advances in next-generation DNA sequencing, the number of newly recognized orphan genetic diseases is growing at a markedly increased pace (~7000 genetic disorders with known disease gene), which is widening the gap between diagnosis of new diseases and discovery of effective cure for them. Currently, repurposing drugs for these orphan diseases is the best practical approach, as it is not only cost-effective but also fast track in the path to finding effective treatment (Beaulieu et al. 2012; Matthews and Hanna 2014; Witherspoon et al. 2015).

As discussed earlier, like many other orphan diseases, there is no effective treatment available for SMA. One translational approach is to target the disease-modifying paralogous gene *SMN2* to produce more SMN protein, which can moderate the disease phenotype. Identification of cellular pathways that can modulate the expression of *SMN2* and matching them with clinic-ready compounds to modulate the identified pathways are essential for repurposing drugs.

15.6   REPURPOSING CLINIC-READY COMPOUNDS FOR SMA

In this section, the past attempts to repurpose clinic-ready compounds as SMA therapeutics are briefly summarized. Promising preclinical therapies (using small drug compounds) are also outlined and reviewed.

15.6.1   Targeting *SMN2* Promoter

1. *Suberanilohydroxamic acid (SAHA)*: *SMN2* is a target of histone deacetylases (HDACs) that through chromatin condensation repress the transcription. Therefore, several HDAC inhibitors (discussed in the following points) have been used both *in vitro* and *in vivo* to increase the transcription of the *SMN2* gene (Avila et al. 2007; Garbes et al. 2009; Mercuri et al. 2007; Narver et al. 2008; Riessland et al. 2010).

SAHA is a Food and Drug Administration (FDA)-approved HDAC inhibitor that is currently used for the treatment of cutaneous T-cell lymphoma. Riessland et al. have tested its efficacy in SMA mouse model where it moderately improved the disease phenotype and survival (Riessland et al. 2010). Further studies are needed to test the toxicity and efficacy of SAHA in clinical settings.

2. *Sodiumbutyrate and phenylbutyrate*: Both are HDAC inhibitors used to treat urea cycle disorders. They showed increase in SMN protein levels in both cell culture and mouse models of SMA. However, in clinical trials, no improvement was seen in SMA patients (Andreassi et al. 2004; Chang et al. 2001; Mercuri et al. 2007).

3. *Valproic acid* (*VPA*): An FDA-approved HDAC inhibitor that is used to treat epilepsy, psychiatric disorders, and migraines. VPA showed great promise in *in vitro* and *in vivo* SMA models and was well tolerated by SMA patients (Brichta et al. 2003; Sumner et al. 2003; Tsai et al. 2008). However, the results from several clinical trials showed no clinical benefit in SMA patients upon treatment with VPA (Mercuri et al. 2007; Swoboda et al. 2009).

4. *Prolactin* (*PRL*): Luteotropic hormone PRL is not commonly used so far in clinical settings. In the past, it has been successfully used to treat lactation-deficient mothers and was proven safe in humans (Powe et al. 2010). It has been reported that the lactation hormone PRL through the activation of the JAK2/STAT5 pathway upregulates *SMN2* gene transcription (Farooq et al. 2011). It has been successfully tested in preclinical studies in SMA mice models. However, the absence of clinical-grade recombinant PRL has put a halt in further testing this hormone in patient population.

5. *Human Growth Hormone* (*HGH*): HGH is an FDA-approved compound with wide clinical use. It has been reported to increase SMN protein levels by targeting the STAT5 pathway and has been successfully tested in preclinical studies in SMA mice models (MacKenzie et al. 2014). However, phase 2 clinical trial in SMA patient population with a single low dose of HGH showed no improvement in muscle strength and function upon HGH treatment (Kirschner et al. 2014). Further studies are needed with higher HGH dose to accurately evaluate the therapeutic benefit of HGH in SMA patients.

15.6.2 Targeting *SMN2*-Derived mRNA and Protein

6. *Hydroxyurea* (*HU*): HU is an antineoplastic drug used for treating different types of cancers (skin, ovary, and myelocytic leukemia). HU has been tested in cultured SMA-patient-derived lymphocytes where it increased *SMN2*-derived full-length SMN transcript (Grzeschik et al. 2005; Liang et al. 2008). These results were, however, not replicated in clinical trials where type II and III SMA patients showed no clinical improvement upon treatment with HU (Chen et al. 2010).

7. *Celecoxib*: The FDA-approved nonsteroidal anti-inflammatory drug celecoxib is commonly used to treat rheumatoid arthritis and osteoarthritis. This drug has been used safely in child population as well.

A low dose of celecoxib activates the p38 mitogen-activated protein kinase pathway that has been shown to play an important role in the post-transcriptional regulation of the *SMN2* gene (Farooq et al. 2009, 2013a; Hadwen et al. 2014). In preclinical SMA mice studies, celecoxib treatment increased levels of SMN protein and ameliorated the disease phenotype (Farooq et al. 2013). A dose response clinical trial is currently underway to investigate the effect of low-dose celecoxib in type II and III SMA patients. This low-risk and cost-effective option offers a lot of promise.

8. *Bortezomib*: Bortezomib (also known as Velcade®) is an FDA-approved ubiquitin–proteasome pathway inhibitor that has been used to treat certain types of cancers. It has been documented that bortezomib has increased SMN levels both *in vitro* and in preclinical SMA mouse model (Kwon et al. 2011). However, its toxicity and its inability to cross the blood–brain barrier are major obstacles to use it in a clinical setting.

9. *Salbutamol*: Salbutamol (also known as Albuterol) is used for the treatment of respiratory problems such as asthma, bronchitis, and emphysema. It has been reported that salbutamol treatment in SMA patients resulted in a significant increase in *SMN2* full-length transcript levels (in patient's leukocytes) (Angelozzi et al. 2008; Tiziano et al. 2010). Further clinical studies are required to evaluate the therapeutic potential of salbutamol in SMA patients.

15.6.3 OTHER PROTECTIVE COMPOUNDS

10. *Fasudil*: Fasudil is an FDA-approved rho-kinase inhibitor that has neuroprotective qualities and is used for treatment of cerebral vasospasm. Fasudil treatment increases muscle fiber and postsynaptic endplate size through rho-kinase inhibition. It also improves the survival of SMA mice, independently of SMN protein upregulation (Bowerman et al. 2012). Further studies are required to understand the role of the rho-kinase pathway and the therapeutic potential of fasudil for SMA.

11. *Gabapentin*: Gabapentin is used as an antiepileptic medicine to treat nerve pain and seizures. Gabapentin has been tested in experimental models of motor neuron disease because of its neuroprotective properties. However, no clinical improvement was observed in SMA patients upon treatment with gabapentin (Merlini et al. 2003; Miller et al. 2001).

15.7 CONCLUSIONS: NEED FOR INTERVENTION

Although several drug therapies have been studied in randomized control trials, no definite improvement has been noted with any of the putative SMA treatments beside ASOs. Several clinic-ready compounds have been demonstrated to increase SMN levels and improve strength and life span in animal models of SMA. Efforts are being made by the international SMA community to have a consensus on designing clinical trials for current and future SMA therapeutics. Repurposed drugs provide an alternative approach to fast track treatments from bench to bedside in the most cost-effective way (Farooq 2016a,b).

REFERENCES

Andreassi, C., C. Angelozzi, F. D. Tiziano, T. Vitali, E. De Vincenzi, A. Boninsegna, M. Villanova et al. 2004. Phenylbutyrate increases SMN expression in vitro: Relevance for treatment of spinal muscular atrophy, *European Journal of Human Genetics*, 12: 59–65.

Angelozzi, C., F. Borgo, F. D. Tiziano, A. Martella, G. Neri, and C. Brahe. 2008. Salbutamol increases SMN mRNA and protein levels in spinal muscular atrophy cells, *Journal of Medical Genetics*, 45: 29–31.

Aoki, Y., T. Yokota, and M. J. Wood. 2013. Development of multiexon skipping antisense oligonucleotide therapy for Duchenne muscular dystrophy, *BioMed Research International*, 2013: 402369.

Avila, A. M., B. G. Burnett, A. A. Taye, F. Gabanella, M. A. Knight, P. Hartenstein, Z. Cizman et al. 2007. Trichostatin A increases SMN expression and survival in a mouse model of spinal muscular atrophy, *The Journal of Clinical Investigation*, 117: 659–671.

Beaulieu, C. L., M. E. Samuels, S. Ekins, C. R. McMaster, A. M. Edwards, A. R. Krainer, G. G. Hicks, B. J. Frey, K. M. Boycott, and A. E. Mackenzie. 2012. A generalizable pre-clinical research approach for orphan disease therapy, *Orphanet Journal of Rare Diseases*, 7: 39.

Bowerman, M., L. M. Murray, J. G. Boyer, C. L. Anderson, and R. Kothary. 2012. Fasudil improves survival and promotes skeletal muscle development in a mouse model of spinal muscular atrophy, *BMC Medicine*, 10: 24.

Brichta, L., Y. Hofmann, E. Hahnen, F. A. Siebzchnrubl, H. Raschke, I. Blumcke, I. Y. Eyupoglu, and B. Wirth. 2003. Valproic acid increases the SMN2 protein level: A well-known drug as a potential therapy for spinal muscular atrophy, *Human Molecular Genetics*, 12: 2481–2489.

Buhler, D., V. Raker, R. Luhrmann, and U. Fischer. 1999. Essential role for the tudor domain of SMN in spliceosomal U snRNP assembly: Implications for spinal muscular atrophy, *Human Molecular Genetics*, 8: 2351–2357.

Campbell, L., A. Potter, J. Ignatius, V. Dubowitz, and K. Davies. 1997. Genomic variation and gene conversion in spinal muscular atrophy: Implications for disease process and clinical phenotype, *American Journal of Human Genetics*, 61: 40–50.

Chang, J. G., H. M. Hsieh-Li, Y. J. Jong, N. M. Wang, C. H. Tsai, and H. Li. 2001. Treatment of spinal muscular atrophy by sodium butyrate, *Proceedings of the National Academy of Sciences of the United States of America*, 98: 9808–9813.

Chen, T. H., J. G. Chang, Y. H. Yang, H. H. Mai, W. C. Liang, Y. C. Wu, H. Y. Wang et al. 2010. Randomized, double-blind, placebo-controlled trial of hydroxyurea in spinal muscular atrophy, *Neurology*, 75: 2190–2197.

Dominguez, E., T. Marais, N. Chataruet, S. Benkhelifa-Ziyyat, S. Duque, P. Ravassard, R. Carcenac et al. 2010. Intravenous scAAV9 delivery of a codon-optimized SMN1 sequence rescues SMA mice, *Human Molecular Genetics*, 20: 681–693.

Dubowitz, V. 1995. Chaos in the classification of SMA: A possible resolution, *Neuromuscular Disorders*, 5: 3–5.

Farooq, F. T. 2012. Role of p38 and STAT5 kinase pathways in the regulation of survival of motor neuron gene expression for development of novel spinal muscular atrophy therapeutics, Doctoral thesis, University of Ottawa, Ottawa, Ontario, Canada. Retrieved from http://www.ruor.uottawa.ca/handle/10393/23090.

Farooq, F. T. 2016a. Re-positioning clinic ready compounds for the treatment of rare genetic diseases. *Frontiers in Molecular Biosciences. Conference Abstract: Middle East Molecular Biology Society Second Annual Conference.*

Farooq, F. T. 2016b. Preclinical assessment of clinic ready compounds for the treatment of Spinal Muscular Atrophy. *Frontiers in Neuroscience. Conference Abstract: International Conference—Educational Neuroscience.* doi:10.3389/conf.fnins.2016.92.00008.

Farooq, F., F. Abadia-Molina, D. MacKenzie, J. Hadwen, F. Shamim, S. O'Reilly, M. Holcik, and A. MacKenzie. 2013a. Celecoxib increases SMN and survival in a severe spinal muscular atrophy mouse model via p38 pathway activation, *Human Molecular Genetics*, 22: 3415–3424.

Farooq, F., S. Balabanian, X. Liu, M. Holcik, and A. MacKenzie. 2009. p38 Mitogen-activated protein kinase stabilizes SMN mRNA through RNA binding protein HuR, *Human Molecular Genetics*, 18: 4035–4045.

Farooq, F., M. Holcik, and A. MacKenzie. 2013b. Spinal muscular atrophy: Classification, diagnosis, background, molecular mechanism and development of therapeutics. In Kishore, U. (ed.), *Neurodegenerative Diseases*, InTech, Rijeka, Croatia.

Farooq, F. and A. MacKenzie. 2015. Current and emerging treatment options for spinal muscular atrophy, *Degenerative Neurological and Neuromuscular Disease*, 5: 75–81.

Farooq, F., F. A. Molina, J. Hadwen, D. MacKenzie, L. Witherspoon, M. Osmond, M. Holcik, and A. MacKenzie. 2011. Prolactin increases SMN expression and survival in a mouse model of severe spinal muscular atrophy via the STAT5 pathway, *The Journal of Clinical Investigation*, 121: 3042–3050.

Feldkotter, M., V. Schwarzer, R. Wirth, T. F. Wienker, and B. Wirth. 2002. Quantitative analyses of SMN1 and SMN2 based on real-time lightCycler PCR: Fast and highly reliable carrier testing and prediction of severity of spinal muscular atrophy, *American Journal of Human Genetics*, 70: 358–368.

Fischer, U., Q. Liu, and G. Dreyfuss. 1997. The SMN-SIP1 complex has an essential role in spliceosomal snRNP biogenesis, *Cell*, 90: 1023–1029.

Foust, K. D., X. Wang, V. L. McGovern, L. Braun, A. K. Bevan, A. M. Haidet, T. T. Le et al. 2010. Rescue of the spinal muscular atrophy phenotype in a mouse model by early postnatal delivery of SMN, *Nature Biotechnology*, 28: 271–274.

Friesen, W. J. and G. Dreyfuss. 2000. Specific sequences of the Sm and Sm-like (Lsm) proteins mediate their interaction with the spinal muscular atrophy disease gene product (SMN), *The Journal of Biological Chemistry*, 275: 26370–26375.

Garbes, L., M. Riessland, I. Holker, R. Heller, J. Hauke, C. Trankle, R. Coras, I. Blumcke, E. Hahnen, and B. Wirth. 2009. LBH589 induces up to 10-fold SMN protein levels by several independent mechanisms and is effective even in cells from SMA patients non-responsive to valproate, *Human Molecular Genetics*, 18: 3645–3658.

Grzeschik, S. M., M. Ganta, T. W. Prior, W. D. Heavlin, and C. H. Wang. 2005. Hydroxyurea enhances SMN2 gene expression in spinal muscular atrophy cells, *Annals of Neurology*, 58: 194–202.

Hadwen, J., D. MacKenzie, F. Shamim, K. Mongeon, M. Holcik, A. MacKenzie, and F. Farooq. 2014. VPAC2 receptor agonist BAY 55-9837 increases SMN protein levels and moderates disease phenotype in severe spinal muscular atrophy mouse models, *Orphanet Journal of Rare Diseases*, 9: 4.

Harada, Y., R. Sutomo, A. H. Sadewa, T. Akutsu, Y. Takeshima, H. Wada, M. Matsuo, and H. Nishio. 2002. Correlation between SMN2 copy number and clinical phenotype of spinal muscular atrophy: Three SMN2 copies fail to rescue some patients from the disease severity, *Journal of Neurology*, 249: 1211–1219.

Hua, Y., K. Sahashi, F. Rigo, G. Hung, G. Horev, C. F. Bennett, and A. R. Krainer. 2011. Peripheral SMN restoration is essential for long-term rescue of a severe spinal muscular atrophy mouse model, *Nature*, 478: 123–126.

Iannaccone, S. T., B. S. Russman, R. H. Browne, C. R. Buncher, M. White, and F. J. Samaha. 2000. Prospective analysis of strength in spinal muscular atrophy. DCN/Spinal Muscular Atrophy Group, *Journal of Child Neurology*, 15: 97–101.

Jones, K. W., K. Gorzynski, C. M. Hales, U. Fischer, F. Badbanchi, R. M. Terns, and M. P. Terns. 2001. Direct interaction of the spinal muscular atrophy disease protein SMN with the small nucleolar RNA-associated protein fibrillarin, *The Journal of Biological Chemistry*, 276: 38645–38651.

Kariya, S., R. Mauricio, Y. Dai, and U. R. Monani. 2009. The neuroprotective factor Wld(s) fails to mitigate distal axonal and neuromuscular junction (NMJ) defects in mouse models of spinal muscular atrophy, *Neuroscience Letters*, 449: 246–251.

Kariya, S., G. H. Park, Y. Maeno-Hikichi, O. Leykekhman, C. Lutz, M. S. Arkovitz, L. T. Landmesser, and U. R. Monani. 2008. Reduced SMN protein impairs maturation of the neuromuscular junctions in mouse models of spinal muscular atrophy, *Human Molecular Genetics*, 17: 2552–2569.

Kirschner, J., D. Schorling, D. Hauschke, C. Rensing-Zimmermann, U. Wein, U. Grieben, G. Schottmann et al. 2014. Somatropin treatment of spinal muscular atrophy: A placebo-controlled, double-blind crossover pilot study, *Neuromuscular Disorders*, 24: 134–142.

Kwon, D. Y., W. W. Motley, K. H. Fischbeck, and B. G. Burnett. 2011. Increasing expression and decreasing degradation of SMN ameliorate the spinal muscular atrophy phenotype in mice, *Human Molecular Genetics*, 20: 3667–3677.

Lefebvre, S., L. Burglen, S. Reboullet, O. Clermont, P. Burlet, L. Viollet, B. Benichou, C. Cruaud, P. Millasseau, and M. Zeviani. 1995. Identification and characterization of a spinal muscular atrophy-determining gene, *Cell*, 80: 155–165.

Liang, W. C., C. Y. Yuo, J. G. Chang, Y. C. Chen, Y. F. Chang, H. Y. Wang, Y. H. Ju, S. S. Chiou, and Y. J. Jong. 2008. The effect of hydroxyurea in spinal muscular atrophy cells and patients, *Journal of the Neurological Sciences*, 268: 87–94.

Lorson, C. L., H. Rindt, and M. Shababi. 2010. Spinal muscular atrophy: Mechanisms and therapeutic strategies, *Human Molecular Genetics*, 19: R111–R118.

MacKenzie, A. 2010. Genetic therapy for spinal muscular atrophy, *Nature Biotechnology*, 28: 235–237.

MacKenzie, D., F. Shamim, K. Mongeon, A. Trivedi, A. MacKenzie, and F. Farooq. 2014. Human growth hormone increases SMN expression and survival in severe spinal muscular atrophy mouse model, *Journal of Neuromuscular Diseases*, 1: 65–74.

Matthews, E. and M. G. Hanna. 2014. Repurposing of sodium channel antagonists as potential new anti-myotonic drugs, *Experimental Neurology*, 261: 812–815.

Meister, G., D. Buhler, B. Laggerbauer, M. Zobawa, F. Lottspeich, and U. Fischer. 2000. Characterization of a nuclear 20S complex containing the survival of motor neurons (SMN) protein and a specific subset of spliceosomal Sm proteins, *Human Molecular Genetics*, 9: 1977–1986.

Mercuri, E., E. Bertini, S. Messina, A. Solari, A. D'Amico, C. Angelozzi, R. Battini et al. 2007. Randomized, double-blind, placebo-controlled trial of phenylbutyrate in spinal muscular atrophy, *Neurology*, 68: 51–55.

Merlini, L., A. Solari, G. Vita, E. Bertini, C. Minetti, T. Mongini, E. Mazzoni, C. Angelini, and L. Morandi. 2003. Role of gabapentin in spinal muscular atrophy: Results of a multicenter, randomized Italian study, *Journal of Child Neurology*, 18: 537–541.

Miller, R. G., D. H. Moore, V. Dronsky, W. Bradley, R. Barohn, W. Bryan, T. W. Prior et al., SMA Study Group. 2001. A placebo-controlled trial of gabapentin in spinal muscular atrophy, *Journal of the Neurological Sciences*, 191: 127–131.

Mourelatos, Z., L. Abel, J. Yong, N. Kataoka, and G. Dreyfuss. 2001. SMN interacts with a novel family of hnRNP and spliceosomal proteins, *The EMBO Journal*, 20: 5443–5452.

Narayanan, U., T. Achsel, R. Luhrmann, and A. G. Matera. 2004. Coupled in vitro import of U snRNPs and SMN, the spinal muscular atrophy protein, *Molecular Cell*, 16: 223–234.

Narver, H. L., L. Kong, B. G. Burnett, D. W. Choe, M. Bosch-Marce, A. A. Taye, M. A. Eckhaus, and C. J. Sumner. 2008. Sustained improvement of spinal muscular atrophy mice treated with trichostatin A plus nutrition, *Annals of Neurology*, 64: 465–470.

O'Hagen, J. M., A. M. Glanzman, M. P. McDermott, P. A. Ryan, J. Flickinger, J. Quigley, S. Riley et al. 2007. An expanded version of the Hammersmith Functional Motor Scale for SMA II and III patients, *Neuromuscular Disorders*, 17: 693–697.

Ogino, S., D. G. Leonard, H. Rennert, W. J. Ewens, and R. B. Wilson. 2002. Genetic risk assessment in carrier testing for spinal muscular atrophy, *American Journal of Medical Genetics*, 110: 301–307.

Ogino, S., R. B. Wilson, and B. Gold. 2004. New insights on the evolution of the SMN1 and SMN2 region: Simulation and meta-analysis for allele and haplotype frequency calculations, *European Journal of Human Genetics*, 12: 1015–1023.

Passini, M. A., J. Bu, E. M. Roskelley, A. M. Richards, S. P. Sardi, C. R. ORiordan, K. W. Klinger, L. S. Shihabuddin, and S. H. Cheng. 2010. CNS-targeted gene therapy improves survival and motor function in a mouse model of spinal muscular atrophy, *The Journal of Clinical Investigation*, 120: 1253–1264.

Pearn, J. 1978. Incidence, prevalence, and gene frequency studies of chronic childhood spinal muscular atrophy, *Journal of Medical Genetics*, 15: 409–413.

Pellizzoni, L., J. Baccon, B. Charroux, and G. Dreyfuss. 2001a. The survival of motor neurons (SMN) protein interacts with the snoRNP proteins fibrillarin and GAR1, *Current Biology*, 11: 1079–1088.

Pellizzoni, L., B. Charroux, and G. Dreyfuss. 1999. SMN mutants of spinal muscular atrophy patients are defective in binding to snRNP proteins, *Proceedings of the National Academy of Sciences of the United States of America*, 96: 11167–11172.

Pellizzoni, L., B. Charroux, J. Rappsilber, M. Mann, and G. Dreyfuss. 2001b. A functional interaction between the survival motor neuron complex and RNA polymerase II, *The Journal of Cell Biology*, 152: 75–85.

Powe, C. E., M. Allen, K. M. Puopolo, A. Merewood, S. Worden, L. C. Johnson, A. Fleischman, and C. K. Welt. 2010. Recombinant human prolactin for the treatment of lactation insufficiency, *Clinical Endocrinology*, 73: 645–653.

Riessland, M., B. Ackermann, A. Forster, M. Jakubik, J. Hauke, L. Garbes, I. Fritzsche et al. 2010. SAHA ameliorates the SMA phenotype in two mouse models for spinal muscular atrophy, *Human Molecular Genetics*, 19: 1492–1506.

Roberts, D. F., J. Chavez, and S. D. Court. 1970. The genetic component in child mortality, *Archives of Disease in Childhood*, 45: 33–38.

Schrank, B., R. Gotz, J. M. Gunnersen, J. M. Ure, K. V. Toyka, A. G. Smith, and M. Sendtner. 1997. Inactivation of the survival motor neuron gene, a candidate gene for human spinal muscular atrophy, leads to massive cell death in early mouse embryos, *Proceedings of the National Academy of Sciences of the United States of America*, 94: 9920–9925.

Simic, G. 2008. Pathogenesis of proximal autosomal recessive spinal muscular atrophy, *Acta Neuropathologica*, 116: 223–234.

Sugarman, E. A., N. Nagan, H. Zhu, V. R. Akmaev, Z. Zhou, E. M. Rohlfs, K. Flynn et al. 2012. Pan-ethnic carrier screening and prenatal diagnosis for spinal muscular atrophy: Clinical laboratory analysis of >72,400 specimens, *European Journal of Human Genetics*, 20: 27–32.

Sumner, C. J., T. N. Huynh, J. A. Markowitz, J. S. Perhac, B. Hill, D. D. Coovert et al. 2003. Valproic acid increases SMN levels in spinal muscular atrophy patient cells, *Annals of Neurology*, 54: 647–654.

Swoboda, K. J., C. B. Scott, S. P. Reyna, T. W. Prior, B. LaSalle, S. L. Sorenson, J. Wood et al. 2009. Phase II open label study of valproic acid in spinal muscular atrophy, *PLoS One*, 4: e5268.

Thomas, N. H. and V. Dubowitz. 1994. The natural history of type I (severe) spinal muscular atrophy, *Neuromuscular Disorders*, 4: 497–502.

Tiziano, F. D., R. Lomastro, A. M. Pinto, S. Messina, A. D'Amico, S. Fiori, C. Angelozzi et al. 2010. Salbutamol increases survival motor neuron (SMN) transcript levels in leucocytes of spinal muscular atrophy (SMA) patients: Relevance for clinical trial design, *Journal of Medical Genetics*, 47: 856–858.

Tsai, L. K., M. S. Tsai, C. H. Ting, and H. Li. 2008. Multiple therapeutic effects of valproic acid in spinal muscular atrophy model mice, *Journal of Molecular Medicine (Berlin, Germany)*, 86: 1243–1254.

Valori, C. F., K. Ning, M. Wyles, R. J. Mead, A. J. Grierson, P. J. Shaw, and M. Azzouz. 2010. Systemic delivery of scAAV9 expressing SMN prolongs survival in a model of spinal muscular atrophy, *Science Translational Medicine*, 2: 35–42.

Wang, C. H., R. S. Finkel, E. S. Bertini, M. Schroth, A. Simonds, B. Wong, A. Aloysius et al., Participants of the International Conference on SMA Standard of Care. 2007. Consensus statement for standard of care in spinal muscular atrophy, *Journal of Child Neurology*, 22: 1027–1049.

Witherspoon, L., S. O'Reilly, J. Hadwen, N. Tasnim, A. MacKenzie, and F. Farooq. 2015. Sodium channel inhibitors reduce DMPK mRNA and protein, *Clinical and Translational Science*, 8: 298–304.

Zerres, K. and K. E. Davies. 1999. 59th ENMC International Workshop: Spinal muscular atrophies: Recent progress and revised diagnostic criteria 17–19 April 1998, Soestduinen, the Netherlands, *Neuromuscular Disorders*, 9: 272–278.

Zerres, K., S. Rudnik-Schoneborn, E. Forrest, A. Lusakowska, J. Borkowska, and I. Hausmanowa-Petrusewicz. 1997a. A collaborative study on the natural history of childhood and juvenile onset proximal spinal muscular atrophy (type II and III SMA): 569 patients, *Journal of the Neurological Sciences*, 146: 67–72.

Zerres, K., B. Wirth, and S. Rudnik-Schoneborn. 1997b. Spinal muscular atrophy—Clinical and genetic correlations, *Neuromuscular Disorders*, 7: 202–207.

16 Rescuing Ischemic Brain Injury by Targeting the Immune Response through Repositioned Drugs

Diana Amantea and Giacinto Bagetta

CONTENTS

16.1 INTRODUCTION

Ischemic stroke is a leading cause of death and long-term disability worldwide, for which the only available therapy consists in blood flow restoration by pharmacological or mechanical lysis (or removal) of the occluding thrombus. Given their narrow therapeutic window and strict eligibility criteria, only a limited number (<10%) of patients can benefit from these emergency procedures, thus leaving ischemic stroke an unmet clinical need (Emberson et al., 2014; Khatri et al., 2014; Fransen et al., 2015). The situation is further aggravated by the fact that virtually all neuroprotective drugs tested to date have failed to reach the clinical setting, because of their lack of efficacy or undesired toxic effects (Ginsberg, 2008; Fagan, 2010; Grupke et al., 2015). Therefore, the identification of novel druggable targets that allow to extend the therapeutic time window while providing little side effects is currently an urgent challenge. In this context, the modulation of the immune response that crucially contributes to both the early and late development of ischemic brain damage has been considered a promising strategy (Amantea et al., 2015b). Recent preclinical work demonstrates the efficacy of immune-polarizing therapies based on repositioning existing drugs characterized by a well-known safety profile in human stroke (Amantea et al., 2015a). The repositioning approach coincides, at least in part, with the view of Sir James Black (winner of the 1988 Nobel Prize in Physiology or Medicine) in that "The most fruitful basis for the discovery of a new drug is to start with an old drug." Nowadays, the latter concept will permit to significantly decrease the risk of clinical failure that has dominated the unsuccessful translation of neuroprotective drugs in ischemic stroke patients during the last decades (Ginsberg, 2008).

16.2 TARGETING THE IMMUNE SYSTEM IN ISCHEMIC STROKE

In the last decades, numerous neuroprotective drugs have been developed in the preclinical setting; however, none of these strategies has reached the clinic providing a discouraging perspective of reducing the burden of the ischemic stroke (Ginsberg, 2008; Grupke et al., 2015). The reasons of clinical trial failure lie in the complexity of human stroke syndromes and in the heterogeneity of patients that cannot be adequately reproduced by the preclinical models, where the drugs usually display beneficial effects. In addition, when tested in patients, most neuroprotective compounds were administered outside their time window of efficacy and, occasionally, in concomitance with thrombolysis (Grupke et al., 2015). Thus, despite the apparently promising preclinical evidence, the feasibility of neuroprotection in patients is disputed, highlighting the necessity of a robust proof of concept supported by proven surrogate measures to predict clinical outcomes (Tymianski, 2013).

In this view, recent gene expression profiling studies performed in human stroke have contributed to increase our understanding of disease pathophysiology for the identification of diagnostic and prognostic biomarkers and for the characterization of novel pharmacological targets. In particular, genomic profiling studies of peripheral blood from ischemic stroke patients have highlighted that the immune system plays a crucial role in the progression of cerebral ischemia (Tang et al., 2006; Barr et al., 2010, 2015; Oh et al., 2012; Brooks et al., 2014a,b; Stamova et al., 2014; Asano et al., 2016). The relevance of the peripheral immune system represents a rather novel and therapeutically useful information, since previous work has mainly focused on the brain parenchyma, demonstrating the contribution of virtually all the components of the neurovascular unit, including neurons, glia, and endothelial cells, to disease pathophysiology. Targeting these cerebral mechanisms implicated in the ischemic cascade (e.g., by glutamate receptors antagonists, calcium channel blockers, free-radical scavengers, etc.) has to date failed to produce effective drugs (Ginsberg, 2008). By contrast, a growing number of recent experimental data suggest that innate and adaptive immune mechanisms may represent promising targets to rescue ischemic brain injury, providing a longer time window than previously expected (Amantea et al., 2015b).

This approach should be rationally designed, bearing in mind that the immune system plays a time-locked dualistic role in the progression of ischemic cerebral injury, providing either detrimental or beneficial effects, depending on the production of specific soluble inflammatory mediators or on the activation of specialized immune cells located in the brain (i.e., microglia) or recruited from the periphery (Amantea et al., 2015b; Gill and Veltkamp, 2015).

The occlusion of a cerebral artery causes a drastic reduction of cerebral blood flow in the ischemic core region, where neuronal death rapidly occurs in concomitance with the release of damage-associated molecular pattern molecules that trigger microglia activation and proliferation (Li et al., 2013; Benakis et al., 2015). As a consequence, microglia shift from a resting ramified phenotype to a phagocytic amoeboid pheno-type that provides debris clearance and contributes to tissue repair (Schilling et al., 2005; Fang et al., 2014; Li et al., 2015). However, upon activation, microglia also pro-duces pro-inflammatory cytokines and reactive oxygen species (ROS) that participate to blood–brain barrier (BBB) rupture, which is followed by the brain infiltration of circulating monocytes, macrophages, neutrophils, and lymphocytes (Gelderblom et al., 2009; Chu et al., 2014; Ritzel et al., 2015). The relevance of blood-borne cells in the development of ischemic brain injury was initially demonstrated by the study of Moore et al. (2005), showing a significantly different RNA expression profile in peripheral blood mononuclear cells between ischemic stroke patients and control subjects. The fact that peripheral blood gene expression profiles correlate with ischemic brain injury was further confirmed by more recent studies (Tang et al., 2006; Sharp and Jickling, 2013). Interestingly, all the stroke-specific profiles reported to date have demonstrated that most of the genes modulated in the acute phase after ischemic stroke in the blood are implicated in the regulation of the innate immune system (Tang et al., 2006; Barr et al., 2010; Oh et al., 2012; Brooks et al., 2014a,b). In fact, the majority of these genes is expressed in circulating neutrophils and, to a lesser extent, in monocytes (Tang et al., 2006). Given the high degree of heterogeneity of innate immune cells, understanding the exact role of each phenotype in disease pathophysiology is crucial for the devel-opment of effective immunomodulatory drugs (Li et al., 2013; Yamasaki et al., 2014; Jickling et al., 2015; Wieghofer et al., 2015). In fact, although microglia and infiltrat-ing myeloid cells have traditionally been considered as pro-inflammatory mediators during ischemic brain injury, more recent evidence highlights their beneficial roles (Lalancette-Hébert et al., 2007; Faustino et al., 2011; Womble et al., 2014; Herz et al., 2015; Jickling et al., 2015; Sippel et al., 2015). In particular, innate immune cells may develop into classic phenotypes that promote ischemic injury or may acquire alter-natively activated phenotypes, namely M2 microglia/macrophages or N2 neutrophils, that provide tissue repair and remodelling.

16.3   THE DUALISTIC ROLE OF INNATE IMMUNITY

Macrophages infiltrating the ischemic brain release a series of pro-inflammatory cyto-kines (Amantea et al., 2010; Ritzel et al., 2015) and display enhanced phagocytic com-petence (Geissmann et al., 2010), thus triggering either pro-inflammatory responses or debris clearance and tissue recovery (Ritzel et al., 2015). A more accurate spatiotem-poral analysis of these reposes has demonstrated that local microglia and infiltrating macrophages display an M2 reparative phenotype during the early stages after the isch-emic insult, whereas, after stimulation by ischemic neurons, these cells gradually shift toward a pro-inflammatory M1 phenotype that prevails days after injury (Perego et al., 2011; Hu et al., 2012; Fumagalli et al., 2015; Ritzel et al., 2015). These phenotypes characterize the detrimental effects of microglia/macrophages during an ischemic insult, through the release of ROS and other neurotoxic mediators (i.e., tumor necrosis factor α, interleukin [IL]-1β, monocyte chemoattractant protein-1, macrophage inflam-matory protein-1α, and IL-6). Interestingly, recent studies have originally reported that in the penumbra, sublethal ischemic neurons may release IL-4, a potent M2-polarizing cytokine that counteracts the M1-like polarization process, thus providing an endogenous repair mechanism (Xiong et al., 2011; Zhao et al., 2015). Moreover, the group of Planas recently demonstrated that early after the ischemic event, Ly6C^{hi} proinflammatory monocytes infiltrate the core of the damage and progressively mature into alternatively activated M2-like macrophages probably involved in the reparative mechanisms occurring in the subacute phases after the insult (Miró-Mur et al., 2016).

Therefore, growing evidence highlights the occurrence of an active interplay between the ischemic milieu and the immune system, where specific microenvironmental stimuli, associated with the spatiotemporal progression of the insult, induce complex and mixed polarization dynamics in microglia and monocytes–macrophages (Fumagalli et al., 2015). In turn, these immune cells strongly affect the development of ischemic brain damage depending on their polarization status, M1 phenotypes being responsible for detrimental effects while M2-like phenotypes aimed at repairing the tissue.

Few years ago, a similar scenario was described for neutrophils whose dualistic nature has contributed to revolutionize the conception of innate immunity in ischemic stroke. In fact, these myeloid cells were classically considered to be merely detrimental through the release of ROS and cytokines and *via* the activation of proteases that cause BBB damage, brain edema, and cerebral damage (Jickling et al., 2015). Accordingly, a positive correlation between brain infiltration of neutrophils and poor neurological outcome was demonstrated in both human stroke and animal models (Matsuo et al., 1994; Garcia-Bonilla et al., 2014; Gelderblom et al., 2014; Neumann et al., 2015). Moreover, higher peripheral neutrophil count and neutrophil to lymphocyte ratio have been correlated with poor outcome in patients (Brooks et al., 2014a,b; Tokgoz et al., 2014; Maestrini et al., 2015). Nonetheless, all the attempts to block the neutrophilic response to treat or prevent ischemic stroke injury have failed in the clinic, very likely because these immune cells may also play beneficial roles (Jickling et al., 2015). In fact, through the release of arginase I, activated neutrophils exert a peripheral immunosuppressant effect by reversibly inhibiting T-cell-mediated responses to stroke injury (Sippel et al., 2015). The ability of neutrophils to shift toward N2 anti-inflammatory phenotypes is strongly dependent on the stimuli they receive from the environment, including those arising following an ischemic stroke (Easton, 2013). Thus, similar to microglia/macrophages, neutrophil polarization may represent a promising therapeutic strategy that should be aimed at inhibiting N1-induced responses, while stimulating N2 shift.

The mechanisms by which M2 or N2 polarized cells provide beneficial effects are not completely understood. In addition to their ability to counteract detrimental inflammatory responses, innate immune cells may also trigger specific regenerative mechanisms. In fact, microglia may release the chemokine CXCL13 that promotes striatal neurogenesis through the activation of CXCR5 (Chapman et al., 2015). Moreover, in mice stroke models, CCR2 recruits a subpopulation of monocytes/macrophages that preserve the neurovascular unit *via* the release of transforming growth factor β1 (Gliem et al., 2015). By contrast, the chemokine fractalkine provides inflammatory responses, as demonstrated by the evidence that deficiency of its receptor increases M2 polarization markers in stroke models and improves outcomes (Tang et al., 2014; Fumagalli et al., 2015). In this context, the identification of novel therapeutic targets for the treatment of ischemic stroke will significantly benefit from a deeper understanding of the molecular mechanisms that regulate the polarization shifts of myeloid immune cells.

16.4 RECENT PROGRESS IN DRUG REPOSITIONING FOR ISCHEMIC STROKE

Drug repositioning has an enormous significance in ischemic stroke where virtually all the neuroprotective drugs tested to date have failed to translate into the clinical setting because of excessive toxicity or lack of efficacy in patients (Ginsberg, 2008). These agents included glutamate antagonists (selfotel and aptiganel), adhesion molecule antagonists (enlimomab), nitric oxide synthase inhibitors (lubeluzole), calcium channel blockers (nimodipine), glycine antagonists (gavestinel), and free-radical scavengers (tirilazad-mesylate and NXY-059); none of these preclinically effective agents resulted in the improved outcome for patients with stroke. Thus, the use of repositioned drugs with an established safety profile and validated targets in human stroke will allow to significantly reduce the risk of clinical trial failure and will open new avenues for the discovery of effective therapeutics for stroke recovery (Jin and Wong, 2014; Strittmatter, 2014; Amantea et al., 2015a). This is an urgent need since the only licenced drug for the acute treatment of ischemic stroke is the tissue plasminogen activator (tPA), a pharmacological tool that is characterized by a very narrow therapeutic window that has a high risk of hemorrhage. In addition, tPA administration has also been associated with an induction of cytotoxic and pro-inflammatory effects, including M1 phenotypes in microglia, when administered 4.5 h after stroke (Won et al., 2015).

A lot of evidence has highlighted that certain antibiotics represent promising neuroprotective drugs in ischemic stroke. Nowadays, there is renewed interest in antibiotics because of their surprisingly heterogeneous ancillary properties not related to their anti-infective activities, including their ability to act as neuroprotectants (Stock et al., 2013). In particular, anti-inflammatory effects have been ascribed to a series of antibacterial and antifungal drugs, namely beta-lactams (Periti, 1998; Wei et al., 2012; Bisht et al., 2014; Lujia et al., 2014), dapsone (Gordon et al., 2012; Kast et al., 2012), fluoroquinolones (Blasi et al., 2012), griseofulvin (Ginsburg et al., 1987; Hussain et al., 1999), macrolides (Amsden, 2005; Cao et al., 2006; Er et al., 2010; Corrales-Medina and Musher, 2011), metronidazole (Rizzo et al., 2010), rifampicin (Gupta et al., 1975; Yulug et al., 2014), and tetracyclines (Gordon et al., 2012; Moon et al., 2012).

Among their ancillary effects, the neuroprotective properties of antibiotics against neurodegenerative (Forloni et al., 2009; Noble et al., 2009; Stoilova et al., 2013; Ruzza et al., 2014) and neuroinflammatory (Noble et al., 2009; Sultan et al., 2013) conditions are of great interest for the development of effective therapies against neurodegenerative diseases such as Parkinson's disease (PD), Alzheimer's disease (AD), human transmissible spongiform encephalopathies, and, most notably, ischemic stroke (Fagan, 2010; Reglodi et al., 2015). In this context, rifampicin, a macrocyclic antibiotic, enhances brain β-amyloid (Aβ) clearance and provides protective functions against chronic neurodegeneration and acute cerebral ischemia (Yulug et al., 2014). Furthermore, minocycline, a semisynthetic tetracycline derivative, prevents Aβ and tau protein accumulation in AD models (Noble et al., 2009), while it reduces apoptosis, neuroinflammation, infarct size, and vascular injury in ischemic stroke (Liao et al., 2013). Recently, the ability of ceftriaxone to increase the expression of glutamate transporter 1 and to attenuate pro-inflammatory responses has been shown to underlie neuroprotection in PD (Bisht et al., 2014; Hsu et al., 2015), amyotrophic lateral sclerosis (Soni et al., 2014; Zhao et al., 2014), and in cerebral ischemia models (Lujia et al., 2014; Hu et al., 2015). Neuroprotection against cerebral ischemia-reperfusion injury has also been recently demonstrated to be produced by the administration of macrolide antibiotics (Katayama et al., 2014; Inaba et al., 2015; Amantea et al., 2016). Fortuitously, the protective effects are observed at doses already approved for the anti-infective activity, characterized by good safety profiles that will likely accelerate translation of these drugs to the clinical setting.

Recent work has also highlighted that various targets for vascular protection in ischemic stroke can be approached with repurposed drugs, including statins, angiotensin receptor blockers (ARBs), minocycline, and melatonin. In fact, atorvastatin, candesartan, and minocycline reduce hemorrhagic transformation and decrease infarct size and neurological deficits in experimental stroke models (Guan et al., 2011). Moreover, melatonin reduced postischemic oxidative/nitrosative damage to the ischemic neurovascular units and improved the preservation of BBB permeability at an early phase following transient focal cerebral ischemia in mice (Chen et al., 2006; Hung et al., 2008; Andrabi et al., 2015). Interestingly, beyond the advantages derived from blood pressure lowering by ARBs, inhibition of the angiotensin II type 1 receptor may reduce inflammation, restore autoregulation, prevent apoptosis, and promote angiogenesis (Kozak et al., 2008, 2009; Fagan, 2010).

Another drug that preserves the integrity of the neurovascular unit in stroke is recombinant human erythropoietin. During the past decade, this growth factor was considered an auspicious therapeutic strategy for various types of brain injuries. However, the promising results obtained in preclinical stroke settings led to a hurried clinical trial that was suddenly aborted in phase II (Souvenir et al., 2015).

In addition to the earlier mentioned approaches, other existing drugs have been tested in ischemic stroke with the aim of improving histological and neurological outcomes but none of them has to date reached the clinic. A major issue that needs to be highlighted is that in most cases the approach used to identify potentially effective existing drugs was not rationally designed. In fact, although most of these drugs showed neuroprotective properties, they were often chosen regardless of their potential molecular target(s). Moreover, very often, multiple mechanisms were described and the exact molecular target(s) remained elusive. At variance with these experiences, the attempts to modulate the inflammatory reaction and the immune system with existing drugs were more rigorously designed around a systematic repositioning approach. In fact, as also detailed in the next paragraph, drugs to be tested were selected on the basis of their ability to exert immunomodulatory effects in different settings or because their target was clearly involved in ischemic stroke pathophysiology.

An original example of this approach was run around the concept that the hallmarks of stroke, namely, vascular impairment, neurodegeneration and, more importantly, neuroinflammation and immune cell recruitment are also found in multiple sclerosis (MS) (Lopes Pinheiro et al., 2016). Preclinical studies have demonstrated that drugs routinely used to mitigate neuroinflammation in MS are also effective neuroprotectants in stroke models. Examples include fingolimod (Kraft et al., 2013; Fu et al., 2014; Zhu et al., 2015), glatiramer acetate (Cruz et al., 2015), and antibodies blocking the leukocyte integrin VLA-4 (Neumann et al., 2015).

16.5 REPOSITIONING EXISTING DRUGS TO POLARIZE INNATE IMMUNITY

In recent years, a number of drugs acting by polarizing either brain or circulating myeloid cells toward beneficial phenotypes have been validated in preclinical stroke models (Table 16.1). Intriguingly, the majority of these studies were (often unintentionally) based on repositioning existing drugs. Based on their anti-inflammatory and immunomodulatory properties, a number of antibacterial drugs have provided promising results in ischemic stroke. By triggering microglia/macrophage polarization toward noninflammatory protective phenotypes, the tetracycline antibiotic minocycline provides neurovascular remodelling during stroke recovery (Liao et al., 2013; Yang et al., 2015). This drug has also shown promising results in early clinical trials involving ischemic stroke patients (Liao et al., 2013). However, the risk of minocycline-induced vasculitis should be carefully pondered before considering this drug for ischemic stroke patients (Baratta et al., 2015; Klaas et al., 2015).

Following the concept of drug repositioning, we have recently investigated the neuroprotective effects of the macrolide antibiotic azithromycin in a mouse model

TABLE 16.1

Immunomodulatory Drugs Showing Neuroprotection in Acute Ischemic Stroke Models

Drug	Mechanism	Effect	Animal Model	References
Azithromycin	Unknown	↑ M2/M1 ratio	Adult male mice subjected to transient MCAo	Amantea et al., 2016
Eplerenone	MR antagonist	↑ M2/M1 ratio	Adult male mice subjected to transient MCAo	Frieler et al., 2011, 2012
Extendin-4	Glucagon-like receptor 1 agonist	↑ M2/M1 ratio	Healthy young adult and aged diabetic/obese mice subjected to permanent MCAo	Darsalia et al., 2014
Metformin	AMPK activator	↑ M2/M1 ratio	Adult male mice subjected to permanent MCAo	Jin et al., 2014
Minocycline	Anti-apoptotic/ anti-inflammatory	↑ M2/M1 ratio	Adult male spontaneously hypertensive rats subjected to transient MCAo	Yang et al., 2015
PHA 568487	α7-nAChR agonist	↑ M2/M1 ratio	Adult male mice subjected to permanent MCAo	Han et al., 2014a,b
Rosiglitazone	PPARγ agonist	↑ M2/M1 ratio	Adult male mice subjected to transient or permanent MCAo	Ballesteros et al., 2014; Han et al., 2015
Bexarotene	RXR agonist	↑ N2/N1 ratio	Adult male mice subjected to transient MCAo	Certo et al., 2015
Rosiglitazone	PPARγ agonist	↑ N2/N1 ratio	Adult mice subjected to permanent MCAo	Cuartero et al., 2013

MCAo, middle cerebral artery occlusion.

of transient middle cerebral artery occlusion. Azithromycin is approved worldwide to treat a variety of community-acquired infections and displays the peculiar pharmacokinetic characteristic of accumulating in circulating leukocytes, mainly macrophages and neutrophils, because of its stability at low pH values in lysosomes (Fieta et al., 1997; Bosnar et al., 2005; Liu et al., 2007). Therefore, in addition to its prolonged antibacterial activity, azithromycin also exerts anti-inflammatory and immunomodulatory effects (Parnham et al., 2014). We originally hypothesized that these properties could confer novel therapeutic potential on azithromycin and, in fact, we demonstrated that this macrolide antibiotic affords neuroprotection in experimental ischemic stroke (Amantea et al., 2016). In fact, azithromycin significantly reduced BBB leakage and brain damage over a prolonged period of time in mice subjected to transient middle cerebral artery occlusion. The beneficial effects of the drug are ascribed to its ability to reduce brain infiltration of circulating neutrophils and to shift polarization of microglia and peripheral macrophages toward the noninflammatory M2 phenotype (Amantea et al., 2016). The low toxicity profile of azithromycin, long proven by its use in humans, meets the need for testing this drug as a novel treatment strategy for ischemic stroke (Sutherland et al., 2012). Indeed, the neuroprotective effects of azithromycin in stroke have been patented (Amantea et al., 2014), and a multicenter placebo-controlled phase IIb clinical trial (ASTRIS) has recently received approval from the ethical committee of the coordinator center.

In order to provide an effective immunomodulation aimed at affording neuroprotection in ischemic stroke, a series of receptors have been validated as promising targets in animal models.

Blockade of the myeloid mineralocorticoid receptor (MR) by the potassium-sparing diuretics, eplerenone and spironolactone, decreases the expression of M1 markers, while partially preserving the ischemia-induced expression of M2 markers. These immunomodulatory properties underlie the amelioration of stroke outcome provided by MR antagonists (Frieler et al., 2011, 2012). On the other hand, agonists of the α-7 nicotinic acetylcholine receptor (α-7 nAChR) have been reported to promote reduction of the M1/M2 macrophage ratio and thus neuroprotection in rodents (Han et al., 2014a,b).

Two drugs used to treat type 2 diabetes, the glucagon-like receptor 1 agonist, exendin-4, and the activator of adenosine 5′-monophosphate-activated protein kinase (AMPK), metformin, provide tissue repair and functional recovery in mice that underwent transient focal ischemia by increasing M2 markers (Darsalia et al., 2014; Jin et al., 2014). These and other findings highlight that, regardless of their molecular target, antidiabetic drugs may represent promising neurotherapeutics for ischemic stroke. Accordingly, the thiazolidinediones rosiglitazone and pioglitazone have also been validated as effective immunomodulators in preclinical stroke models. Rosiglitazone facilitates M2 polarization of microglia and promotes resolution of inflammation after focal cerebral ischemia (Ballesteros et al., 2014; Han et al., 2015). In turn, pioglitazone favors an anti-inflammatory milieu through the activation of peroxisome proliferator-activated receptor (PPAR)γ (Gliem et al., 2015). PPARγ have been addressed as promising candidate targets, since their activation by rosiglitazone promotes polarization of neutrophils toward the N2 phenotype that may contribute to resolution of inflammation in ischemic stroke models (Cuartero et al., 2013). More specifically, we have recently demonstrated that the heterodimer retinoid

X receptor (RXR)–PPARγ mediates the neuroprotective effects of the antineoplastic drug bexarotene (Certo et al., 2015). Acute administration of this rexinoid reduces BBB rupture, brain infarct damage, and neurological deficit produced by transient middle cerebral artery occlusion in mice. Both the amelioration of histological outcome and the ability of bexarotene to revert ischemia-induced spleen atrophy are dependent on the activation of the RXR/PPARγ heterodimer. Moreover, bexarotene elevates Ym1-immunopositive N2 neutrophils both in the ipsilateral hemisphere and in the spleen of mice subjected to transient middle cerebral artery occlusion, pointing to a major role for peripheral neutrophil polarization in neuroprotection (Certo et al., 2015). Together with similar evidence (Pan et al., 2015), these studies emphasize the therapeutic potential of repurposing RXR–PPARγ agonists to trigger polarization of innate immune cells toward protective phenotypes in stroke.

16.6   CONCLUDING REMARKS

Repositioning existing drugs will open new avenues for the development of effective therapeutic strategies for ischemic stroke. Several drugs already approved for other disease conditions have offered promising results in animal models of ischemic stroke where their efficacy as neuroprotectants has been ascribed to distinct and very heterogeneous mechanisms. Among these preclinical successes, the most promising approaches consist in the use of M2- or N2-polarizing agents. In fact, although the exact molecular mechanisms involved in immune cell shift toward beneficial phenotypes have not been completely understood, boosting reparative innate responses allows to significantly reduce the risk of toxicity. This, together with the fact that repositioned drugs are characterized by a well-established safety profile in humans, will hustle clinical translation after three decades of disappointing attempts (Ginsberg, 2008; Fagan, 2010; Grupke et al., 2015).

REFERENCES

Amantea D, Bagetta G, Caltagirone C, Corasaniti MT, Nappi G. Azithromycin, its pharmaceutically acceptable salts or solvates for use as neuroprotectors. Patent number IT1405314-B; 2014.

Amantea D, Bagetta G, Tassorelli C, Mercuri NB, Corasaniti MT. Identification of distinct cellular pools of interleukin-1β during the evolution of the neuroinflammatory response induced by transient middle cerebral artery occlusion in the brain of rat. *Brain Res.* 2010;1313:259–269.

Amantea D, Certo M, Bagetta G. Drug repurposing and beyond: The fundamental role of pharmacology. *Funct Neurol.* 2015a;30(1):79–81.

Amantea D, Certo M, Petrelli F, Tassorelli C, Micieli G, Corasaniti MT, Puccetti P, Fallarino F, Bagetta G. Azithromycin protects mice against ischemic stroke injury by promoting macrophage transition toward M2 phenotype. *Exp Neurol.* 2016;275(Pt 1):116–125.

Amantea D, Micieli G, Tassorelli C, Cuartero MI, Ballesteros I, Certo M, Moro MA, Lizasoain I, Bagetta G. Rational modulation of the innate immune system for neuroprotection in ischemic stroke. *Front Neurosci.* 2015b;9:147.

Amsden G. Anti-inflammatory effects of macrolides e an underappreciated benefit in the treatment of community-acquired respiratory tract infections and chronic inflammatory pulmonary conditions? *J Antimicrob Chemother.* 2005;55:10e21.

Andrabi SS, Parvez S, Tabassum H. Melatonin and ischemic stroke: Mechanistic roles and action. *Adv Pharmacol Sci*. 2015;2015:384750.

Asano S, Chantler PD, Barr TL. Gene expression profiling in stroke: Relevance of blood-brain interaction. *Curr Opin Pharmacol*. 2016;26:80–86.

Ballesteros I, Cuartero MI, Pradillo JM et al. Rosiglitazone-induced CD36 up-regulation resolves inflammation by PPARγ and 5-LO-dependent pathways. *J Leukoc Biol*. 2014;95(4):587–598.

Baratta JM, Dyck PJ, Brand P, Thaisetthawatkul P, Dyck PJ, Engelstad JK, Goodman B, Karam C. Vasculitic neuropathy following exposure to minocycline. *Neurol Neuroimmunol Neuroinflamm*. November 12, 2015;3(1):e180.

Barr TL, Conley Y, Ding J, Dillman A, Warach S, Singleton A, Matarin M. Genomic bio-markers and cellular pathways of ischemic stroke by RNA gene expression profiling. *Neurology*. 2010;75(11):1009–1014.

Barr TL, VanGilder R, Rellick S et al. A genomic profile of the immune response to stroke with implications for stroke recovery. *Biol Res Nurs*. 2015;17:248–256.

Benakis C, Garcia-Bonilla L, Iadecola C, Anrather J. The role of microglia and myeloid immune cells in acute cerebral ischemia. *Front Cell Neurosci*. 2015;8:461.

Bisht R, Kaur B, Gupta H, Prakash A. Ceftriaxone mediated rescue of nigral oxidative dam-age and motor deficits in MPTP model of Parkinson's disease in rats. *Neurotoxicology*. September 2014;44:71–79.

Blasi F, Mantero M, Aliberti S. Antibiotics as immunomodulant agents in COPD. *Curr Opin Pharmacol*. 2012;12:293–299.

Bosnar M, Kelnerić Z, Munić V, Eraković V, Parnham MJ. Cellular uptake and efflux of azithro-mycin, erythromycin, clarithromycin, telithromycin, and cethromycin. *Antimicrob Agents Chemother*. 2005;49:2372–2377.

Brooks SD, Spears C, Cummings C et al. Admission neutrophil-lymphocyte ratio predicts 90 day outcome after endovascular stroke therapy. *J Neurointerv Surg*. 2014a;6(8):578–583.

Brooks SD, Van Gilder R, Frisbee JC, Barr TL. Genomics for the advancement of clinical translation in stroke. In *Rational Basis for Clinical Translation in Stroke Therapy*. Micieli G, Amantea D (Eds.). CRC Press, Boca Raton, FL, 2014b, pp. 123–136.

Cao X, Dong M, Shen J, Wu B, Wu C, Du X, Wang Z, Qi Y, Li B. Tilmicosin and tylosin have anti-inflammatory properties via modulation of COX-2 and iNOS gene expression and production of cytokines in LPS-induced macrophages and monocytes. *Int J Antimicrob Agents*. 2006;27:431–438.

Certo M, Endo Y, Ohta K, Sakurada S, Bagetta G, Amantea D. Activation of RXR/PPARγ underlies neuroprotection by bexarotene in ischemic stroke. *Pharmacol Res*. 2015;102:298–307.

Chapman KZ, Ge R, Monni E, Tatarishvili J, Ahlenius H, Arvidsson A, Ekdahl CT, Lindvall O, Kokaia Z. Inflammation without neuronal death triggers striatal neurogenesis com-parable to stroke. *Neurobiol Dis*. 2015;83:1–15.

Chen HY, Chen TY, Lee MY et al. Melatonin decreases neurovascular oxidative/nitrosative damage and protects against early increases in the blood-brain barrier permeability after transient focal cerebral ischemia in mice. *J Pineal Res*. 2006;41:175–182.

Chu HX, Kim HA, Lee S et al. Immune cell infiltration in malignant middle cerebral artery infarction: Comparison with transient cerebral ischemia. *J Cereb Blood Flow Metab*. 2014;34(3):450–459.

Corrales-Medina V, Musher D. Immunomodulatory agents in the treatment of community-acquired pneumonia: A systematic review. *J Infect*. 2011;63:187–199.

Cruz Y, Lorea J, Mestre H, Kim-Lee JH, Herrera J, Mellado R, Gálvez V, Cuellar L, Musri C, Ibarra A. Copolymer-1 promotes neurogenesis and improves functional recovery after acute ischemic stroke in rats. *PLoS One*. 2015;10(3):e0121854.

Cuartero MI, Ballesteros I, Moraga A, Nombela F, Vivancos J, Hamilton JA, Corbí ÁL, Lizasoain I, Moro MA. N2 neutrophils, novel players in brain inflammation after stroke: Modulation by the PPARγ agonist rosiglitazone. *Stroke*. 2013;44:3498–3508.

Darsalia V, Hua S, Larsson M, Mallard C, Nathanson D, Nyström T, Sjöholm Å, Johansson ME, Patrone C. Exendin-4 reduces ischemic brain injury in normal and aged type 2 diabetic mice and promotes microglial M2 polarization. *PLoS One*. 2014;9(8): e103114.

Easton AS. Neutrophils and stroke—Can neutrophils mitigate disease in the central nervous system? *Int Immunopharmacol*. 2013;17(4):1218–1225.

Emberson J, Lees KR, Lyden P et al., Stroke Thrombolysis Trialists' Collaborative Group. Effect of treatment delay, age, and stroke severity on the effects of intravenous thrombolysis with alteplase for acute ischaemic stroke: A meta-analysis of individual patient data from randomised trials. *Lancet* 2014;384(9958):1929–1935.

Er A, Yazar E, Uney K, Elmas M, Altan F, Cetin G. Effects of tylosin on serum cytokine levels in healthy and lipopolysaccharide-treated mice. *Acta Vet Hung*. 2010;58:75–81.

Fagan SC. Drug repurposing for drug development in stroke. *Pharmacotherapy*. 2010;30(7 Pt 2):51S–54S.

Fang H, Chen J, Lin S, Wang P, Wang Y, Xiong X, Yang Q. CD36-mediated hematoma absorption following intracerebral hemorrhage: Negative regulation by TLR4 signaling. *J Immunol*. 2014;192(12):5984–5992.

Faustino JV, Wang X, Johnson CE, Klibanov A, Derugin N, Wendland MF, Vexler ZS. Microglial cells contribute to endogenous brain defenses after acute neonatal focal stroke. *J Neurosci*. 2011;31(36):12992–13001.

Fieta A, Merlini C, Grassi GC. Requirements for intracellular accumulation and release of clarithromycin and azithromycin by human phagocytes. *J Chemother*. 1997;9:23–31.

Forloni G, Salmona M, Marcon G, Tagliavini F. Tetracyclines and prion infectivity. *Infect Disord Drug Targets*. 2009;9:23–30.

Fransen PS, Berkhemer OA, Lingsma HF et al. Multicenter randomized clinical trial of endovascular treatment of acute ischemic stroke in the Netherlands (MR CLEAN) investigators. Time to reperfusion and treatment effect for acute ischemic stroke: A randomized clinical trial. *JAMA Neurol*. December 2015;21:1–7.

Frieler RA, Meng H, Duan SZ, Berger S, Schütz G, He Y, Xi G, Wang MM, Mortensen RM. Myeloid-specific deletion of the mineralocorticoid receptor reduces infarct volume and alters inflammation during cerebral ischemia. *Stroke*. 2011;42:179–185.

Frieler RA, Ray JJ, Meng H et al. Myeloid mineralocorticoid receptor during experimental ischemic stroke: Effects of model and sex. *J Am Heart Assoc*. 2012;1(5):e002584.

Fu Y, Zhang N, Ren L et al. Impact of an immune modulator fingolimod on acute ischemic stroke. *Proc Natl Acad Sci USA*. 2014;111(51):18315–18320.

Fumagalli S, Perego C, Pischiutta F, Zanier ER, De Simoni MG. The ischemic environment drives microglia and macrophage function. *Front Neurol*. 2015;6:81.

Garcia-Bonilla L, Moore JM, Racchumi G, Zhou P, Butler JM, Iadecola C, Anrather J. Inducible nitric oxide synthase in neutrophils and endothelium contributes to ischemic brain injury in mice. *J Immunol*. 2014;193(5):2531–2537.

Geissmann F, Manz MG, Jung S, Sieweke MH, Merad M, Ley K. Development of monocytes, macrophages, and dendritic cells. *Science*. 2010;327:656–661.

Gelderblom M, Leypoldt F, Steinbach K et al. Temporal and spatial dynamics of cerebral immune cell accumulation in stroke. *Stroke*. 2009;40(5):1849–1857.

Gelderblom M, Melzer N, Schattling B et al. Transient receptor potential melastatin subfamily member 2 cation channel regulates detrimental immune cell invasion in ischemic stroke. *Stroke*. 2014;45(11):3395–3402.

Gill D, Veltkamp R. Dynamics of T cell responses after stroke. *Curr Opin Pharmacol*. 2015;26:26–32.

Ginsberg MD. Neuroprotection for ischemic stroke: Past, present and future. *Neuropharmacology*. 2008; 55:363–389.

Ginsburg C, Gan V, Petruska M. Randomized controlled trial of intralesional corticosteroid and griseofulvin vs. griseofulvin alone for treatment of kerion. *Pediatr Infect Dis J.* 1987;6:1084–1087.

Gliem M, Klotz L, van Rooijen N, Hartung HP, Jander S. Hyperglycemia and PPARγ antagonistically influence macrophage polarization and infarct healing after ischemic stroke. *Stroke*. 2015;46(10):2935–2942.

Gordon R, Mays R, Sambrano B, Mayo T, Lapolla W. Antibiotics used in nonbacterial dermatologic conditions. *Dermatol Ther*. 2012;25:38e54.

Grupke S, Hall J, Dobbs M, Bix GJ, Fraser JF. Understanding history, and not repeating it. Neuroprotection for acute ischemic stroke: From review to preview. *Clin Neurol Neurosurg*. 2015;129:1–9.

Guan W, Kozak A, Fagan SC. Drug repurposing for vascular protection after acute ischemic stroke. *Acta Neurochir Suppl*. 2011;111:295–298.

Gupta S, Grieco M, Siegel I. Suppression of T-lymphocyte rosettes by rifampin. Studies in normals and patients with tuberculosis. *Ann Intern Med*. 1975;82:484–488.

Han L, Cai W, Mao L, Liu J, Li P, Leak RK, Xu Y, Hu X, Chen J. Rosiglitazone promotes white matter integrity and long-term functional recovery after focal cerebral ischemia. *Stroke*. 2015;46(9):2628–2636.

Han Z, Li L, Wang L, Degos V, Maze M, Su H. Alpha-7 nicotinic acetylcholine receptor agonist treatment reduces neuroinflammation, oxidative stress, and brain injury in mice with ischemic stroke and bone fracture. *J Neurochem*. 2014a;131(4):498–508.

Han Z, Shen F, He Y, Degos V, Camus M, Maze M, Young WL, Su H. Activation of α-7 nicotinic acetylcholine receptor reduces ischemic stroke injury through reduction of proinflammatory macrophages and oxidative stress. *PLoS One*. 2014b;9(8):e105711.

Herz J, Sabellek P, Lane TE, Gunzer M, Hermann DM, Doeppner TR. Role of neutrophils in exacerbation of brain injury after focal cerebral ischemia in hyperlipidemic mice. *Stroke*. 2015;46(10):2916–2925.

Hsu CY, Hung CS, Chang HM, Liao WC, Ho SC, Ho YJ. Ceftriaxone prevents and reverses behavioral and neuronal deficits in an MPTP-induced animal model of Parkinson's disease dementia. *Neuropharmacology* April 2015;91:43–56.

Hu X, Li P, Guo Y et al. Microglia/macrophage polarization dynamics reveal novel mechanism of injury expansion after focal cerebral ischemia. *Stroke*. 2012;43:3063–3070.

Hu YY, Xu J, Zhang M, Wang D, Li L, Li WB. Ceftriaxone modulates uptake activity of glial glutamate transporter-1 against global brain ischemia in rats. *J Neurochem*. January 2015;132(2):194–205.

Hung YC, Chen TY, Lee EJ, Chen WL, Huang SY, Lee WT, Lee MY, Chen HY, Wu TS. Melatonin decreases matrix metalloproteinase-9 activation and expression and attenuates reperfusion-induced hemorrhage following transient focal cerebral ischemia in rats. *J Pineal Res*. November 2008;45(4):459–467.

Hussain I, Muzaffar F, Rashid T, Ahmad T, Jahangir M, Haroon T. A randomized, comparative trial of treatment of kerion celsi with griseofulvin plus oral prednisolone vs. griseofulvin alone. *Med Mycol*. 1999;37:97–99.

Inaba T, Katayama Y, Ueda M, Nito C. Neuroprotective effects of pretreatment with macrolide antibiotics on cerebral ischemia reperfusion injury. *Neurol Res*. June 2015;37(6):514–524.

Jickling GC, Liu D, Ander BP, Stamova B, Zhan X, Sharp FR. Targeting neutrophils in ischemic stroke: Translational insights from experimental studies. *J Cereb Blood Flow Metab*. 2015;35(6):888–901.

Jin G, Wong STC. Toward better drug repositioning: Prioritizing and integrating existing methods into efficient pipelines. *Drug Discov Today*. 2014;19:637–644.

Jin Q, Cheng J, Liu Y, Wu J, Wang X, Wei S, Zhou X, Qin Z, Jia J, Zhen X. Improvement of functional recovery by chronic metformin treatment is associated with enhanced alternative activation of microglia/macrophages and increased angiogenesis and neurogenesis following experimental stroke. *Brain Behav Immun.* 2014;40:131–142.

Kast R, Lefranc F, Karpel-Massler G, Halatsch M. Why dapsone stops seizures and may stop neutrophils' delivery of VEGF to glioblastoma. *Br J Neurosurg.* 2012;26:813–817.

Katayama Y, Inaba T, Nito C, Ueda M, Katsura K. Neuroprotective effects of erythromycin on cerebral ischemia reperfusion-injury and cell viability after oxygen-glucose deprivation in cultured neuronal cells. *Brain Res.* November 7, 2014;1588:159–167.

Khatri P, Yeatts SD, Mazighi M et al., IMS III Trialists. Time to angiographic reperfusion and clinical outcome after acute ischaemic stroke: An analysis of data from the Interventional Management of Stroke (IMS III) phase 3 trial. *Lancet Neurol.* 2014;13(6):567–574.

Klaas JP, Matzke T, Makol A, Fulgham JR. Minocycline-induced polyarteritis nodosa-like vasculitis presenting as brainstem stroke. *J Clin Neurosci.* 2015;22(5):904–907.

Kozak A, El-Remessy AB, Ergul A et al. Candesartan augments ischemia-induced proangiogenic state and results in sustained functional improvement after stroke. *Stroke.* 2009;40:1870–1876.

Kozak W, Kozak A, Elewa HF et al. Vascular protection with candesartan after experimental acute stroke in hypertensive rats: A dose-response study. *J Pharmacol Exp Ther.* 2008;326:773–782.

Kraft P, Göb E, Schuhmann MK et al. FTY720 ameliorates acute ischemic stroke in mice by reducing thrombo-inflammation but not by direct neuroprotection. *Stroke.* 2013;44(11):3202–3210.

Lalancette-Hébert M, Gowing G, Simard A, Weng YC, Kriz J. Selective ablation of proliferating microglial cells exacerbates ischemic injury in the brain. *J Neurosci.* 2007;27(10):2596–2605.

Li F, Faustino J, Woo MS, Derugin N, Vexler ZS. Lack of the scavenger receptor CD36 alters microglial phenotypes after neonatal stroke. *J Neurochem.* 2015;135(3):445–452.

Li T, Pang S, Yu Y, Wu X, Guo J, Zhang S. Proliferation of parenchymal microglia is the main source of microgliosis after ischaemic stroke. *Brain.* 2013;136:3578–3588.

Liao TV, Forehand CC, Hess DC, Fagan SC. Minocycline repurposing in critical illness: Focus on stroke. *Curr Top Med Chem.* 2013;13(18):2283–2290.

Liu P, Allaudeen H, Chandra R, Phillips K, Jungnik A, Breen JD, Sharma A. Comparative pharmacokinetics of azithromycin in serum and white blood cells of healthy subjects receiving a single-dose extended-release regimen versus a 3-day immediate-release regimen. *Antimicrob Agents Chemother.* 2007;51:103–109.

Lopes Pinheiro MA, Kooij G, Mizee MR, Kamermans A, Enzmann G, Lyck R, Schwaninger M, Engelhardt B, de Vries HE. Immune cell trafficking across the barriers of the central nervous system in multiple sclerosis and stroke. *Biochim Biophys Acta.* 2016, 1862(3):461–471. doi:10.1016/j.bbadis.2015.10.018.

Lujia Y, Xin L, Shiquan W, Yu C, Shuzhuo Z, Hong Z. Ceftriaxone pretreatment protects rats against cerebral ischemic injury by attenuating microglial activation-induced IL-1β expression. *Int J Neurosci.* September 2014;124(9):657–665.

Maestrini I, Strbian D, Gautier S et al. Higher neutrophil counts before thrombolysis for cerebral ischemia predict worse outcomes. *Neurology.* 2015;85:1408–1416.

Matsuo Y, Onodera H, Shiga Y, Nakamura M, Ninomiya M, Kihara T, Kogure K. Correlation between myeloperoxidase-quantified neutrophil accumulation and ischemic brain injury in the rat. Effects of neutrophil depletion. *Stroke.* 1994;25:1469–1475.

Miró-Mur F, Pérez-de-Puig I, Ferrer-Ferrer M, Urra X, Justicia C, Chamorro A, Planas AM. Immature monocytes recruited to the ischemic mouse brain differentiate into macrophages with features of alternative activation. *Brain Behav Immun.* 2016;53:18–33. doi:10.1016/j.bbi.2015.08.010.

Moon A, Gil S, Gill S, Chen P, Matute-Bello G. Doxycycline impairs neutrophil migration to the airspaces of the lung in mice exposed to intratracheal lipopolysaccharide. *J Inflamm.* 2012;9:31.

Moore DF, Li H, Jeffries N et al. Using peripheral blood mononuclear cells to determine a gene expression profile of acute ischemic stroke: A pilot investigation. *Circulation.* 2005;111:212–221.

Neumann J, Riek-Burchardt M, Herz J et al. Very-late-antigen-4 (VLA-4)-mediated brain invasion by neutrophils leads to interactions with microglia, increased ischemic injury and impaired behavior in experimental stroke. *Acta Neuropathol.* 2015;129(2):259–277.

Noble W, Garwood CJ, Hanger DP. Minocycline as a potential therapeutic agent in neurodegenerative disorders characterised by protein misfolding. *Prion.* 2009;3:78–83.

Oh SH, Kim OJ, Shin DA, Song J, Yoo H, Kim YK, Kim JK. Alteration of immunologic responses on peripheral blood in the acute phase of ischemic stroke: Blood genomic profiling study. *J Neuroimmunol.* 2012;249:60–65.

Pan J, Jin JL, Ge HM, Yin KL, Chen X, Han LJ, Chen Y, Qian L, Li XX, Xu Y. Malibatol A regulates microglia M1/M2 polarization in experimental stroke in a PPARγ-dependent manner. *J Neuroinflammation.* 2015;12:51.

Parnham MJ, Haber VE, Giamarellos-Bourboulis EJ, Perletti G, Verleden GM, Vos R. Azithromycin: Mechanisms of action and their relevance for clinical applications. *Pharmacol Ther.* 2014;143:225–245.

Perego C, Fumagalli S, De Simoni MG. Temporal pattern of expression and colocalization of microglia/macrophage phenotype markers following brain ischemic injury in mice. *J Neuroinflammation.* 2011;8:174.

Periti P. Immunopharmacology of oral betalactams. *J Chemother.* 1998;10:91–96.

Reglodi D, Renaud J, Tamas A, Tizabi Y, Socías SB, Del-Bel E, Raisman-Vozari R. Novel tactics for neuroprotection in Parkinson's disease: Role of antibiotics, polyphenols and neuropeptides. *Prog Neurobiol.* November 2, 2015, pii: S0301-0082(15)00128-8. doi:10.1016/j.pneurobio.2015.10.004.

Ritzel RM, Patel AR, Grenier JM, Crapser J, Verma R, Jellison ER, McCullough LD. Functional differences between microglia and monocytes after ischemic stroke. *J Neuroinflammation.* 2015;12:106.

Rizzo A, Paolillo R, Guida L, Annunziata M, Bevilacqua N, Tufano M. Effect of metronidazole and modulation of cytokine production on human periodontal ligament cells. *Int Immunopharmacol.* 2010;10:744–750.

Ruzza P, Siligardi G, Hussain R et al. Ceftriaxone blocks the polymerization of a-synuclein and exerts neuroprotective effects in vitro. *ACS Chem Neurosci.* 2014;5:30–38.

Schilling M, Besselmann M, Müller M, Strecker JK, Ringelstein EB, Kiefer R. Predominant phagocytic activity of resident microglia over hematogenous macrophages following transient focal cerebral ischemia: An investigation using green fluorescent protein transgenic bone marrow chimeric mice. *Exp Neurol.* 2005;196(2):290–297.

Sharp FR, Jickling GC. Whole genome expression of cellular response to stroke. *Stroke.* 2013 Jun;44(6 Suppl. 1):S23–S25.

Sippel TR, Shimizu T, Strnad F, Traystman RJ, Herson PS, Waziri A. Arginase I release from activated neutrophils induces peripheral immunosuppression in a murine model of stroke. *J Cereb Blood Flow Metab.* 2015;35(10):1657–1663.

Soni N, Reddy BV, Kumar P. GLT-1 transporter: An effective pharmacological target for various neurological disorders. *Pharmacol Biochem Behav.* December 2014;127:70–81.

Souvenir R, Doycheva D, Zhang JH, Tang J. Erythropoietin in stroke therapy: Friend or foe. *Curr Med Chem.* 2015;22(10):1205–1213.

Stamova B, Jickling GC, Ander BP et al. Gene expression in peripheral immune cells following cardioembolic stroke is sexually dimorphic. *PLoS One.* 2014;9(7):e102550.

Stock ML, Fiedler KJ, Acharya S et al. Antibiotics acting as neuroprotectants via mechanisms independent of their anti-infective activities. *Neuropharmacology*. October 2013;73:174–182.

Stoilova T, Colombo L, Forloni G, Tagliavini F, Salmona M. A new face for old antibiotics: Tetracyclines in treatment of amyloidoses. *J Med Chem*. 2013;56:5987–6006.

Strittmatter SM. Old drugs learn new tricks. *Nat Med*. 2014;20:590–591.

Sultan S, Gebara E, Toni N. Doxycycline increases neurogenesis and reduces microglia in the adult hippocampus. *Front Neurosci*. 2013;7:131.

Sutherland BA, Minnerup J, Balami JS, Arba F, Buchan AM, Kleinschnitz C. Neuroprotection for ischaemic stroke: Translation from the bench to the bedside. *Int J Stroke*. 2012;7(5):407–418.

Tang Y, Xu H, Du X, Lit L, Walker W, Lu A, Ran R, Gregg JP, Reilly M, Pancioli A. Gene expression in blood changes rapidly in neutrophils and monocytes after ischemic stroke in humans: A microarray study. *J Cereb Blood Flow Metab*. 2006;26:1089–1102.

Tang Z, Gan Y, Liu Q, Yin JX, Liu Q, Shi J, Shi FD. CX3CR1 deficiency suppresses activation and neurotoxicity of microglia/macrophage in experimental ischemic stroke. *J Neuroinflammation*. 2014;11:26.

Tokgoz S, Keskin S, Kayrak M, Seyithanoglu A, Ogmegul A. Is neutrophil/lymphocyte ratio predict to short-term mortality in acute cerebral infarct independently from infarct volume? *J Stroke Cerebrovasc Dis*. 2014;23(8):2163–2168.

Tymianski M. Novel approaches to neuroprotection trials in acute ischemic stroke. *Stroke*. 2013;44:2942–2950.

Wei J, Pan X, Pei Z, Wang W, Qiu W, Shi Z, Xiao G. The beta-lactam antibiotic, ceftriaxone, provides neuroprotective potential via antiexcitotoxicity and anti-inflammation response in a rat model of traumatic brain injury. *J Trauma Acute Care Surg*. 2012;73:654–660.

Wieghofer P, Knobeloch KP, Prinz M. Genetic targeting of microglia. *Glia*. 2015;63(1):1–22.

Womble TA, Green S, Shahaduzzaman M, Grieco J, Sanberg PR, Pennypacker KR, Willing AE. Monocytes are essential for the neuroprotective effect of human cord blood cells following middle cerebral artery occlusion in rat. *Mol Cell Neurosci*. 2014;59:76–84.

Won S, Lee JK, Stein DG. Recombinant tissue plasminogen activator promotes, and progesterone attenuates, microglia/macrophage M1 polarization and recruitment of microglia after MCAO stroke in rats. *Brain Behav Immun*. 2015;49:267–279.

Xiong X, Barreto GE, Xu L, Ouyang YB, Xie X, Giffard RG. Increased brain injury and worsened neurological outcome in interleukin-4 knockout mice after transient focal cerebral ischemia. *Stroke*. 2011;42:2026–2032.

Yamasaki R, Lu H, Butovsky O et al. Differential roles of microglia and monocytes in the inflamed central nervous system. *J Exp Med*. 2014;211(8):1533–1549.

Yang Y, Salayandia VM, Thompson JF, Yang LY, Estrada EY, Yang Y. Attenuation of acute stroke injury in rat brain by minocycline promotes blood-brain barrier remodeling and alternative microglia/macrophage activation during recovery. *J Neuroinflammation*. 2015;12:26.

Yulug B, Hanoglu L, Kilic E, Schabitz WR. RIFAMPICIN: An antibiotic with brain protective function. *Brain Res Bull*. 2014;107C:37–42.

Zhao X, Wang H, Sun G, Zhang J, Edwards NJ, Aronowski J. Neuronal Interleukin-4 as a modulator of microglial pathways and ischemic brain damage. *J Neurosci*. 2015;35(32):11281–11291.

Zhao Y, Cudkowicz ME, Shefner JM et al. Systemic pharmacokinetics and cerebrospinal fluid uptake of intravenous ceftriaxone in patients with amyotrophic lateral sclerosis. *J Clin Pharmacol*. October 2014;54(10):1180–1187.

Zhu Z, Fu Y, Tian D et al. Combination of the immune modulator fingolimod with alteplase in acute ischemic stroke: A pilot trial. *Circulation*. 2015;132(12):1104–1112.

Index

9 780367 869076